S0-AKX-931

Stochastic Quantum Mechanics and Quantum Spacetime

Fundamental Theories of Physics

A New International Book Series on The Fundamental Theories of Physics: Their Clarification, Development and Application

Stochastic Quantum Mechanics and Quantum Spacetime

A Consistent Unification of Relativity and Quantum Theory Based on Stochastic Spaces

by

Eduard Prugovečki

Department of Mathematics, University of Toronto, Canada

D. Reidel Publishing Company

A MEMBER OF THE KLUWER ACADEMIC PUBLISHERS GROUP

Dordrecht / Boston / Lancaster

Library of Congress Cataloging in Publication Data

CIP

Prugovečki, Eduard.
Stochastic quantum mechanics and quantum spacetime.

(Fundamental theories of physics)
Bibliography: p.
Includes index.
1. Quantum field theory. 2. Stochastic processes. I. Title.
II. Series.
QC174.45.P78 1984 530.1′32 83–11075
ISBN 90–277–1617–X

Published by D. Reidel Publishing Company,
P.O. Box 17, 3300 AA Dordrecht, Holland.

Sold and distributed in the U.S.A. and Canada
by Kluwer Academic Publishers,
190 Old Derby Street, Hingham, MA 02043, U.S.A.

In all other countries, sold and distributed
by Kluwer Academic Publishers Group,
P.O. Box 322, 3300 AH Dordrecht, Holland.

Printed in The Netherlands.

To MARGARET
for her spiritual support,
dedication and understanding

Table of Contents

* The asterisks indicate sections which can be skipped at a first reading.

Preface

The principal intent of this monograph is to present in a systematic and self-contained fashion the basic tenets, ideas and results of a framework for the consistent unification of relativity and quantum theory based on a quantum concept of spacetime, and incorporating the basic principles of the theory of stochastic spaces in combination with those of Born's reciprocity theory.

In this context, by the physicial consistency of the present framework we mean that the advocated approach to relativistic quantum theory relies on a consistent probabilistic interpretation, which is proven to be a *direct extrapolation* of the conventional interpretation of nonrelativistic quantum mechanics. The central issue here is that we can derive conserved and relativistically convariant *probability* currents, which are shown to merge into their nonrelativistic counterparts in the nonrelativistic limit, and which at the same time explain the physical and mathematical reasons behind the basic fact that no probability currents that consistently describe pointlike particle localizability exist in conventional relativistic quantum mechanics. Thus, it is not that we dispense with the concept of locality, but rather the advanced central thesis is that the classical concept of locality based on pointlike localizability is inconsistent in the realm of relativistic quantum theory, and should be replaced by a concept of *quantum locality* based on stochastically formulated systems of covariance and related to the aforementioned currents.

By the mathematical consistency of the framework we mean that all its main features are derived from basic mutually consistent physical principles by methods that meet contemporary standards of mathematical rigor. Thus, in presenting this framework it can be said that we echo Dirac's sentiments, expressed in many of the statements made by him since the inception of the renormalization program, and in particular that "one must seek a new relativistic quantum mechanics and one's prime concern must be to base it on sound mathematics" (Dirac, 1978b, p. 5). Naturally, in order to establish the link with conventional relativistic quantum theory, we had to resort in the last chapter of this book to the kind of formal manipulations that have become the trademark of most work in so-called 'local' quantum field theory. However, the main purpose of such diversions is to establish that conventional reuslts can be recovered, even though the theoretical machinery originally used in their derivation is dispensed with partly or in toto.

The present approach to relativistic quantum theory assigns great prominence to Born's (1938, 1949) reciprocity principle, which emerges as a basic tenet in the later development of the theory of quantum spacetime. Indeed, it can be truly maintained that the basic physical and mathematical concepts underlying the theoretical framework in this monograph are embodiments of philosophical ideas advocated by M. Born (and partly by A. Landé) as long ago as forty-odd years – albeit the present author had not been aware of this fact until rather late in the development of the present program. In fact, the very concept of stochastic value, which constitutes the cornerstone of the stochastic quantum mechanics described in Part I, was vigorously advocated by Born (1955) in some of his later writings – as illustrated by the quotes heading Chapters 1 and 3. It is also clear from these writings (Born, 1956), and from the published Born–Einstein correspondence (Born, 1971), that a single central philosophical idea dominated his foundational research in physics – an epistemological idea which was even more diametrically opposed to Einstein's determinism (Einstein, 1949; Pais, 1979) than Bohr's ideology revealed in the well-known Bohr–Einstein debate (Bohr, 1961). In essence, this idea is that *all* our physical knowledge – including that pertaining to classical phenomena – is probabilistic rather than deterministic in nature, and consequently physical theories should be reformulated in exclusively probabilistic terms even at their most fundamental structural level (Born, 1956, pp. 167–186). As a logical corollary of this *Weltanschauung*, one should strive to formulate even the most basic physical concepts, such as that of spacetime, in a purely probabilistic manner.

The present work describes in detail the outcome of a systematic endeavor of this nature. Considering the difficulties and the controversies surrounding the long-outstanding issues it is dealing with, it would be imprudent to try to provide the reader with final answers. Rather, the hope is that this work might stimulate foundational research based on clearly stated principles that combine physical, mathematical as well as philosophical ideas in a *consistent* manner.

In this respect our basic premise is that, *assuming* the uncertainty principle is as fundamental as generally believed, it then has to be consistently combined with the principles of relativity theory. Indeed, as we shall remind the reader in the Introduction, as well as in the beginning of each chapter, by direct quotes from the published utterances by Einstein, Dirac, Heisenberg, Born and many other well-known researchers in relativity and quantum theory, these leading physicists were not only keenly aware that this issue was central to the further development of quantum theory, but also that the various approaches to relativistic quantum physics that became popular during the second half of this century were leading us astray from any cogent attempt ever to try to systematically answer all the foundational questions related to this issues.

Thus, in a spirit of cautious but consistent extrapolation from the physically and philosophical sound ideas advocated by the founders of quantum mechanics and relativity theory, we have opted for an inductive rather than deductive method of exposition. Hence, instead of deducing all results from a series of axioms for quantum spacetime, we first introduce nonrelativistic quantum mechanics on stochastic phase space as the first extrapolation of conventional quantum theory. Then we move on to the relativistic regime, by allowing the Poincaré group to take over the role which the Galilei group has played in this extrapolation. We follow that by introducing a concept of quantum spacetime whose detailed structure we eventually pinpoint by systematically using Born's reciprocity principle. Consequently, in addition to forming a cohesive group, each of the introduced basic principles can be also considered separately on its own terms. This method has the advantage that if any of these principles is found wanting, the faulty underlying idea can be revised or altogether eliminated from the present framework, without thereby having to disassemble the remaining structure.

To make the text accessible not only to mathematical physicists but also to the entire community of physicists, we have avoided as much as possible the use of mathematical concepts not generally familiar to theoretical physicists. Furthermore, we have also avoided the methodology of stating and then proving theorems – although, as demonstrated even in some intermediate-level textbooks (Prugovečki, 1981a; Schechter, 1981), as a mathematical discipline, quantum theory can be made to match the standards of rigor prevalent in all of contemporary mathematics. Naturally, a certain minimum of functional-analytic concepts and techniques had to be retained in order to substantiate the claim of mathematical consistency, but all concepts and basic theorems not commonly encountered in textbooks on quantum mechanics have been described or explained in the two appendices at the end of this monograph.

I wish to thank all those physicists and mathematicians (A. O. Barut, L. C. Biedenharn, A. Böhm, P. Busch, E. Caianiello, H. Ekstein, R. Giles, R. L. Ingraham, Y. S. Kim, A. Lonke, R. E. O'Connell, S. Roy, M. Sachs, D. Sen, D. Spring, J. C. Taylor, A. S. Wightman, E. P. Wigner and others) who have commented, verbally or in writing, on the various aspects of the program presented in this monograph, and who have at times supplied very useful references to the tremendously rich and varied literature pertaining to the topics with which this monograph is concerned. Special thanks are due to S. T. Ali, J. A. Brooke, R. Gagnon, D. Greenwood, W. Guz and F. E. Schroeck for their valuable comments on the first draft of the manuscript, and to S. M. Shute for typing the manuscript promptly and accurately. Research grants from the National Research Council of Canada (NSERC of late) for the duration of the project are gratefully acknowledged. Last, but certainly not least, an unremittable spiritual debt belongs to those giants of twentieth-century physics, whose uncompromising scientific integrity, as reflected in the subsequently cited critical appraisals of the past accomplishments in relativistic quantum physics, has lent the moral support needed for the kind of systematic reconsideration of the foundations of this subject undertaken in the present monograph.

Department of Mathematics
University of Toronto
Toronto, Canada

EDUARD PRUGOVEČKI
January 1983

Introduction

The two principal theories that revolutionized physics in the twentieth century – namely, relativity and quantum mechanics – display some profound similarities on the most basic epistemological level, as well as equally striking disparities on the structural, mathematical level. The present approach to their unification is based on the premise that a mathematical framework capable of embracing them both into a harmonious whole can be achieved only by concentrating on those structural features which are the direct progeny of their common epistemic heritage.

To single out the key features that allow themselves to be incorporated into a new superstructure devoid of internal inconsistencies, one has to examine the historical development of the basic physical ideas on which these two fundamental theories had been founded, and to scrutinize the reasons behind the failure of past attempts at their unification.

The origin of the common epistemic heritage of relativity and quantum mechanics can be traced to the firm belief of their respective founders in the necessity of basing fundamental physical concepts on operational considerations. In other words, fundamental physical ideas should be theoretical idealizations anchored in actual laboratory procedures rather than in preconceived frameworks arrived at on the basis of ontological arguments. A striking example of the second type of approach is provided by Poincaré, who had discussed, prior to all Einstein's papers on relativity theory, the possibility of adopting Riemannian geometry for the description of three-dimensional physical space. However, he eventually dismissed this possibility on the grounds that "experience . . . tells us not what is the truest, but what is the most convenient geometry" (Poincaré, 1905, pp. 70–71), and then he settled on Euclidean geometry as 'the most convenient'. On the other hand, in 1923, in direct response to Poincaré's arguments, Einstein declared: "The question whether the structure of this [spacetime] continuum is Euclidean, or in accordance with Riemann's general scheme, or otherwise, is . . . properly speaking a physical question which must be answered by experience and not a question of a mere convention to be selected on practical grounds" (Einstein, 1953, p. 193). One paragraph later in the same publication, Einstein also made a statement which is crucial to the central thesis of this monograph, namely that his "proposed physical interpretation of geometry breaks down when applied immediately to spaces of submolecular order of magnitude". When one juxtaposes this statement with one made by Heisenberg only four years later – namely, that "if one wants to clarify the meaning of the expression 'position of the object', e.g. of an electron (in relation to a reference frame), one must provide certain experiments by means of which one can envisage the measurement of 'the position of the electron'" (Heisenberg, 1927, p. 174) – one catches the first glimpse of the main stumbling block standing in the path of the harmonious unification of relativity and quantum mechanics. The essence of the encountered difficulty lies in the question: How can we reconcile the classical relativistic concept of localizability of a macroscopic object with the quantum mechanical concept of localizability of a micro-particle?

The answer to how to treat the localizability problem of a macroparticle in spacetime had already been provided in unambiguous terms by Einstein in his famed 1905 paper that launched special relativity: spacetime coordinates were to be operationally defined by means of rigid rods and standard clocks in relation to inertial frames envisaged as "three rigid material lines perpendicular to one another and issuing from one point" (Einstein *et al.*, 1923, p. 43). Furthermore, test particles (envisaged as material points) were to be used as markers of space-time 'events' representing point-collisions between these test particles, or between test particles and light signals. These light signals represented the cornerstone of Einstein's operational definitions of spacetime distances and of simultaneity of events with respect to any given inertial frame.

A decade later, as he arrived after a prolonged search at his final formulation of general relativity, Einstein (1916) relaxed somewhat his strict adherence to operational principles when he dropped the requirement that the spacetime coordinates should be themselves operationally defined. In all other respects he retained, however, his basic operational outlook. In particular, in the words of Wigner (1955, p. 219), general as well as special relativity were to be based on the postulate "that only coincidences in space-time can be observed directly and only these should be the subject of physical theory". As a matter of fact, this postulate had provided Einstein (1916) with the physical justification of the principle of covariance, which demanded that the formulation of all basic laws in relativistic theories should not be form-dependent on the choice of coordinates. However, with the advent of modern quantum mechanics, Einstein became keenly aware of the difficulties his approach encountered on the microscopic level, as illustrated by the following remarks that he made in the last decade of his life: "If one disregards the quantum structure, one can justify the introduction of the [metric tensor] g_{ik} 'operationally' by pointing to the fact that one can hardly doubt the physical reality of the elementary light cone which belongs to a point." On the other hand, "that kind of physical justification for the introduction of the g_{ik} falls by the wayside unless one limits oneself to the 'macroscopic'. The application of the formal basis of the general theory of relativity to the 'microscopic' can, therefore, based only upon the fact that that tensor is the formally simplest covariant structure which can come under consideration. Such argumentation, however, carries no weight with anyone who doubts that we have to adhere to the continuum at all. All honor to his doubt – but where else is there a passable road?" (Einstein, 1949, p. 676).

Our intention is to demonstrate in this monograph that 'passable roads' actually do exist, and that they can be clearly perceived once we eliminate from orthodox quantum mechanics all the paradoxical assumptions that have already been the focus of much criticism in the past, and then contemplate the task of building a relativistic quantum mechancis from this fresh perspective. To achieve that, we first have to remind ourselves that the orthodox interpretation of quantum mechanics, as advocated by the Copenhagen school, was also operationally based (Heisenberg, 1930), but that the descriptions of operational procedures display a peculiar dichotomy in their treatment of 'system' versus 'apparatus'. Indeed, the former is supposed to abide by quantum mechanical laws, and therefore is subject to the uncertainty principle, whereas the latter is to be described, in some sense, in classical terms. Thus inertial frames and detection devices are, by decree, to receive a classical treatment in the operational analysis of the measurement process. The main justification for this strange kind of duality is, according to Bohr (1961, p. 39), "that by the word 'experiment' we refer to a situation where we can tell others what we have done and what we have learned and that, therefore, the account of the experimental arrangement and of the results of observations must be expressed in unambiguous language with suitable application of the terminology of classical physics".

The above quote constitutes an essential part of the doctrine of quantum mechanical duality that has been severely criticized by many outstanding physicists, the most poignant of this criticism originating in he second half of this century with Landé (1960, 1965). In the case of Landé, the criticism is not aimed at reinstating determinism in physics at the micro-level, but on the contrary, at establishing a *consistently* statistical point of view which the doctrine of dualism has failed in achieving. This lack of consistency is reflected in countless 'paradoxes' – some of which are, however, actual contradictions. They range from the well-known ones described in books on the philosophical foundation of quantum mechanics (Jammer, 1974; d'Espagnat, 1976), to some not widely discussed but nevertheless significant – such as the ambiguous role played by the uncertainty principle when it is applied to the center-of-mass of a system as one makes the transition from micro- to macro-systems via macromolecules, molecular chains and Brownian particles (Prugovečki, 1967, p. 2175). As the system becomes progressively larger, so that purportedly the classical realm is approached, at which stage exactly does the validity of the uncertainty principle stop, and the center-of-mass position and momentum become simultaneously measurable? And as the size of the 'system' approaches macroscopic dimensions, are its physical attributes still subject to the complementarity principle or not? If not, where exactly is the demarcation line? In fact, if we go to the other extreme and adopt a cosmological perspective, the duality of roles of 'system' and 'apparatus' (detectors, inertial frames, etc.) leads to downright absurdities (De Witt and Graham, 1973), since the rather artificial dividing line between 'system' and 'apparatus' melts away when we include the entire Universe in our considerations.

In face of the innumerable repetitions in many textbooks on quantum mechanics of the need to describe the apparatus 'classically' in the *traditional sense*, it comes as a startling realization that the main justification for the contention lies in the naive interpretation of certain ambiguous[1] quotes from Bohr (1928, 1934, 1961), such as the one presented above. Indeed, on reflection, for a precise account of any experimental arrangement, not only is the 'terminology of classical physics' not essential, but even when it is used, it has to be supplemented by notions that lie outside the *deterministic superstructure* of the classical framework. Due to a deeply ingrained force of habit rather than to logical necessity, this superstructure nevertheless insinuates itself into quantum mechanics via the purely continuum description of spacetime, via deterministic descriptions of inertial frames of reference and of fundamental kinematical operations with such frames (spacetime translations, rotations, boosts, etc.), via the notion that 'position' can be defined and measured with arbitrary precision, etc. The inconsistency in this hidden acceptance of determinism in quantum mechanics in the guise of a 'classical description of the apparatus' had been actually brought to the surface by one of the main founders of quantum mechanics and of the orthodox point of view, namely M. Born, who set out to "demonstrate the fallacy of [the] deterministic interpretation of classical mechanics and of the whole physics derived from it", and who eventually concluded that "determinism is out of the question, in the original sense of the word, even in the case of the simplest classical science, that of mechanics" (Born, 1962, p. 34). In other words, whereas the language that theoretical physicists use in everyday communication reinforces the fiction of exactly measurable values for observables, the actual experimental praxis has routinely to cope – classically as well as quantum mechanically – with the kind of uncertainty intrinsic in any measurement due to the inherent lack of precision of *all* instruments. Yet the epistemological implications of that basic fact appear to have been overlooked by most theoreticians of this century, with some notable exceptions (such as Born), who will be mentioned in some of the introductory paragraphs of the five chapters of this monograph.

In contrast, the point of view which we shall advocate is epistemologically a modified[2] version of Landé's (1960, 1965) 'unitary' approach: there is only one reality, namely *quantum* reality, and that physical reality manifests itself in the stochastic behavior of quantum particles constituting all matter in the Universe. *Ergo*, the description of this behavior should be purely *stochastic* (rather than deterministic, i.e. 'classical' in the naive sense of the word) regardless of whether the part of the Universe on which the observer has focused his attention plays the role of 'system' or of 'apparatus'. Thus, 'classical' (i.e. deterministic) descriptions of any physical entity, including spacetime, should be relegated to the role of an approximation liable to be satisfactory only when treating matter in bulk at macroscopic orders of magnitude and for spacetime distances well beyond 10^{-13} cm. As a corollary, the famed wave–particle duality (originating from the insistence on trying to visualize quantum behavior either from the perspective of classical waves or from that of classical point particles), transcends into a principle of quantum-mechanical unity. This unity is based on the notion of *stochastically extended quantum particles*, which are neither waves nor particles in the classical sense of the word, but rather a spatio-temporal phenomenon whose relation to the Universe is described by a number of intrinsic attributes (mass, spin, etc.), and which prominently include that of *proper wave function* (cf. Section 1.2).

The concept of proper wave function had already been introduced by Landé in 1939, but was not pursued by him. It was, however, promptly incorporated by Born (1939) into the reciprocity theory that he had launched the previous year with the avowed aim of unifying relativity and quantum theory into a consistent framework (Born, 1938). During the subsequent decade, Landé (1965, p. 164), as well as Born (1949), working with various collaborators, used the reciprocity principle in attempting to predict the mass spectrum of elementary particles. These attempts, however, apparently failed at providing results in sufficiently good agreement with experiment to attract other adherents who would further develop and verify the theory. Nevertheless, in Part II of this monograph we shall demonstrate not only that Born's reciprocity theory can be incorporated into the present framework and that it leads to experimentally already well-verified results (namely, linear Regge trajectories for mesons and baryons), but that in this framework the concept of proper wave function emerges in a most natural manner by a straightforward extrapolation of conventional quantum theory that is described in Part I.

It is in this last direction that the principal thrust of the present endeavor at a consistent unification of relativity and quantum theory is to be found. Thus, on the one hand, the aim is *not* to make mere cosmetic changes in the present formulations of those two fundamental theories, that might be justified by computational expediency, but would leave inherent contradictions unresolved. On the other hand, albeit the changes introduced might be radically new on the conceptual level, they should be implemented by standard techniques, so as not to render irrelevant to the present framework the great wealth of models and computational results arrived at during the past five decades of developments in relativistic quantum mechanics and quantum field theory.

With these aims in mind, in Chapter 1 we present a careful extrapolation of the conventional formalism of nonrelativistic quantum mechanics based on representations of the Galilei group $\mathcal{G}$ in phase space. Thus we demonstrate on familiar ground that the concept of sharp localizability of quantum particles is unnecessarily restrictive, and can be generalized to that of a stochastic localizability that is both internally consistent, as well as consistent with orthodox quantum mechanics. Mathematically, this amounts to replacing projector-valued (PV) systems of imprimitivity for unitary ray representations of $\mathcal{G}$ with positive operator-valued (POV) systems of covariance for the Galilei group $\mathcal{G}$.

It is then shown in Chapter 2 that when the Poincaré group $\mathcal{P}$ is allowed to take over the role of the Galilei group $\mathcal{G}$, relativistic systems of covariance are obtained that supply a resolution of the long-outstanding problems of particle localizability in relativistic quantum mechanics. Thus, relativistically covariant and conserved *probability* (and not only charge) 4-currents are obtained for relativistic quantum particles that are stochastically extended, i.e. possess a nontrivial proper wave function.

In Chapter 3 (most of which can be skipped at first reading) we show that the extended particle formalism of Chapter 1 can be used to arrive at unified frameworks for classical and quantum statistical mechanics. Thus, on a mathematical level, the derived results vindicate Born's (1955, 1956) contention that, despite formal appearances, there are no fundamental epistemic differences between classical and quantum theories when they are viewed from the perspective of application to the surrounding physical reality rather than as abstract mathematical disciplines. A number of key new concepts (such as informational completeness) that underline this parallelism emerge when stochastic (rather than deterministic) values are introduced at both classical and quantum levels.

In Chapter 4 we re-evaluate Einstein's epistemological analysis of the concept of (classical) spacetime from a purely quantum point of view. The outcome is a concept of *quantum spacetime* in which relativistic propagators describing quantum test particles in free-fall take over the role played by world lines of classical test particles. The result is a geometrodynamics of evolving *stochastic* 3-geometries that replace the pseudo-Riemannian geometries of (classical) special and general relativity.

As a natural byproduct, the considerations of Chapter 4 give rise to a concept of *quantum locality* that is purely stochastic. In Chapter 5 we show that this enables the development of local field theories on quantum spacetime (FTQS), which in some formal aspects resemble their counterparts in local quantum field theory (LQFT), but are not plagued by divergences. In FTQS the rather *ad hoc* rules of 'second quantization' on classical spacetime are replaced by very natural rules for constructing fields on quantum spacetime that act in master Fock spaces of states with variable numbers of quantum particles. Specific models are formulated, such as that of a reciprocally and relativistically invariant electrodynamics on quantum spacetime, whose formal S-matrix expansion merges into that of quantum electrodynamics if Born's fundamental length l_0 is allowed to approach zero (in which limit divergences, of course, re-emerge). However, the absence of singularities in FTQS also permits the development of nonperturbational techniques – and one feasible approach to this task is described in Chapter 5. It is hoped that the present monograph will stimulate research that would implement this or other techniques in the context of actual models of fields on quantum spacetime. For those models that are counterparts of LQFT models currently in fashion, any such enterprise could also throw some light on the validity of the mathematical heuristics presently in use in LQFT computations.

Notes

[1] In the words of Stapp (1972, p. 1098): "The writings of Bohr are extraordinarily elusive. They rarely seem to say what you want to know. They weave a web of words around the Copenhagen interpretation but do not say exactly what it is."

[2] Many aspects of Landé's (1960, 1965) unitary approach, such as his advocacy of Duane's (1923) 'quantum rule' as the 'missing link between wavelike appearances and particle reality', have proved (Shimony, 1966) unfruitful, and should be viewed only in the light of his maxim of "deducing the quantum rules themselves from a nonquantal basis of familiar postulates of symmetry and invariance" (Landé, 1965b, p 127); cf. also the debate between Landé (1969) and Born (1969b) – Heisenberg (1969).

Part I

Stochastic Quantum Mechanics

"Statements like 'A quantity x has a completely definite value' (expressed by a real number and represented by a point in the physical continuum) seem to me to have no physical meaning. Modern physics has achieved its greatest successes by applying a principle of methodology that concepts whose application requires distinctions that cannot in principle be observed are meaningless and must be eliminated."

BORN (1956, p. 167)

"One must seek a new relativistic quantum mechanics and one's prime concern must be to base it on sound mathematics."

DIRAC (1978b, p. 6)

Chapter 1

Nonrelativistic Stochastic Quantum Mechanics

"As an instrument of mathematics the concept of a real number represented by a nonterminating decimal is extremely important and fruitful. As a measure of a physical quantity, the concept is nonsensical . . . According to the heuristic principle employed by Einstein in the theory of relativity and by Heisenberg in quantum theory, concepts that correspond to no conceivable observation ought to be eliminated from physics. This is possible without difficulty in the present case also; we have only to replace statements like $x = \pi$ cm by: the probability of the distribution of values of x has a sharp maximum at $x = \pi$ cm."

BORN (1956, p. 186)

Conventional quantum mechanics is predicated on the assumption that exact, deterministic values of observables are in principle accessible to physical theory despite the limitations imposed on the measurement process by the very nature of quantum theory, which is statistical and subject to the uncertainty principle. This tacit but basic postulate can be traced to the fundamental dichotomy imposed on the physical world by the orthodox theory of measurement, which splits the physical environment into two sharply separated sectors: 'system' and 'apparatus'. Although the dividing line is admittedly arbitrary and chosen by the 'observer' so as to meet his convenience, it is nevertheless requested that, once it is drawn, the 'system' should be considered as subject to quantum mechanical laws, whereas the 'apparatus' should be described 'classically' (Bohr, 1934, 1961; Messiah, 1962).

In this context, it is usually taken for granted that 'classical values' are to be treated as exact and deterministic. However, all experimental praxis has to cope routinely with the fact that all readings of any instrument measuring a quantity with a continuous spectrum are subject to margins of uncertainty, and therefore, are stochastic in nature. It is for this reason that Max Born, although one of the founders of the orthodox interpretation of quantum mechanics, nevertheless argued in his later writings (Born, 1955, 1956), as well as in his correspondence with Einstein (Born, 1971), in favor of banishing exact values from all of physics. In fact, he was the first physicist to point out that, strictly speaking, not even classical mechanics is deterministic when viewed as a physical theory rather than as an abstract mathematical framework (Born, 1955).

The methodological and epistemological implications of this observation are far-reaching: they point the way towards an eventual conceptual unification of all classical and quantum physics, by founding both on the concept of *stochastic* (rather than sharp) values for physical quantities. In the quantum realm, they result in an extrapolation of conventional quantum mechanics which we shall call *stochastic quantum mechanics*. This name reflects the fact that the present approach introduces an extra element of stochasticity in quantum theory via the mathematical formulation – where positive-operator valued (POV) measures take over the role played in the conventional framework (Prugovečki, 1981a) by projector-valued (PV) measures (Section 1.3) – as well as in the physical interpretation, where only stochastic values (as opposed to sharp, deterministic values) are assumed to be accessible to the measurement process.

In its treatment of classical mechanics *vis-à-vis* quantum mechanics, the present stochastic approach is the antithesis of hidden-variable approaches (Bohm, 1952; De Broglie, 1964; Bohm and Bub, 1966) and of some previous stochastic approaches (Nelson 1966, 1967; De la Pena-Auerbach, 1969, 1971) to the epistemic relation of these two fundamental disciplines: instead of trying to reduce quantum theory to classical concepts supplemented by some fundamental stochastic hypothesis, the present approach recasts (Sections 3.4–3.7) classical theory into a quantum mechanical framework by explicitly displaying in it the stochastic nature inherent in all actual measurement processes – regardless of whether they are viewed classically or quantum mechanically. For example, in the stochastic mechanics of Nelson (1966), De la Pena-Auerbach and Ceto (1975), Guerra and Ruggiero (1978), Vigier (1979), Davidson (1979, 1980) and others (cf. De Witt and Elsworthy, 1981, for reviews), the dynamics of the stochastic particle is obtained essentially by using stochastic (forward and backward) derivatives in Newton's second law, whereas the hidden-variable approach of Bohm (1952) or De Broglie (1964), as extended by Kershaw (1964) or Lehr and Park (1977), is based on particle probability derivatives and mean (forward and backward) velocities that obey Smoluchowski-type equations. However, in both instances, the net result is that systems of equations for these quantities can be obtained which, upon making certain assumptions and identifications (e.g. diffusion coefficients are identified with $\hbar/2m$, where m is the mass of the particle), can be shown to be equivalent to the Schrödinger equation. On the other hand, in the present stochastic approach the general foundation of conventional quantum mechanics is preserved from the outset, the only basic difference in the nonrelativistic realm being that instead of dealing with wave functions on configuration space, momentum space, or some other spectral representation spaces (cf. Appendix B) of sharp values of some complete set of observables, in stochastic quantum mechanics one deals with wave functions on *stochastic* phase spaces (cf. Sections 1.1–1.5).

Thus, the basic premise of stochastic quantum mechanics is that quantum theory should be formulated in terms of stochastic values of observables, which in the presence of continuous spectra are necessarily nonsharp. Once such a formulation is achieved, we can investigate the existence of *sharp-point limits*, in which parameters characterizing the widths of such spread-out stochastic values are allowed to approach zero. In this chapter, after formulating nonrelativistic stochastic quantum mechanics – which involves spread-out position values – we shall show that the sharp-point limits do exist, and that their outcome coincides with conventional nonrelativistic quantum mechanics. This result implicitly demonstrates that the concept of sharp localizability does not contradict the combined outcome of the uncertainty principle and of Galilei covariance that is mathematically embodied in Galilei systems of imprimitivity (Section 1.2). As we shall see in the next chapter, this is in sharp contrast with the relativistic case – as exemplified by the non-existence of analogous Poincaré systems of imprimitivity discussed in Section 2.2.

1.1. Stochastic Values and Stochastic Phase Spaces

The calibration of any instrument used in the process of measurement is aimed at achieving concurrence between distinct measurements carried out on the same quantity under operationally identical conditions. This concurrence is essential to the concept of scientific objectivity: without it the criterion of reproducibility of experimental results obtained by various experimenters would be vacuous. It is, however, a basic observational fact that such a concurrence cannot be achieved if the result of each measurement is described by a single number α. Rather, it is essential that via a calibration procedure a confidence interval be assigned to each reading or, more generally, that a probability distribution χ_α peaked at α and describing the margin of uncertainty in the reading be specified together with α (Dietrich, 1973). In that case, we can introduce probability measures[1]

$$\mu_\alpha(\Delta) = \int_\Delta \chi_\alpha(x)\,dx, \tag{1.1}$$

which can be called *confidence measures* since they describe the probability that when a reading α is obtained the actually prepared[2] value had been within Δ. If the reading α were absolutely accurate, then μ_α would be a δ-measure with support at α, and therefore its density χ_α would equal, formally speaking, a δ-'function' centered at α, provided that x varies over a nondegenerate interval containing α. Since, however, this is never the case in practice, it is not only desirable but also necessary to introduce the *stochastic value* (or *point*) $\underset{\sim}{\alpha}$, obtained by adjoining μ_α to α, whenever we relate theoretically predicted values to observed values.

Sets of stochastic values can display various properties which characterize them as members of specific categories of stochastic spaces. The stochastic spaces with which we shall be dealing in this monograph shall be stochastic phase spaces of various kinds, but in order to establish the link with the theory of statistical metric spaces (Menger 1942, 1951; Wald, 1943, 1955), it is desirable to formulate this notion at the most general mathematical level. (The reader uninterested in mathematical abstractions can skip the next five paragraphs, and continue with the one containing Equation (1.6).)

Given a measurable space $(\mathscr{X}, \mathscr{A})$ consisting (cf. Appendix A) of a set $\mathscr{X}$ and a

Boolean σ-algebra $\mathscr{A}$ of subsets of $\mathscr{X}$, we can define a *stochastic space* $\underset{\sim}{\mathscr{X}}$ over $(\mathscr{X}, \mathscr{A})$ by assigning to each $\alpha \in \mathscr{X}$ a probability measure μ_α on $(\mathscr{X}, \mathscr{A})$, and then setting

$$\underset{\sim}{\mathscr{X}} = \{\underset{\sim}{\alpha} = (\alpha, \mu_\alpha) \mid \alpha \in \mathscr{X}\}, \tag{1.2}$$

i.e. identifying $\underset{\sim}{\mathscr{X}}$ with the set of all *stochastic values* $\underset{\sim}{\alpha} = (\alpha, \mu_\alpha)$. In this context, a stochastic value (α, μ_α) will be called *sharp* or *pointlike* if its confidence measure μ_α is the δ-measure at α, i.e. if $\mu_\alpha(\Delta) = 1$ for all $\Delta \in \mathscr{A}$ that contain α, and $\mu_\alpha(\Delta) = 0$ if $\alpha \notin \Delta$. A stochastic value which is not sharp will be called *spread-out*, or *extended*, or *fuzzy.*[3] $\underset{\sim}{\mathscr{X}}$ is taken to constitute a measurable space, its measurable sets $\underset{\sim}{\Delta} = \{\underset{\sim}{\alpha} \mid \alpha \in \Delta\}$ being those for which Δ is measurable, i.e. $\Delta \in \mathscr{A}$; if $\mathscr{X}$ is also a topological space (Appendix A), then $\underset{\sim}{\Delta}$ is open in $\underset{\sim}{\mathscr{X}}$ iff (i.e. if and only if) Δ is open in $\mathscr{X}$.

In this chapter we shall be dealing almost exclusively with the case where $\mathscr{X}$ is an n-dimensional real Euclidean space $\mathbb{R}^n$, where $\mathscr{A}$ coincides with the family $\mathscr{B}^n$ of all Borel sets (cf. Appendix A), and where the confidence measures μ_α possess *confidence functions* $\chi_\alpha(x)$, i.e. can be written in the form (1.1) for all $\Delta \in \mathscr{B}^n$, where the integration is with respect to the Lebesgue measure in $\mathscr{X} = \mathbb{R}^n$. In such cases we shall identify the stochastic value $\underset{\sim}{\alpha}$ with the pair (α, χ_α). Furthermore, in these cases we shall also require that α should be the location of the absolute maximum of the confidence function χ_α, so that the most likely value x corresponding to the reading α equals α itself.

If the set $\mathscr{X}$ is a metric space, so that $d^2(\alpha_1, \alpha_2)$ denotes the (positive definite or indefinite) square of the 'distance' between any two of its elements α_1 and α_2, then the stochastic space $\underset{\sim}{\mathscr{X}}$ of stochastic values $\underset{\sim}{\alpha}$ will become a *stochastic metric space* in a sense that generalizes the concept of statistical metric space introduced by Menger (1942, 1951) and Wald (1943, 1955) to the extent of relaxing some of their axioms (Guz, 1983b), and of allowing for square-metric distribution functions

$$M_{\underset{\sim}{\alpha}_1, \underset{\sim}{\alpha}_2}(\lambda) = \int_{d^2(x_1, x_2) \leqslant \lambda} \mathrm{d}\mu_{\alpha_1}(x_1)\, \mathrm{d}\mu_{\alpha_2}(x_2) \tag{1.3}$$

between two stochastic points $\underset{\sim}{\alpha}_1 = (\alpha_1, \mu_{\alpha_1})$ and $\underset{\sim}{\alpha}_2 = (\alpha_2, \mu_{\alpha_2})$ even if the metric is indefinite. Indeed, in the case of $\mathscr{X} = \mathbb{R}^n$ and of extended stochastic values $\underset{\sim}{\alpha} = (\alpha, \chi_\alpha)$, with χ_α related to μ_α by (1.1), the above distribution function assumes the form

$$M_{\underset{\sim}{\alpha}_1, \underset{\sim}{\alpha}_2}(r^2) = \int_{|x_1 - x_2|^2 \leqslant r^2} \chi_{\alpha_1}(x_1)\chi_{\alpha_2}(x_2)\, \mathrm{d}x_1\, \mathrm{d}x_2, \tag{1.4}$$

so that $M_{\underset{\sim}{\alpha}_1, \underset{\sim}{\alpha}_2}(r^2)$ as a function of $r = \lambda^{1/2}$ is a generalization of the Menger–Wald distribution function $F_{\underset{\sim}{\alpha}_1, \underset{\sim}{\alpha}_2}(r)$ (Schweizer and Sklar, 1962) expressing the probability that for the readings α_1 and α_2 the actual distance between the actually prepared exact values x_1 and x_2 does not exceed r.

An even more general concept combining geometric and stochastic features is arrived at when $\mathscr{X}$ is a Riemannian or pseudo-Riemannian manifold. This case will be discussed in Chapter 4 in the context of quantum spacetime based on *stochastic geometries* (Blokhintsev, 1973, 1975) that generalize conventional geometries in the sense that the latter is recovered if $\underset{\sim}{\mathscr{X}}$ consists exclusively of sharp values.

We can study stochastic (metric) spaces in the abstract by defining for them various operations and concepts in analogy with corresponding operations and concepts for ordinary (metric) spaces, and then investigating the consequences. However, the only major concept of this nature that we shall require in the sequel is that of Cartesian product $\underset{\sim}{\mathscr{X}}' \times \underset{\sim}{\mathscr{X}}''$ of two stochastic spaces $\underset{\sim}{\mathscr{X}}'$ and $\underset{\sim}{\mathscr{X}}''$. We shall say that $\underset{\sim}{\mathscr{X}}$ is the *Cartesian product* of $\underset{\sim}{\mathscr{X}}'$ and $\underset{\sim}{\mathscr{X}}''$, i.e. $\underset{\sim}{\mathscr{X}} = \underset{\sim}{\mathscr{X}}' \times \underset{\sim}{\mathscr{X}}''$, if, first of all, the underlying measurable space $(\mathscr{X}, \mathscr{A})$ is a Cartesian product of $(\mathscr{X}', \mathscr{A}')$ and $(\mathscr{X}'', \mathscr{A}'')$, so that $\mathscr{X} = \mathscr{X}' \times \mathscr{X}''$ and $\mathscr{A} = \mathscr{A}' \times \mathscr{A}''$, and second, for each $\underset{\sim}{\alpha} = (\alpha, \mu_\alpha) \in \mathscr{X}$ the confidence measure μ_α is the Cartesian product $\mu'_{\alpha'} \times \mu''_{\alpha''}$ of two confidence measures $\mu'_{\alpha'}$ and $\mu''_{\alpha''}$, where $\alpha = (\alpha', \alpha'')$. In that case we shall write

$$\underset{\sim}{\alpha} = (\alpha, \mu_\alpha) = \underset{\sim}{\alpha}' \times \underset{\sim}{\alpha}'' = (\alpha', \mu'_{\alpha'}) \times (\alpha'', \mu''_{\alpha''}). \tag{1.5}$$

The stochastic spaces of central interest in nonrelativistic (classical or quantum) mechanics are those defined over $(\Gamma^N, \mathscr{B}^{2N})$, where

$$\Gamma^N = \mathbb{R}^{2N} = \{(q, p) \mid q = (q_1, \ldots, q_N) \in \mathbb{R}^N, p = (p_1, \ldots, p_N) \in \mathbb{R}^N\} \tag{1.6}$$

is the phase space for N degrees of freedom encountered in classical mechanics, and $\mathscr{B}^{2N}$ is the family of all Borel set in Γ^N. We shall call such spaces

$$\underset{\sim}{\Gamma}^N = \{\underset{\sim}{\zeta} = (\zeta, \chi_\zeta) \mid \zeta = (q, p) \in \Gamma^N\} \tag{1.7}$$

nonrelativistic stochastic phase spaces if their stochastic points $\underset{\sim}{\zeta}$ have confidence functions χ_ζ which are congruent under translations in the phase space Γ^N, i.e. if

$$\chi_{q,p}(q', p') = \chi_{0,0}(q' - q, p' - p) \tag{1.8}$$

for all $(q, p) \in \Gamma^N$. In addition, the nonrelativistic stochastic phase spaces with which we shall be dealing will be factorizable in the Cartesian product $\underset{\sim}{\mathbb{R}}^N_{\text{conf}} \times \underset{\sim}{\mathbb{R}}^N_{\text{mom}}$ of a *stochastic configuration space*

$$\underset{\sim}{\mathbb{R}}^N_{\text{conf}} = \{\underset{\sim}{q} = (q, \chi'_q) \mid q \in \mathbb{R}^N\} \tag{1.9}$$

with $\chi'_q(q') = \chi_0(q' - q)$, and a *stochastic momentum space*

$$\underset{\sim}{\mathbb{R}}^N_{\text{mom}} = \{\underset{\sim}{p} = (p, \chi''_p) \mid p \in \mathbb{R}^N\} \tag{1.10}$$

with $\chi''_p(p') = \chi''_0(p' - p)$, so that for all $\underset{\sim}{\zeta} \in \underset{\sim}{\Gamma}^N$

$$\underset{\sim}{\zeta} = (\zeta, \chi_\zeta) = (q, \chi'_q) \times (p, \chi''_p), \quad \zeta = (q, p), \tag{1.11}$$

$$\chi_{q,p}(q', p') = \chi'_q(q')\,\chi''_p(p'). \tag{1.12}$$

Stochastic phase spaces can be jointly considered in both the classical and quantum context. We shall actually do that in Chapter 3, while developing a unified framework for classical quantum statistical mechanics. However, their primary significance to us shall be in the quantum realm, where stochastic phase spaces prove indispensible in solving the following fundamental problems in quantum mechanics.

(1) the description of quantum ensembles in terms of positive-definite distribution functions in phase-space variables $\zeta = (\underset{\sim}{q}, \underset{\sim}{p})$;

(2) the informationally complete description of quantum states in terms of probability measures, i.e. of positive-definite and normalized measures over sample spaces of observable quantities to which correspond *unique* pure or mixed states;

(3) the formulation of Poincaré covariant probability densities and probability currents describing the localization in spacetime of relativistic quantum particles.

Problems 1 and 2 will be dealt with in Chapter 3, whereas problem 3 will be considered in Chapter 2. However, for the understanding of the nature of this problem as well as of its solution in the relativistic regime, it is necessary to first understand the localization problem and all its ramifications in the nonrelativistic context, where conventional quantum mechanics supplies a mathematically consistent solution in the context of sharp localization. It is this problem that constitutes the central issue in this chapter, and we turn to it next.

1.2. Sharp Localizability and Systems of Imprimitivity

In conventional quantum mechanics the concept of localizability at a point $\mathbf{x}$ has to be treated as a limiting case of localizability in arbitrarily small regions R_n,

$$R_1 \supset R_2 \supset \ldots \supset \{\mathbf{x}\}, \qquad \bigcap_{n=1}^{\infty} R_n = \{\mathbf{x}\}, \tag{2.1}$$

if a mathematically as well as physically consistent formulation is to be achieved (Wightman, 1962). Indeed, from the mathematical point of view, there are no normalized elements $\psi(\mathbf{x})$ of a spectral representation space (Appendix B) for the position operators X^j

$$(X^j \psi)(\mathbf{x}) = x^j \psi(\mathbf{x}), \quad j = 1, 2, 3, \tag{2.2}$$

for which the probability measure $P_\psi^{\mathbf{X}}(B)$ of, say, a spin-zero particle could be a δ-measure at $\mathbf{x}_0 \in \mathbb{R}^3$, if we request that X^j together with self-adjoint operators P^j, $j = 1, 2, 3$, playing the role of momentum operators satisfy the canonical commutation relations (CCRs)

$$[X^i, P^j] = i\hbar\delta^{ij}, \qquad [X^i, X^j] = [P^i, P^j] = 0. \tag{2.3}$$

This is due to Von Neumann's theorem,[4] which states that any representation of the CCRs is unitarily equivalent to either the Schrödinger representation, corresponding to

$$P_\psi^{\mathbf{X}}(B) = \int_B |\psi(\mathbf{x})|^2 \, d\mathbf{x}, \quad \langle\psi \mid \psi\rangle = 1, \tag{2.4a}$$

$$\langle\psi_1|\psi_2\rangle = \int_{\mathbb{R}^3} \psi_1^*(\mathbf{x})\,\psi_2(\mathbf{x})\, d\mathbf{x}, \tag{2.4b}$$

or to a direct sum of such representations. In turn, in the Schrödinger representation the inner product in (2.4a) equals zero whenever B is a one-point set $\{\mathbf{x}_0\}$. The δ-'functions' $\delta_{\mathbf{x}_0}(\mathbf{x}) = \delta(\mathbf{x} - \mathbf{x}_0)$, heuristically purported to supply the required solutions, clearly do not fulfill that role in a mathematically meaningful manner since they are actually singular distributions for which $|\psi_{\mathbf{x}_0}(\mathbf{x})|^2$ in (2.4) is not defined. In subsequent sections it will become clear that this mathematical impossibility is a reflection of the physical impossibility of actually measuring an *exact* real value. In fact, this provides the first hint as to the basic need for implementing Born's advice that "statements like 'A quantity x has a completely definite value' . . . are meaningless and must be eliminated [from physics]" (Born, 1956, p. 167).

However, in case of regions (understood as open sets or closures of open sets in Euclidean spaces $\mathbb{R}^n$) we can arrive at a mathematically meaningful concept of sharp localizability in a probabilistic sense: a particle in a state ψ is *sharply localized* in space at time t if there is a bounded region B for which the probability $P_\psi(B; t)$ of finding the particle within B equals 1. If that is the case, we shall also say that the state ψ is *localized within* B at time t. A quantum mechanical theory for a given particle will be said to be *local in the strict sense* if for any (however small) region B and any instant t there is a state ψ localized within B.

Due to the aforementioned Von Neumann theorem on CCRs, all nonrelativistic theories of quantum particles are local in the strict sense. However, we shall see in Section 2.2 that precisely the opposite is true in the relativistic case, since no *consistent* relativistic theory of quantum particles is local in the strict sense.

On the pure mathematical level this is due to basic kinematical features that distinguish Newtonian from relativistic mechanics. Thus, the most general transformation relating Cartesian coordinates x^j, $j = 1, 2, 3$, of the same particles in two distinct inertial frames can be decomposed into a space translation by the 3-vector $\mathbf{a}$, a velocity boost by the 3-velocity $\mathbf{v}$, and a rotation that can be described by 3×3 real matrix $R \in SO(3)$. If we also allow for the possibility that standard clocks tied to the two frames and running at the same rate might not have been synchronized but rather display a difference b in the time values they register, then the general coordinate transformation $g = (b, \mathbf{a}, \mathbf{v}, R)$ constitutes an element of the (restricted) Galilei group $\mathcal{G}$. This group element acts on the position vector $\mathbf{x}$ with coordinates x^j, on the 3-momentum $\mathbf{p}$, and on the time t measured with respect to an inertial frame $\mathscr{L}$ in the following well-known manner (Levy-Leblond, 1971),

$$\mathbf{x} \longmapsto \mathbf{x}' = R\mathbf{x} + \mathbf{v}t + \mathbf{a}, \tag{2.5a}$$

$$\mathbf{p} \longmapsto \mathbf{p}' = R\mathbf{p} + m\mathbf{v}, \tag{2.5b}$$

$$t \longmapsto t' = t + b, \tag{2.5c}$$

in order to reproduce the corresponding coordinate values in the new frame $\mathscr{L}' = g^{-1}\mathscr{L}$ for a point particle of mass m under the assumption that Newtonian mechanics is strictly valid. In the quantum mechanical context, for a particle without spin, this transformation law translates into the requirement that the wave functions $\psi(\mathbf{x}, t)$ in the configuration representation and in the Schrödinger picture transform in accordance to an irreducible unitary ray representation $\hat{U}(b, \mathbf{a}, \mathbf{v}, R)$ of $\mathscr{G}$, such as (Levy-Leblond, 1971):

$$\hat{U}(b, \mathbf{a}, \mathbf{v}, R) : \psi(\mathbf{x}, t) \longmapsto \psi'(\mathbf{x}, t)$$

$$= \exp\left\{\frac{i}{\hbar}\left[-\frac{m\mathbf{v}^2}{2}(t-b) + m\mathbf{v}\cdot(\mathbf{x}-\mathbf{a})\right]\right\}\psi(R^{-1}[\mathbf{x}-\mathbf{a}-\mathbf{v}(t-b)], t-b). \tag{2.6}$$

The reason for choosing a ray representation, for which the multiplication law is

$$\hat{U}(g_1 g_2) = \exp\left[-\frac{im}{\hbar}\left(\frac{1}{2}\mathbf{v}_1^2 b_2 + \mathbf{v}_1 \cdot R_1 \mathbf{a}_2\right)\right]\hat{U}(g_1)\hat{U}(g_2), \tag{2.7}$$

$$g_1 g_2 = (b_1, \mathbf{a}_1, \mathbf{v}_1, R_1) \cdot (b_2, \mathbf{a}_2, \mathbf{v}_2, R_2)$$

$$= (b_1 + b_2, \mathbf{a}_1 + R_1\mathbf{a}_2 + b_2\mathbf{v}_1, \mathbf{v}_1 + R_1\mathbf{v}_2, R_1 R_2), \tag{2.8}$$

rather than a vector one, stems from the requirement that the infinitesimal generators of space translations and of velocity boosts,

$$\hat{U}(0, \mathbf{0}, \mathbf{v}, I) = \exp\left[\frac{i}{\hbar} m\mathbf{v} \cdot \mathbf{X}\right], \tag{2.9a}$$

$$\hat{U}(0, \mathbf{a}, \mathbf{0}, I) = \exp\left[-\frac{i}{\hbar}\mathbf{a} \cdot \mathbf{P}\right], \tag{2.9b}$$

should give rise to a representation (2.3) of the CCRs. Furthermore, ray representations are required for the Galilei invariance of the Schrödinger equation (Bargmann, 1954, Section 6g).

Since $\{X^j\}$ and $\{P^j\}$ constitute sets of commuting self-adjoint operators, they possess joint spectral measures (cf. Appendix B) $E^{\mathbf{X}}(B)$ and $E^{\mathbf{P}}(B)$, respectively. From (2.6) we immediately get

$$(E^{\mathbf{X}}(B)\psi)(\mathbf{x}, t) = \chi_B(\mathbf{x})\psi(\mathbf{x}, t), \tag{2.10}$$

where $\chi_S(\mathbf{x})$ denotes in general the characteristic function of a set S, i.e. $\chi_S(\mathbf{x}) = 1$ for $\mathbf{x} \in S$ and $\chi_S(\mathbf{x}) = 0$ for $\mathbf{x} \notin S$. To compute $E^{\mathbf{P}}(B)$, we apply to $\psi(\mathbf{x}, t)$ the Fourier–Plancherel transform[5] U_{F},

$$U_{\mathrm{F}} : \psi(\mathbf{x}, t) \longmapsto \tilde{\psi}(\mathbf{k}, t) = (2\pi\hbar)^{-3/2} \int_{\mathbb{R}^3} \exp\left(-\frac{i}{\hbar}\mathbf{k}\cdot\mathbf{x}\right) \psi(\mathbf{x}, t)\, d\mathbf{x}, \tag{2.11a}$$

$$U_{\mathrm{F}}^{-1} : \tilde{\psi}(\mathbf{k}, t) \longmapsto \psi(\mathbf{x}, t) = (2\pi\hbar)^{-3/2} \int_{\mathbb{R}^3} \exp\left(\frac{i}{\hbar}\mathbf{x}\cdot\mathbf{k}\right) \tilde{\psi}(\mathbf{k}, t)\, d\mathbf{k}, \tag{2.11b}$$

and observe that since

$$(P^j\psi)(\mathbf{x}, t) = -i\hbar \frac{\partial}{\partial x^j} \psi(\mathbf{x}, t), \tag{2.12}$$

we shall have

$$(P^j\psi)^\sim(\mathbf{k}, t) = k^j \tilde{\psi}(\mathbf{k}, t), \tag{2.13}$$

Consequently, $\tilde{\psi}(\mathbf{k}, t)$ belongs to the momentum representation, and we have

$$(E^{\mathbf{P}}(B)\psi)^\sim(\mathbf{k}, t) = \chi_B(\mathbf{k})\tilde{\psi}(\mathbf{k}, t), \tag{2.14}$$

Furthermore, if we introduce the free Hamiltonian H_0 by setting

$$\hat{U}(-t, \mathbf{0}, \mathbf{0}, I) = \exp\left(-\frac{i}{\hbar} H_0 t\right), \tag{2.15}$$

then by (2.6)

$$(H_0\psi)^\sim(\mathbf{k}, t) = \frac{\mathbf{k}^2}{2m} \tilde{\psi}(\mathbf{k}, t), \quad H_0 = \frac{P^2}{2m}. \tag{2.16}$$

From (2.6) and (2.11) we also obtain:

$$U_{\mathrm{F}}\hat{U}(b, \mathbf{a}, \mathbf{v}, R)U_{\mathrm{F}}^{-1} : \tilde{\psi}(\mathbf{k}, t) \longmapsto \tilde{\psi}'(\mathbf{k}, t)$$

$$= \exp\left\{\frac{i}{\hbar}\left[\frac{m\mathbf{v}^2}{2}(t-b) - \mathbf{k}\cdot(\mathbf{a} + \mathbf{v}(t-b))\right]\right\} \tilde{\psi}(R^{-1}(\mathbf{k} - m\mathbf{v}), t-b), \tag{2.17}$$

Hence, combining (2.6) with (2.10), and (2.14) with (2.17), we obtain

$$\hat{U}(g)E^{\mathbf{X}}(B_1)\hat{U}^{-1}(g) = E^{\mathbf{X}}(gB_1), \tag{2.18a}$$

$$gB_1 = \{R\mathbf{x} + \mathbf{v}t + \mathbf{a} \mid \mathbf{x} \in B_1\}, \tag{2.18b}$$

$$\hat{U}(g)E^{\mathbf{P}}(B_2)\hat{U}^{-1}(g) = E^{\mathbf{P}}(gB_2), \tag{2.19a}$$

$$gB_2 = \{R\mathbf{k} + m\mathbf{v} \mid \mathbf{k} \in B_2\}, \tag{2.19b}$$

for the general element $g = (0, \mathbf{a}, \mathbf{v}, R)$ of the *isochronous* (Levy-Leblond, 1971, p. 229) *Galilei group* $\mathcal{G}'$. This signifies that $\{E^{\mathbf{X}}(B) \mid B \in \mathcal{B}^3\}$ and $\{E^{\mathbf{P}}(B) \mid B \in \mathcal{B}^3\}$ constitute (transitive) systems of imprimitivity for the representation $\hat{U}(0, \mathbf{a}, \mathbf{v}, R)$ of $\mathcal{G}'$ with the base $\mathbb{R}^3$ of configuration or momentum space of values, respectively.

To define in general a system of imprimitivity (Mackey, 1968), consider a group G of homeomorphisms (i.e. one-to-one continuous mappings) of a topological space $\mathcal{X}$ onto itself. Let $\mathcal{B}$ denote the family of Borel sets on $\mathcal{X}$ (i.e. the Boolean σ-algebra generated by all open sets in $\mathcal{X}$ – cf. Appendix A). A *projector-valued (PV) measure* over $(\mathcal{X}, \mathcal{B})$ is a family of orthogonal projection operators $E(B)$, $B \in \mathcal{B}^n$, for which $E(\mathcal{X})$ equals the identity operator $\mathbb{1}$ on $\mathcal{H}$, and

$$E\left(\bigcup_{i=1}^{\infty} B_i\right) = \sum_{i=1}^{\infty} E(B_i), \quad B_i \cap B_j = \emptyset, i \neq j, \tag{2.20}$$

for any sequence $B_1, B_2 \ldots$ of disjoint Borel sets. Such a PV measure is said to be a *system of imprimitivity* for a unitary representation $\mathcal{U}(g)$ of G with base $\mathcal{X}$ if

$$\mathcal{U}(g)E(B)\mathcal{U}^{-1}(g) = E(gB), \quad gB = \{g(x) \mid x \in B\} \tag{2.21}$$

for all $g \in G$ and $B \in \mathcal{B}$. A system of imprimitivity is *transitive* if G is transitive, i.e., if for any two $x, x' \in \mathcal{X}$ there is a homeomorphism $g \in G$ such that $x' = g(x)$.

Early axiomatizations of the concept of quantum locality (Wightman, 1962; Kálnay, 1971) adopted systems of imprimitivity as fundamental to the concept of localizability. However, albeit precise, we shall see in Section 2.2 that this mathematical tool is too restrictive to cope with the problem of localizability of quantum particles in relativistic spacetime. In fact, as opposed to $\mathcal{G}$, the Poincaré group $\mathcal{P}$ does not preserve simultaneity, so that there is no relativistic counterpart of $\mathcal{G}'$, and therefore (2.18) and (2.19) possess no relativistic counterparts. For the momentum representation that is not a serious limitation (at least not in the special relativity case – cf. Sections 2.1 and 2.2) since by working in the Heisenberg picture with time-independent wave functions $\tilde{\psi}(\mathbf{k})$, we can replace $\hat{U}(g)$ by the following representation (which coincides with the relation (2.17) taken at $t = 0$ in accordance with (2.16)):

$$\tilde{U}(0, \mathbf{a}, \mathbf{v}, R) : \tilde{\psi}(\mathbf{k}) \mapsto \tilde{\psi}'(\mathbf{k})$$

$$= \exp\left[\frac{i}{\hbar}\left(\frac{\mathbf{k}^2}{2m} b - \mathbf{k} \cdot \mathbf{a}\right)\right] \tilde{\psi}(R^{-1}(\mathbf{k} - m\mathbf{v})). \tag{2.22}$$

This representation of $\mathcal{G}$ in $L^2(\mathbb{R}^3)$ does possess a system of imprimitivity $\tilde{E}^{\mathbf{P}}(B)$ with base over the 3-momentum space $\mathbb{R}^3$,

$$\tilde{U}(g)\tilde{E}^{\mathbf{P}}(B_2)\tilde{U}^{-1}(g) = \tilde{E}^{\mathbf{P}}(gB_2), \quad g \in \mathcal{G}, \tag{2.23a}$$

$$(\tilde{E}^{\mathbf{P}}(B)\tilde{\psi})(\mathbf{k}) = \chi_B(\mathbf{k})\tilde{\psi}(\mathbf{k}), \tag{2.23b}$$

which has the relativistic counterpart (2.1.15) for the representation of $\mathcal{P}$ in (2.1.9) (i.e., in Eq. (1.9) of Chapter 2). However, due to a theorem by Hegerfeldt (1974) (cf. the last four paragraphs in Section 2.2), we can state with certainty that no consistent solution of the localizability problem exists in the relativistic configuration space as long as we restrict ourselves to systems of imprimitivity and to sharp localizability.

As a consequence of (2.20) we find for PV-measures (Prugovečki, 1981a, p. 232) that

$$E(R \cap S) = E(R)E(S) \tag{2.24}$$

for any two Borel sets R and S. Hence, there has been an attempt by Jauch and Piron (1967) (and further developed by Amrein (1969)) to replace the equality sign in (2.20) by $\leqslant$, demanding only that for $B_1 \cap B_2 = \emptyset$ the projectors $E(B_1)$ and $E(B_2)$ should project onto subspaces of $\mathcal{H}$ which have only the zero vector in common, and to replace (2.24) by

$$E(R \cap S) = \operatorname*{s-lim}_{n \to \infty} (E(R)E(S))^n. \tag{2.25}$$

However, it turns out that Hegerfeldt's no-go theorem applies (Skagerstam, 1976) equally well to the somewhat weaker notion of localizability formulated within such a framework. In fact, as long as localizability in a bounded region B is associated with a projector $E(B)$, then it leads to sharp localizability in the earlier defined sense, since for any orthogonal projection operators $E(B) \neq 0$ there will be states ψ for which

$$E(B)\psi = \psi, \quad \|\psi\| = 1, \tag{2.26}$$

and for which, therefore,

$$P_\psi(B) = \langle \psi \mid E(P)\psi \rangle \tag{2.27}$$

equals 1. The essence of Hegerfeldt's no-go theorem, as well as of the operational analysis discussed at the end of Section 2.2, is that there cannot be a consistent notion of sharp localizability in the relativistic mechanics of quantum particles. Hence, we turn next to the task of formulating the weakest conditions under which a sensible notion of particle localizability is still available in a (nonrelativistic or relativistic) quantum theory.

1.3. Stochastic Localizability and Systems of Covariance

A general notion of quantum localizability is arrived at if the localization in a space region B is not considered in isolation from all other regions, but rather holistically, i.e. as part of all spatial relationships. Thus, if we discard the naive notion of sharp geometric points as an extreme mathematical idealization which has no literal counterpart in the actual physical world of material objects, then the entire concept of a spatial region with sharp boundaries that differentiate it unambiguously from the rest of the universe becomes a gratuitous assumption, to be discarded if it cannot be fitted within a consistent framework of ideas underlying a physical theory. Hence, if we consider as primary the formalism of quantum mechanics, with its accepted probabilistic interpretation, then we observe that in its most basic aspects this framework requires only a stochastic notion of localizability, which should associate a probability $P_\psi(B)$ with every state vector ψ and every region B. The only other prerequisite that the conventional Hilbert space formulation of quantum mechanics imposes is that this probability should be the expectation value of some self-adjoint operator $F(B)$ associated with that region:

$$P_\psi(B) = \langle \psi | F(B) \psi \rangle, \quad \| \psi \| = 1. \tag{3.1}$$

Since (3.1) is supposed to hold for any state vector $\psi \in \mathcal{H}$, by the Heillinger–Toeplitz theorem (cf. Appendix B) the operators $F(B)$ must be bounded.

Viewed in this light, all the remaining set-theoretical properties of $F(B)$ are not a consequence of quantum mechanics *per se*, but rather of general probability theory. Thus, since $P_\psi(B) \geqslant 0$ and $P_\psi(\emptyset) = 0$ for every state vector, we conclude that each $F(B)$ is a positive-definite operator:

$$F(B) \geqslant 0, \qquad F(\emptyset) = 0. \tag{3.2}$$

If we assume that $P_\psi(B)$ is to be part of a probability measure on Borel sets, then it follows that the operator-valued set-theoretical function $F(B)$ can be extended to all Borel sets B in such a manner that

$$F\left(\bigcup_{n=1}^{\infty} B_n\right) = \sum_{n=1}^{\infty} F(B_n), \quad B_i \cap B_j = \emptyset, i \neq j, \tag{3.3}$$

due to (3.1) and the fact that for any ϕ, $\psi \in \mathcal{H}$ and any linear operator F defined on the entire Hilbert space:

$$\langle \phi | F \psi \rangle = \frac{1}{2i} \langle \phi + i\psi | F(\phi + i\psi) \rangle + \frac{1}{2} \langle \phi + \psi | F(\phi + \psi) \rangle + $$
$$+ \frac{i-1}{2} (\langle \phi | F\phi \rangle + \langle \psi | F\psi \rangle). \tag{3.4}$$

In fact, (3.4) also implies that the infinite sum in (3.3) should converge weakly, and since $F(B_n) \geqslant 0$, we conclude by (B.6) and (B.7) in Appendix B that

$$\sum_{n=1}^{\infty} F(B_n) = \underset{N \to \infty}{\text{s-lim}} \sum_{n=1}^{\infty} F(B_n). \tag{3.5}$$

Finally, since $P_\psi(B) = 1$ for all ψ when B is the entire 'space' $\mathscr{X}$ (i.e. $\mathscr{X} = \mathbb{R}^3$ in nonrelativistic theory, and $\mathscr{X}$ is equal to a hypersurface of operational simultaneity in the relativistic case), we should have, by (3.1),

$$F(\mathscr{X}) = I. \tag{3.6}$$

An extended operator-valued measure on Borel sets, which assumes finite values on compact (cf. Appendix A) sets and satisfies (3.2) and (3.3), is called a *positive operator-value (POV) measure*; if in addition (3.6) is also satisfied, then we shall say that $F(B)$ is a *normalized POV measure.*[6] The PV measures $E(B)$ defined in Section 1.2 are obviously special instances of normalized POV measures in which $F(B)$ for each Borel set B is an orthogonal projection operator. In fact, it is easily seen that a normalized POV measure $E(B)$ which satisfies (2.24) is necessarily a PV measure, since by setting in (2.24) $R = S$ we conclude that $E(R) = E^2(R)$, i.e. that the operators $E(R)$ are projectors. Thus, (2.24) emerges as a necessary and sufficient condition for a POV measure to be a PV measure. However, in Section 1.4 we shall encounter instances of POV measures which are neither PV measures nor derived from PV measures (as is the case with those discussed later in this section). Rather, they embody a notion of localizability in spatial regions R which are not the seats of pointlike test particles, but rather of stochastically extended test particles, i.e. mathematically they are sets of stochastic values.

These considerations suggest that we should extend the concept of a quantum theory local in the strict sense defined in Section 1.2 into a concept of a theory *local in the stochastic sense,*[7] in which a positive-definite operator $F(B)$ is associated with every region B in space, and in which the family of all these operators generate a POV measure $F(B)$ over the Borel sets in the space $\mathscr{X}$. Whether $\mathscr{X}$ itself is to be visualized as a set of 'sharp points' giving rise to conventional (i.e. deterministic) geometries, or as a set of 'hazy clouds' giving rise to the stochastic geometries briefly mentioned in Section 1.1, is a question which, for the time being, we leave open. Heuristically, however, we expect that in the first case stochastically local theories would be local in the strict sense, whereas in the second case sharp locality would be unrealizable for any bounded region B, i.e. that $P_\psi(B) < 1$ for all state vector ψ, so that we can never assert with total certainty that a particle is to be found in a bounded region B.

The above crucial points will be fully discussed in Chapter 4, when we shall re-examine, with the uncertainty principle in mind, the epistemic meaning of spacetime by using the operational method originally deployed by Einstein while formulating special and general relativity in the years prior to the advent of quantum mechanics. For the time being we note that the underlying attitude, which *ipso facto* brings into question the naive notions of localizability that can be traced to the ancient origins[8] of geometry, is an operational one: it views geometry as a reflection of measurement procedures dealing with spatio-temporal relationships, rather than as a body of *a priori* given disciplines to be judged exclusively on basis of the internal consistency of their axioms.

In this context we observe that the separation of spatial from temporal degrees of freedom in the discussion of localizability is foreign to the spirit of relativity theory. In fact, if we consider spatial and temporal relationships from an operational point of view, then the totally unsymmetric treatment of these two aspects – whereby time is looked upon as a sharply defined parameter whereas spatial localizability is treated in terms of regions with possibly nonsharp boundaries – does not accurately reflect experimental praxis: uncertainties in the measurement of time are as all-pervading as they are in measurements of position. Hence, a discussion of localizability in spacetime – be it Newtonian (cf. Ehlers, 1973), Minkowskian or Einsteinian, namely in the spacetime of general relativity – appears mandatory for the attainment of a physically accurate description of particle localizability in quantum mechanics.

Spacetime localizability can be consistently defined only in relation to certain spacetime domains $\mathring{D}$, i.e. we can refer only to a probability $P(\mathring{B}/\mathring{D})$, such as for $\psi(\mathbf{x}, t)$ in (2.10)

$$P_\psi(\mathring{B}/\mathring{D}) = \int_{\mathring{B}} |\psi(\mathbf{x}, t)|^2 \, d\mathbf{x} \, dt \bigg/ \int_{\mathring{D}} |\psi(\mathbf{x}, t)|^2 \, d\mathbf{x} \, dt, \tag{3.7}$$

for observing a particle in a spacetime region $\mathring{B}$ in relation to the domain $\mathring{D}$ in which $\mathring{B}$ is contained. The immediate mathematical reason why $\mathring{D}$ cannot be taken to be all of (in the present case Newtownian) spacetime is obvious from (2.4): we have

$$P_\psi^{\mathbf{X}}(\mathbb{R}^3) = \int_{\mathbb{R}^3} |\psi(\mathbf{x}, t)|^2 \, d\mathbf{x} = 1, \tag{3.8a}$$

and therefore if we set $\mathring{D} = \mathbb{R}^4$ in (3.7) we get a divergent expression. Thus, since the Schrödinger law of time-evolution expressed by the unitary operator (2.15) is meant to insure conservation of probability over all of space, this feature in turn prohibits a notion of probability over all of spacetime.

Exactly the same situation is encountered in the relativistic context. It has nevertheless been suggested by Stueckelberg (1941), and more recently by others (Horwitz and Piron, 1973; Reuse, 1979, 1980), that in order to solve the relativistic particle localizability problem we should introduce wave functions ψ that are square integrable over all Minkowski spacetime. It has to be realized, however, that in addition to having to express 'evolution' in terms of an unphysical 'evolution parameter' τ which itself does not represent time, such an approach also imposes a totally unphysical behavior on the system: if $|\psi|^2$ is integrable over all spacetime, then the probability of observing the particle in a spacetime slice or in a spacetime region converges to zero as that slice or region is translated in a timelike direction corresponding to $t \to \pm\infty$. Thus, in such a framework, an otherwise isolated particle would exhibit a stochastic on–off existence as time flows by, and would eventually fade away in the distant past or future. Such behavior is totally at odds with intuitive notions as well as with the experimental observations, which show that particles are created or annihilated only as a result of interaction with antiparticles, and not as a steady, ever-present process. Hence, the introduction of probabilities (3.7) only in relation to domains D which, if infinite in spacelike directions, are finite in timelike directions, appears mandatory to the concept of spacetime localization of quantum particles.

On the other hand, relative probabilities like $P_\psi(\mathring{B}/\mathring{D})$ can be expressed as quotients of expectation values of operators:

$$P_\psi(\mathring{B}/\mathring{D}) = \langle\psi|F^{\mathbf{X}}(\mathring{B})\psi\rangle \,/\, \langle\psi|F^{\mathbf{X}}(\mathring{D})\psi\rangle. \tag{3.8b}$$

To achieve such a formulation, we have to work in the Heisenberg picture, where the state vectors $\psi(\mathbf{x})$ are time-independent, whereas observable quantities, such as $E^{\mathbf{X}}(B)$ appearing in (2.10), are represented by time-dependent operators:

$$E^{\mathbf{X}}(B, t) = \exp\left(\frac{i}{\hbar} H_0 t\right) E^{\mathbf{X}}(B) \exp\left(-\frac{i}{\hbar} H_0 t\right). \tag{3.9}$$

To arrive at (3.8b) we introduce the Bochner integrals (cf. Appendix B)

$$F^{\mathbf{X}}(\mathring{B}) = \int_{-\infty}^{+\infty} E^{\mathbf{X}}(B_t, t)\, dt, \tag{3.10}$$

for Borel sets $\mathring{B}$ in $\mathbb{R}^4$, where B_t is the cross-section of $\mathring{B}$ at time t:

$$B_t = \{\mathbf{x} \,|\, (t, \mathbf{x}) \in \mathring{B}\} \subset \mathbb{R}^3. \tag{3.11}$$

In view of (2.15), i.e. of the Schrödinger equation

$$i\hbar \frac{\partial}{\partial t} \psi(\mathbf{x}, t) = -\frac{\hbar^2}{2m} \Delta\psi(\mathbf{x}, t), \tag{3.12}$$

and of (2.10), we obtain upon setting in general $\psi(\mathbf{x}, 0) = \psi(\mathbf{x})$,

$$\int_{\mathbb{R}^3} |(E^{\mathbf{X}}(B)\psi)(\mathbf{x}, t)|^2 \, d\mathbf{x} = \int_{\mathbb{R}^3} |(E^{\mathbf{X}}(B, t)\psi)(\mathbf{x})|^2 \, d\mathbf{x}, \tag{3.13}$$

and consequently (3.8b) immediately follows.

We note that whereas $E^{\mathbf{X}}(B)$ and $E^{\mathbf{X}}(B, t)$ are projectors, that is not the case with $F^{\mathbf{X}}(\mathring{B})$. Indeed,

$$F^{\mathbf{X}}(\mathring{B}) = (F^{\mathbf{X}}(\mathring{B}))^* \geqslant 0, \tag{3.14}$$

but $F^{\mathbf{X}}(\mathring{B})$ is not necessarily idempotent, i.e. by (3.9)–(3.11) we can have:

$$(F^{\mathbf{X}}(\mathring{B}))^2 \neq F^{\mathbf{X}}(\mathring{B}). \tag{3.15}$$

Hence $F^{\mathbf{X}}(\mathring{B})$ constitutes a POV measure which is not a PV measure – and in fact is not even normalized over any spacetime domain $\mathring{D}$. This provides an indirect indication that no time operator T can be introduced in nonrelativistic quantum mechanics. As a matter of fact, such a time operator would be expected to satisfy, by analogy with (2.3),

$$[T, H_0] = i\hbar. \tag{3.16}$$

On account of Von Neumann's theorem, which classifies all representations of the CCRs (based on the Weyl form) for finitely many degrees of freedom, this would imply that H_0 is not bounded from below, which by (2.16) is false. Hence, as opposed to the case of relativistic quantum mechanics considered in the next chapter (see also Section 4.6), nonrelativistic quantum mechanics at best allows for the existence of time superoperators (Misra *et al.*, 1979), or for nonself-adjoint choices for T (Recami, 1977).

The mathematical reason why (3.15) impedes the association of a time operator T with the POV measure $F^{\mathbf{X}}(\mathring{B})$ is a general one: whereas to any PV measure $E(B)$ in $\mathbb{R}^n$ we can associate uniquely (cf. Appendix B) a family $\{A_1, \ldots, A_n\}$ of commuting self-adjoint operators whose joint spectral measure is then $E(B)$, that statement is no longer true of POV measures which are not PV measures. In the latter case operators representing generalized observables in the mean stochastic sense can still be associated (Davies and Lewis, 1969; Neumann, 1972; Ali and Emch, 1974; Ali and Prugovečki, 1977; Schroeck, 1981) with a POV measure, but the association is not unique and therefore requires physical justification based on the interpretation of each POV measure under consideration; in any event, the resulting mean stochastic observables will certainly not relate to measurements of sharp values, as will be seen in the specific instances considered in Sections 1.8 and 1.10.

On the other hand, POV measures can be used to generalize in a straightforward manner the concept of system of imprimitivity defined in the preceding section. Thus, in complete analogy with (2.21), we shall say that a POV measure $F(B)$ over $(\mathscr{X}, \mathscr{A})$ is a *system of covariance* for a unitary representation $\mathscr{U}(g)$ of a group G of homeomorphisms on the topological space $\mathscr{X}$ if

$$\mathscr{U}(g)F(B)\mathscr{U}^{-1}(g) = F(gB), \quad g \in G. \tag{3.17}$$

If we now reinterpret the representation (2.6) of the Galilei group $\mathscr{G}$ in the Heisenberg picture, where all observables are represented by time-dependent operators $A(t)$, by setting

$$\check{U}(b, \mathbf{a}, \mathbf{v}, R) : \psi(\mathbf{x}) \mapsto \psi'(\mathbf{x})$$
$$= \exp\left\{\frac{i}{\hbar}\left[\frac{m\mathbf{v}^2}{2}\, b + m\mathbf{v}\cdot(\mathbf{x}-\mathbf{a})\right]\right\} \psi(R^{-1}[\mathbf{x}-\mathbf{a}+b\mathbf{v}]), \tag{3.18a}$$

$$\check{U}(b, \mathbf{a}, \mathbf{v}, R) : A(t) \mapsto A'(t) = A(t-b), \tag{3.18b}$$

then we immediately see that the POV measure $F^{\mathbf{X}}(\mathring{B})$ in (3.10) constitutes a system of covariance for this 'representation':

$$\check{U}(b, \mathbf{a}, \mathbf{v}, R)\, F^{\mathbf{X}}(\mathring{B})\, \check{U}^{-1}(b, \mathbf{a}, \mathbf{v}, R) = F^{\mathbf{X}}(g\mathring{B}), \tag{3.19a}$$

$$g\mathring{B} = \{(t + b, R\mathbf{x} + \mathbf{v}t + \mathbf{a}) \mid (t, \mathbf{x}) \in \mathring{B}\}. \tag{3.19b}$$

Thus, whereas the PV measure $E^{\mathbf{X}}(B)$ in (2.10), that pertains to sharp localizability in the absolute space of nonrelativistic mechanics, determined a system of imprimitivity which, by (2.18), was related only to the isochronous Galilei group $\mathscr{G}'$, the POV measure $F^{\mathbf{X}}(\mathring{B})$, that pertains to sharp measurements in Newtonian spacetime, determines a system of covariance which, by (3.19), is related to the Galilei group $\mathscr{G}$ in its entirety. Hence the introduction of POV measures appears to be mandatory already at the nonrelativistic level for the formulation of Galilei covariance properties in full. And understanding Galilei covariance, with its relationship to nonrelativistic localizability in Newtonian spacetime, paves the way to understanding Poincaré covariance, with its analogous relationship to relativistic localizability in Minkowskian spacetime.

1.4. Phase-Space Representations of the Galilei Group

Mathematically, the localizability problem that we face in relativistic quantum mechanics is: given the fact that, by Hegerfeldt's (1974) theorem, there are no PV measures which correspond to sharply localized quantum particles and at the same time are consistent with relativistic covariance and causality, construct instead relativistically covariant POV measures corresponding to the kind of stochastic localizability discussed in the preceding section. In addition to relativistic covariance, such a construction has also to satisfy other consistency conditions imposed by physical circumstances and by the multitude of already existing and partly successful models in conventional relativistic quantum mechanics. Among all those criteria, the most crucial is the one which relativity theory itself had to face when it was originally formulated by Einstein (1905): at velocities $v \ll c$, the nonrelativistic theory has to be recovered as a first-order approximation in v/c. Hence, corresponding POV measures that constitute a system of covariance obeying (3.17) for the Galilei group $\mathscr{G}$ have to be found, and these nonrelativistic systems of covariance have to be in turn consistent with all the central postulates and results of conventional nonrelativistic quantum mechanics. As we shall see in Chapter 3, as a byproduct of successfully completing this search, new results, hitherto underived by conventional methods, will be arrived at in the context of statistical mechanics.

The method of constructing Galilei systems of covariance corresponding to nonsharp stochastic localizability was originally developed (Prugovečki, 1976a, b; Ali and Prugovečki, 1977a) in the configuration representation,[9] where preliminary measurement–theoretical analyses (Prugovečki, 1967, 1973, 1974, 1975; Ali and Emch, 1974; Ali and Doebner, 1976) indicated that the conventional interpretation of $\psi(\mathbf{x})$ in nonrelativistic quantum mechanics could be extrapolated to stochastic configuration space. However, it eventually became clear (Ali and Prugovečki, 1977b; Prugovečki, 1978f) that such a construction could be carried out by a method independent of any of the conventional representations of nonrelativistic quantum mechanics, and that in fact these new systems of covariance are at least as fundamental as the familiar configuration or momentum systems of imprimitivity (2.10) and (2.14), respectively.

Indeed, as is the case with the latter, the Galilei systems of covariance for nonsharp localizability are based on realizations of the CCRs (2.3), such as

$$Q^j = q^j + i\hbar \frac{\partial}{\partial p^j}, \qquad P^j = -i\hbar \frac{\partial}{\partial q^j}. \tag{4.1}$$

In (4.1), q^j and p^j are to be considered as independent variables, which can be thought of formally as labels in the six-dimensional classical phase space $\Gamma = \mathbb{R}^6$. As a consequence, Q^j and P^j are self-adjoint operators in the Hilbert space $L^2(\Gamma)$ of square-integrable functions $\psi(\mathbf{q}, \mathbf{p})$, with inner product

$$\langle \psi_1 | \psi_2 \rangle_\Gamma = \int_\Gamma \psi_1^*(\mathbf{q}, \mathbf{p})\, \psi_2(\mathbf{q}, \mathbf{p})\, d\mathbf{q}\, d\mathbf{p}. \tag{4.2}$$

Furthermore, due to the presence of q^j in (4.1), Q^j and P^j do not commute, but rather satisfy the CCRs:

$$[Q^i, P^j] = i\hbar\delta^{ij}, \qquad [Q^i, Q^j] = [P^i, P^j] = 0. \tag{4.3}$$

In Section 1.7 it will be seen that (4.1) is a special case of the more general form

$$Q^j = f^j(\mathbf{q}, \mathbf{p}) + i\hbar \frac{\partial}{\partial p^j}, \qquad P^j = g^j(\mathbf{q}, \mathbf{p}) - i\hbar \frac{\partial}{\partial q^j} \tag{4.4}$$

that can be arrived at by gauge transformations of the first kind. In turn, in Section 2.7 (4.4) will emerge as a very special case of a general realization of (relativistic) CCRs in terms of covariant derivatives, with f^j and g^j being related to affine connections (cf. Sections 4.5 and 4.6).

For the time being, however, the realization (4.1) turns out to be very well suited to the task of relating the new formalism on $L^2(\Gamma)$ to the conventional formalism on $L^2(\mathbb{R}^3)$. Indeed, its Q^j and P^j operators are the infinitesimal generators for velocity boosts and space translations, respectively, of the following counterpart of $\hat{U}(g)$ in (2.6):

$$U(b, \mathbf{a}, \mathbf{v}, R) : \psi(\mathbf{q}, \mathbf{p}; t) \longmapsto \psi'(\mathbf{q}, \mathbf{p}; t)$$

$$= \exp\left\{\frac{i}{\hbar}\left[-\frac{m\mathbf{v}^2}{2}(t-b) + m\mathbf{v}\cdot(\mathbf{q}-\mathbf{a})\right]\right\} \times$$

$$\times\, \psi(R^{-1}(\mathbf{q} - \mathbf{v}(t-b) - \mathbf{a}), R^{-1}(\mathbf{p} - m\mathbf{v}); t - b). \tag{4.5a}$$

For time translations, we get, in terms of **P** in (4.1), the following counterpart of (2.15):

$$U(-t, \mathbf{0}, \mathbf{0}, I) = \exp\left(-\frac{i}{\hbar}\frac{\mathbf{P}^2}{2m}\, t\right). \tag{4.5b}$$

In view of (2.5), it is immediately obvious that $U(b, \mathbf{a}, \mathbf{v}, R)$ preserves the inner product (4.2), and that in fact it is a unitary operator on $L^2(\Gamma)$ for each $g = (0, \mathbf{a}, \mathbf{v}, R) \in \mathscr{G}$. Furthermore, (2.7) remains true when $U(g)$ in (4.5a) replaces $\hat{U}(g)$ in (2.6), so that we are dealing with a unitary ray representation of $\mathscr{G}$. However, in contradistinction to $\hat{U}$, this new representation is highly reducible.

To find its irreducible subrepresentatives for spin zero, and at the same time confirm the physical significance of those representations by relating them to the conventional configuration representation $\hat{U}(g)$ in (2.6), we shall construct unitary transformations W_ξ that map the space $L^2(\mathbb{R}^3)$ in which $\hat{U}(g)$ acts onto subspaces $W_\xi L^2(\mathbb{R}^3)$ of $L^2(\Gamma)$ that are left invariant by $U(g)$ in (4.5). Thus, by analogy with (2.11a), let us examine mappings W_ξ which act on $\psi \in L^2(\mathbb{R}^3)$ as follows:

$$W_\xi : \psi(\mathbf{x}) \longmapsto \psi(\mathbf{q}, \mathbf{p}) = \int_{\mathbb{R}^3} \xi^*_{\mathbf{q},\mathbf{p}}(\mathbf{x})\, \psi(\mathbf{x})\, d\mathbf{x}, \tag{4.6a}$$

$$\xi_{\mathbf{q},\mathbf{p}}(\mathbf{x}) = \left(\hat{U}\left(0, \mathbf{q}, \frac{\mathbf{p}}{m}, I\right) \xi\right)(\mathbf{x}) = \exp\left[\frac{i}{\hbar}\, \mathbf{p} \cdot (\mathbf{x} - \mathbf{q})\right] \xi(\mathbf{x} - \mathbf{q}). \tag{4.6b}$$

The function $\xi(\mathbf{x})$ can be regarded as a parametric quantity which, when given, completely determines the transformation W_ξ. Its basic properties follow from the basic requirement that W_ξ should be an isometric mapping of $L^2(\mathbb{R}^3)$ into $L^2(\Gamma)$:

$$\langle W_\xi \psi_1 \mid W_\xi \psi_2 \rangle_\Gamma = \langle \psi_1 \mid \psi_2 \rangle. \tag{4.7}$$

Replacing in (4.2) ψ_i, $i = 1, 2$, by $W_\xi \psi_i$ given in (4.6), and using the unitarity property of Fourier–Plancherel transforms, we get

$$\langle W_\xi \psi \mid W_\xi \psi \rangle_\Gamma = (2\pi\hbar)^3 \int_{\mathbb{R}^3} d\mathbf{x}\, |\xi(\mathbf{x})|^2 \int_{\mathbb{R}^3} d\mathbf{q}\, |\psi(\mathbf{x} + \mathbf{q})|^2. \tag{4.8}$$

Hence, we conclude that (4.7) is true if $\xi(\mathbf{x})$ belongs to $L^2(\mathbb{R}^3)$ and

$$\|\xi\| = \langle \xi | \xi \rangle^{1/2} = (2\pi\hbar)^{-3/2}. \tag{4.9}$$

Since for such $\xi(\mathbf{x})$ the transformation W_ξ is isometric, its range $W_\xi L^2(\mathbb{R}^3)$ is a closed subspace of $L^2(\Gamma)$, and on this range W_ξ has an inverse W_ξ^{-1} (cf. Appendix B). Upon expressing in (4.7) $W_\xi \psi_2$ in terms of (4.6a) and then using Fubini's theorem (Appendix A) to reverse orders of integration, we arrive at the conclusion that for $\psi(\mathbf{q},\mathbf{p})$ in $W_\xi L^2(\mathbb{R}^3)$

$$W_\xi^{-1}: \psi(\mathbf{q},\mathbf{p}) \longmapsto \psi(\mathbf{x}) = \int_\Gamma \xi_{\mathbf{q},\mathbf{p}}(\mathbf{x})\, \psi(\mathbf{q},\mathbf{p})\, d\mathbf{q}\, d\mathbf{p}. \tag{4.10}$$

Thus, for each $\xi(\mathbf{x})$ satisfying (4.9), W_ξ as well as W_ξ^{-1} are unitary integral transforms. Using (4.6a), we immediately get that (4.7) is equivalent to

$$\int_\Gamma |\xi_{\mathbf{q},\mathbf{p}}\rangle\, d\mathbf{q}\, d\mathbf{p}\, \langle \xi_{\mathbf{q},\mathbf{p}}| = \mathbb{1}, \tag{4.11}$$

where $\mathbb{1}$ stands for the identity operator on $L^2(\mathbb{R}^3)$. Furthermore, since by (4.6b) $\xi_{\mathbf{q},\mathbf{p}}$ is norm-continuous in $(\mathbf{q},\mathbf{p}) \in \Gamma$, the family $\{\xi_{\mathbf{q},\mathbf{p}}\}$ constitutes a continuous resolution of the identity in $L^2(\mathbb{R}^3)$ (Davies, 1976). Since each $\xi_{\mathbf{q},\mathbf{p}}$ is obtained from $\xi(\mathbf{x})$ by applying $\hat{U}(0, \mathbf{q}, \mathbf{p}/m, I)$ to it, the function $\xi(\mathbf{x})$ is called a *resolution generator*. In Section 1.7 we shall see that each resolution generator is uniquely determined by its irreducible representation, and that under certain conditions it physically represents the proper wave function of an extended test particle placed at the origin of an inertial frame.

To ascertain some of these conditions, we first have to establish when $W_\xi L^2(\mathbb{R}^3)$ is a subspace of $L^2(\Gamma)$ left invariant by $U(g)$ in (4.5). If we set in (3.18) $\mathbf{x}' = g^{-1}\mathbf{x}$, $g = (b, \mathbf{a}, \mathbf{v}, R)$, we observe that by (4.6) at $t = 0$,

$$(W_\xi \hat{U}(g)\psi)(\mathbf{q},\mathbf{p}) = \int \xi^*_{\mathbf{q},\mathbf{p}}(R\mathbf{x}' - b\mathbf{v} + \mathbf{a}) \exp\left\{\frac{i}{\hbar}\left[\frac{m\mathbf{v}^2}{2}\, b + m\mathbf{v}\cdot(R\mathbf{x}' - b\mathbf{v})\right]\right\} \psi(\mathbf{x}')\, d\mathbf{x}', \tag{4.12}$$

$$\xi_{\mathbf{q},\mathbf{p}}(R\mathbf{x}' - b\mathbf{v} + \mathbf{a}) = \exp\left\{\frac{i}{\hbar}\, \mathbf{p}\cdot(R\mathbf{x}' - b\mathbf{v} + \mathbf{a} - \mathbf{q})\right\} \xi(R\mathbf{x}' - \mathbf{q} - b\mathbf{v} + \mathbf{a}), \tag{4.13}$$

since the Jacobian for the change of variables $\mathbf{x} \longmapsto \mathbf{x}'$ equals 1. On the other hand, by (4.5) and (4.6), at $t = 0$

$$(U(g)W_\xi\psi)(\mathbf{q},\mathbf{p}) = \exp\left\{\frac{i}{\hbar}\left[\frac{m\mathbf{v}^2}{2}\, \mathbf{b} + m\mathbf{v}\cdot(\mathbf{q} - \mathbf{a})\right]\right\} \times$$

$$\times \int_{\mathbb{R}^3} \xi^*_{g^{-1}(\mathbf{q},\mathbf{p})}(\mathbf{x}')\, \psi(\mathbf{x}')\, d\mathbf{x}', \tag{4.14}$$

$$\xi_{g^{-1}(\mathbf{q},\mathbf{p})}(\mathbf{x}') = \exp\left[\frac{i}{\hbar}\, (\mathbf{p} - m\mathbf{v})\cdot(R\mathbf{x}' - \mathbf{q} - b\mathbf{v} + \mathbf{a})\right] \times$$

$$\times\, \xi(\mathbf{x}' - R^{-1}(\mathbf{q} + b\mathbf{v} - \mathbf{a})). \tag{4.15}$$

Upon comparing (4.13) with (4.15) we immediately see that (4.12) and (4.14) are equal if and only if

$$\xi(\mathbf{x}) = \xi(R\mathbf{x}), \quad R \in \mathrm{SO}(3), \tag{4.16}$$

i.e. iff $\xi(\mathbf{x})$ is rotationally invariant and therefore a function of $|\mathbf{x}|$ only. When that is the case the above argument establishes not only that $U(g)$ leaves the subspace $W_\xi L^2(\mathbb{R}^3) \subset L^2(\Gamma)$ invariant, but also that the restriction of $U(g)$ to this subspace coincides with $W_\xi \hat{U}(g) W_\xi^{-1}$, and therefore it is an irreducible representation of the Galilei group $\mathcal{G}$.

In view of Von Newmann's theorem on the CCRs, this result is of interest in itself. By that theorem $L^2(\Gamma)$ is decomposable into a direct sum of irreducible subspaces, and since the Hilbert space $L^2(\Gamma)$ is separable, such a direct sum contains a countable number of elements. On the other hand, by the above construction, to each rotationally invariant $\xi(\mathbf{x})$ of norm $(2\pi\hbar)^{-3/2}$ we can assign a distinct irreducible subrepresentation of $U(g)$, and there is an uncountable number of such ξ. This indicates that the aforementioned decomposition is highly nonunique. In fact, all such direct sum decompositions have been derived (Ali and Prugovečki, 1983), and in addition to nonuniqueness they reveal that $U(g)$ contains irreducible representations for all integer spin values.

We can also work on $L^2(\Gamma)$ in the Heisenberg picture, in which states and observables are represented, respectively, by $\psi(\mathbf{q}, \mathbf{p}) = \psi(\mathbf{q}, \mathbf{p}; 0)$ and

$$A(t) = \exp\left(\frac{i}{\hbar}\frac{\mathbf{P}^2}{2m}t\right) A \exp\left(-\frac{i}{\hbar}\frac{\mathbf{P}^2}{2m}t\right), \tag{4.17}$$

where $\mathbf{P}$ has the components in (4.1). Then

$$\begin{aligned}\bar{U}(b, \mathbf{a}, \mathbf{v}, R)\colon \psi(\mathbf{q}, \mathbf{p}) &\mapsto \psi'(\mathbf{q}, \mathbf{p}) \\ &\doteq \exp\left\{\frac{i}{\hbar}\left[\frac{m\mathbf{v}^2}{2} b + m\mathbf{v}\cdot(\mathbf{q}-\mathbf{a})\right]\right\} \psi(R^{-1}[\mathbf{q}+b\mathbf{v}-\mathbf{a}], R^{-1}[\mathbf{p}-m\mathbf{v}]),\end{aligned} \tag{4.18a}$$

$$\bar{U}(b, \mathbf{a}, \mathbf{v}, R)\colon A(t) \mapsto A'(t) = A(t-b), \tag{4.18b}$$

is the $L^2(\Gamma)$ counterpart of $\check{U}(g)$ in (3.18). In this chapter we shall often work at $t = 0$, in which case for $g \in \mathcal{G}'$ the effects of U and $\bar{U}$ on the state vector are indistinguishable, and therefore we shall not differentiate between them notationwise.

1.5. Localizability in Nonrelativistic Stochastic Phase Spaces

In the preceding sections, after having started with the reducible representation $U(g)$ of the Galilei group $\mathcal{G}$ (where $U(g)$ acts on time-dependent elements of $L^2(\Gamma)$ in accordance with (4.5)), we have constructed subspaces $W_\xi L^2(\mathbb{R}^3)$ which reduce

$U(g)$. These irreducible subspaces were the ranges of isometric transformations W_ξ of $L^2(\mathbb{R}^3)$ into $L^2(\Gamma)$, and were determined by resolution generators $\xi(\mathbf{x})$ satisfying (4.9) and (4.16). Each transformation W_ξ maps the configuration space wave function $\psi(\mathbf{x}, t)$ of a spinless particle into an element $\psi(\mathbf{q}, \mathbf{p}; t)$ of $W_\xi L^2(\mathbb{R}^3)$ in accordance with (4.6).

The first indication as to the physical meaning of such a $\psi(\mathbf{q}, \mathbf{p})$ (at any instant t) is obtained from the *marginality properties*

$$\int_{\mathbb{R}^3} |\psi(\mathbf{q}, \mathbf{p})|^2 \, d\mathbf{p} = \int_{\mathbb{R}^3} \chi^\xi_{\mathbf{q}}(\mathbf{x}) \, |\psi(\mathbf{x})|^2 \, d\mathbf{x}, \tag{5.1a}$$

$$\int_{\mathbb{R}^3} |\psi(\mathbf{q}, \mathbf{p})|^2 \, d\mathbf{q} = \int_{\mathbb{R}^3} \bar{\chi}^\xi_{\mathbf{p}}(\mathbf{k}) \, |\tilde{\psi}(\mathbf{k})|^2 \, d\mathbf{k}, \tag{5.1b}$$

$$\chi^\xi_{\mathbf{q}}(\mathbf{x}) = (2\pi\hbar)^3 \, |\xi(\mathbf{x} - \mathbf{q})|^2, \qquad \bar{\chi}^\xi_{\mathbf{p}}(\mathbf{k}) = (2\pi\hbar)^3 \, |\tilde{\xi}(\mathbf{k} - \mathbf{p})|^2, \tag{5.2}$$

which are easily derived by inserting the expressions (4.6) for $\psi(\mathbf{q}, \mathbf{p})$ into the left-hand sides of (5.1), and then using the standard properties of Fourier–Plancherel transforms. In view of (4.9) the functions $\chi^\xi_{\mathbf{q}}(\mathbf{x})$ and $\bar{\chi}^\xi_{\mathbf{p}}(\mathbf{k})$ in (5.2) are normalized probability distributions on $\mathbb{R}^3$. Hence, they can be interpreted, in accordance with (1.9) and (1.10), as the confidence functions of the respective stochastic points

$$\underset{\sim}{q} = (\mathbf{q}, \chi^\xi_{\mathbf{q}}) \in \underset{\sim}{\mathbb{R}}^3_{\text{conf}}, \qquad \underset{\sim}{p} = (\mathbf{p}, \bar{\chi}^\xi_{\mathbf{p}}) \in \underset{\sim}{\mathbb{R}}^3_{\text{mom}}, \tag{5.3a}$$

in stochastic configuration and momentum spaces, respectively.

In view of these facts, and of the conventional interpretation of $|\psi(\mathbf{x})|^2$ and $|\tilde{\psi}(\mathbf{k})|^2$ as probability densities, the interpretation of the right-hand sides of (5.1) becomes self-evident: if $|\psi(\mathbf{x})|^2$ is the probability density for a point particle to be detected at $\mathbf{x}$ by means of a perfectly accurate position indicator, and if a realistic position-measuring apparatus is used which produces with probability density $\chi^\xi_0(\mathbf{x} - \mathbf{q})$ the reading $\mathbf{q}$ when a particle is present at $\mathbf{x}$, then since these two probability distributions are independent, by elementary probability calculus

$$\int_{\mathbb{R}^3} \chi^\xi_0(\mathbf{x} - \mathbf{q}) \, |\psi(\mathbf{x})|^2 \, d\mathbf{x}, \qquad \chi^\xi_0(\mathbf{x} - \mathbf{q}) = \chi^\xi_{\mathbf{q}}(\mathbf{x}) \tag{5.3b}$$

is the probability that such a position indicator will provide the response $\mathbf{q}$ when a point particle in the state $\psi(\mathbf{x})$ is present. In the *sharp-point limit*

$$\chi^\xi_{\mathbf{q}}(\mathbf{x}) = \chi^\xi_0(\mathbf{x} - \mathbf{q}) \longrightarrow \delta^3(\mathbf{x} - \mathbf{q}) \tag{5.4}$$

of a perfectly accurate position indicator the right-hand side of (5.1a) merges into $|\psi(\mathbf{q})|^2$, i.e. the conventional interpretation is recovered. All these observations also hold true of (5.1b), with the proviso that in this latter instance momentum takes over the role of position. Hence, (5.1a) and (5.1b) are indeed the probability densities on the stochastic configuration space $\underset{\sim}{\mathbb{R}}^3_{\text{conf}}$ and stochastic momentum space $\underset{\sim}{\mathbb{R}}^3_{\text{mom}}$, respectively.

All this suggests that $|\psi(\mathbf{q},\mathbf{p})|^2$ itself possesses a physical interpretation, namely that of a probability density on the *stochastic phase space*

$$\Gamma_\xi = \underset{\sim}{\mathbb{R}}^3_{\text{conf}} \times \underset{\sim}{\mathbb{R}}^3_{\text{mom}} = \{\underset{\sim}{\zeta} = (\mathbf{q}, \chi^\xi_{\mathbf{q}}) \times (\mathbf{p}, \bar{\chi}^\xi_{\mathbf{p}}) \,|\, (\mathbf{q},\mathbf{p}) \in \Gamma\}, \tag{5.5}$$

defined in accordance with (1.7)–(1.12). However, whereas mathematically we could also write

$$\underset{\sim}{\zeta} = (\zeta, \chi^\xi_\zeta), \qquad \chi^\xi_\zeta(\mathbf{x},\mathbf{k}) = \chi^\xi_{\mathbf{q}}(\mathbf{x})\,\bar{\chi}^\xi_{\mathbf{p}}(\mathbf{k}), \tag{5.6}$$

and from the classical point of view $\chi^\xi_\zeta(\mathbf{x},\mathbf{k})$ could be indeed operationally defined as a confidence function for measurements of phase space values $(\mathbf{x},\mathbf{k}) \in \Gamma$, the uncertainty principle forbids such a straightforward operational interpretation in the quantum realm. On the other hand, the very method of constructing the irreducible spaces

$$L^2(\Gamma_\xi) \overset{\text{def}}{=} W_\xi L^2(\mathbb{R}^3) \subset L^2(\Gamma) \tag{5.7}$$

by means of W_ξ in (4.6) suggests an alternative operational interpretation for the normalized wave function

$$\psi(\mathbf{q},\mathbf{p}) = (W_\xi \psi)(\mathbf{q},\mathbf{p}) = \langle \xi_{\mathbf{q},\mathbf{p}} | \psi \rangle \tag{5.8}$$

assigned by W_ξ to the normalized configuration space wave function $\psi(\mathbf{x})$.

Indeed, the operational interpretation of $\xi_{\mathbf{q},\mathbf{p}}$ is self-evident: by (4.6b), $\xi_{\mathbf{q},\mathbf{p}}(\mathbf{x})$ is obtained from the configuration space resolution generator $\xi(\mathbf{x})$ by the operations of translating $\xi(\mathbf{x})$ in space by the amount $\mathbf{q}$ and then boosting it to the 3-velocity $\mathbf{p}/m$. Thus, if we think of $\xi(\mathbf{x})$ as representing the *configuration space proper wave function*[10] of an extended test particle of mass m that is used as a microdetector placed at the origin of the laboratory (inertial) frame, then $\xi_{\mathbf{q},\mathbf{p}}(\mathbf{x})$ is the proper wave function of that test particle after it has been submitted to the kinematical procedures of space translation by $\mathbf{q}$ and velocity boost by $\mathbf{u} = \mathbf{p}/m$. Naively, we can think of $\xi_{\mathbf{q},\mathbf{p}}$ as the state of such a test particle of stochastic position $\underset{\sim}{\mathbf{q}}$ and stochastic momentum $\underset{\sim}{\mathbf{p}}$ given in (5.3). This does not violate the uncertainty principle, since the spreads Δx^j and Δk^j

$$(\Delta x^j)^2 = \int_{\mathbb{R}^3} (x^j - q^j)^2\, \chi^\xi_{\mathbf{q}}(\mathbf{x})\, d\mathbf{x}, \qquad (\Delta k^j)^2 = \int_{\mathbb{R}^3} (k^j - p^j)^2\, \bar{\chi}^\xi_{\mathbf{p}}(\mathbf{k})\, d\mathbf{k}, \tag{5.9}$$

satisfy that principle on account of (4.9) and (5.2):

$$(\Delta x^j)(\Delta k^j) \geqslant \hbar/2, \qquad j = 1, 2, 3. \tag{5.10}$$

In view of (5.8), or equivalently (4.6a), and by analogy with (2.4a), we interpret

$$P_\psi^\xi(B) = \int_B |(W_\xi \psi)(\mathbf{q}, \mathbf{p})|^2 \, d\mathbf{q} \, d\mathbf{p} \tag{5.11}$$

as the probability of observing stochastic values $(\mathbf{q}, \mathbf{p})$ within **B** when a simultaneous measurement (cf. Section 1.12) of stochastic position $\underset{\sim}{\mathbf{q}}$ and stochastic momentum $\underset{\sim}{\mathbf{p}}$ is performed on a particle with state vector $\psi(\mathbf{x})$. The invokes analogy becomes even more striking if we recall that in the momentum representation, where the wave function is

$$\tilde{\psi}(\mathbf{k}) = (U_F\psi)(\mathbf{k}) = (2\pi\hbar)^{-3/2} \int_{\mathbb{R}^3} \exp\left(-\frac{i}{\hbar}\mathbf{k}\cdot\mathbf{x}\right) \psi(\mathbf{x}) \, d\mathbf{x}, \tag{5.12}$$

we work with the following realization of the CCRs,

$$\tilde{X}^j = U_F X^j U_F^{-1}, \qquad \tilde{P}^j = U_F P^j U_F^{-1}, \tag{5.13a}$$

$$(\tilde{X}^j \tilde{\psi})(\mathbf{k}) = i\hbar \frac{\partial}{\partial k^j} \tilde{\psi}(\mathbf{k}), \qquad (\tilde{P}^j \tilde{\psi})(\mathbf{k}) = k^j \tilde{\psi}(\mathbf{k}), \tag{5.13b}$$

and the probability for observing sharp momentum values $\mathbf{k} \in B \subset \mathbb{R}^3$ equals

$$P_\psi^{\mathbf{P}}(B) = \int_B |(U_F\psi)(\mathbf{k})|^2 \, d\mathbf{k}. \tag{5.14}$$

Comparing (5.12) and (4.6a) as well as (5.14) and (5.11), we see that, in a sense, the transition from the configuration to the (sharp) momentum representation can be regarded as an extremal case of transitions to stochastic phase space representations. This conclusion will receive further support from considerations in the next section.

Despite all these clear indications as to the meaning of $P_\psi^\xi(B)$ in (5.11), this interpretation cannot be deduced from the conventional one (as in case of (5.1)), but merely inferred from it. Hence, additional corroborative evidence is required to verify both the internal consistency of this extrapolation of the conventional interpretation, as well as its ability to provide new physical evidence that agrees with experimental data.

The second type of supporting evidence will be gradually accumulated as we describe in later chapters actual models yielding experimental predictions. On the other hand, the first type of evidence is more readily derived, and it will emerge from the considerations in this chapter, starting with those in the next section.

1.6. Stochastic Phase-Space Systems of Covariance

The probability measure $P_\psi^\xi(B)$ in (5.11) can be written as an expectation value

$$P_\psi^\xi(B) = \langle \psi | \hat{\mathbb{P}}_\xi(B) \psi \rangle \tag{6.1}$$

if we take note of (5.8) and introduce the Bochner integral

$$\hat{\mathbb{P}}_\xi(B) = \int_B |\xi_{\mathbf{q},\mathbf{p}}\rangle \, d\mathbf{q} \, d\mathbf{p} \, \langle \xi_{\mathbf{q},\mathbf{p}}|. \tag{6.2}$$

It is evident that $\hat{\mathbb{P}}_\xi(B)$ constitutes a POV measure in $L^2(\mathbb{R}^3)$ over the Borel sets of Γ, and by (4.11) $\hat{\mathbb{P}}_\xi(B)$ is normalized on Γ. From (4.12) to (4.15) we obtain that

$$\hat{U}(g)\hat{\mathbb{P}}_\xi(B)\hat{U}^*(g) = \int_B |\xi_{g(\mathbf{q},\mathbf{p})}\rangle \, d\mathbf{q} \, d\mathbf{p} \, \langle \xi_{g(\mathbf{q},\mathbf{p})}|, \tag{6.3}$$

so that $\hat{\mathbb{P}}_\xi(B)$ constitutes a system of covariance for the representation $\hat{U}(g)$ in (2.6) restricted to the isochronous Galilei group $\mathscr{G}'$:

$$\hat{U}(g)\hat{\mathbb{P}}_\xi(B)\hat{U}^{-1}(g) = \hat{\mathbb{P}}_\xi(gB), \quad g \in \mathscr{G}'. \tag{6.4}$$

In order to introduce the stochastic phase-space systems of covariance in a generally familiar context, we have expressed these systems of covariance on the spectral representation space $L^2(\mathbb{R}^3)$ for the position observables $\{X^j\}$, i.e. in the configuration representation. However, we could have carried out our considerations as well in the momentum representation, or for that matter directly in the space $L^2(\Gamma_\xi)$ defined by (5.7), in which $U(g)$ of (4.5) has been shown to induce an irreducible representation of the Galilei group $\mathscr{G}$. In fact, since W_ξ is a unitary transformation of $L^2(\mathbb{R}^3)$ onto $L^2(\Gamma_\xi)$, $W_\xi \hat{\mathbb{P}}_\xi(B) W_\xi^{-1}$ is a POV measure in $L^2(\Gamma_\xi)$. Hence by (6.2) and (4.7)

$$\mathbb{P}_\xi(B) = \int_B |W_\xi \xi_{\mathbf{q},\mathbf{p}}\rangle \, d\mathbf{q} \, d\mathbf{p} \, \langle W_\xi \xi_{\mathbf{q},\mathbf{p}}| \tag{6.5}$$

is also a POV measure when $\mathbb{P}_\xi(B)$ are considered to be operators on $L^2(\Gamma_\xi)$. Moreover, since the restriction of $U(g)$ to $L^2(\Gamma_\xi)$ was shown in Section 1.4 to coincide with $W_\xi \hat{U}(g) W_\xi^{-1}$, we deduce from (6.4) that

$$U(g)\,\mathbb{P}_\xi(B)U^{-1}(g) = \mathbb{P}_\xi(gB), \quad g \in \mathscr{G}'. \tag{6.6}$$

Hence, we are justified in calling $\mathbb{P}_\xi(B)$ and $U(g)$ an *isochronous Galilei system of covariance on the stochastic phase space* Γ_ξ.

The vectors $W_\xi \xi_{\mathbf{q},\mathbf{p}}$ that determine $\mathbb{P}_\xi(B)$ give rise by (4.11) to a continuous resolution of the identity on $L^2(\Gamma_\xi)$. They also possess additional remarkable properties. By (4.6a) or (5.8)

$$(W_\xi \xi_{\mathbf{q},\mathbf{p}})(\mathbf{q}',\mathbf{p}') = \int_{\mathbb{R}^3} \xi^*_{\mathbf{q}',\mathbf{p}'}(\mathbf{x})\,\xi_{\mathbf{q},\mathbf{p}}(\mathbf{x})\,d\mathbf{x}, \tag{6.7}$$

and due to the isometry of W_ξ we can write:

$$(W_\xi \xi_{\mathbf{q},\mathbf{p}})(\mathbf{q}',\mathbf{p}') = \langle W_\xi \xi_{\mathbf{q}',\mathbf{p}'} | W_\xi \xi_{\mathbf{q},\mathbf{p}}\rangle_\Gamma. \tag{6.8}$$

In terms of the function K_ξ defined by

$$K_\xi(\mathbf{q}', \mathbf{p}'; \mathbf{q}, \mathbf{p}) = (W_\xi \xi_{\mathbf{q},\mathbf{p}})(\mathbf{q}', \mathbf{p}') = \langle \xi_{\mathbf{q}',\mathbf{p}'} | \xi_{\mathbf{q},\mathbf{p}} \rangle, \tag{6.9}$$

and in view of (4.2), the equality (6.8) assumes the form

$$K_\xi(\mathbf{q}', \mathbf{p}'; \mathbf{q}, \mathbf{p}) = \int_\Gamma K_\xi(\mathbf{q}', \mathbf{p}'; \mathbf{q}'', \mathbf{p}'') K_\xi(\mathbf{q}'', \mathbf{p}''; \mathbf{q}, \mathbf{p})\, d\mathbf{q}''\, d\mathbf{p}'', \tag{6.10}$$

which characterizes K_ξ in (6.9) as a *reproducing kernel* (Aronsajn, 1950). In fact, we can indeed consider (6.9) to be the kernel of an integral operator $\mathbb{P}_\xi$,

$$(\mathbb{P}_\xi \psi)(\mathbf{q}, \mathbf{p}) = \int_\Gamma K_\xi(\mathbf{q}, \mathbf{p}; \mathbf{q}', \mathbf{p}')\, \psi(\mathbf{q}', \mathbf{p}')\, d\mathbf{q}'\, d\mathbf{p}', \tag{6.11}$$

acting[11] on $\psi \in L^2(\Gamma)$. Since

$$K_\xi^*(\mathbf{q}', \mathbf{p}'; \mathbf{q}, \mathbf{p}) = \langle \xi_{\mathbf{q}',\mathbf{p}'} | \xi_{\mathbf{q},\mathbf{p}} \rangle^* = K_\xi(\mathbf{q}, \mathbf{p}; \mathbf{q}', \mathbf{p}'), \tag{6.12}$$

$\mathbb{P}_\xi$ is self-adjoint, and by (6.10) it is also idempotent, i.e. $\mathbb{P}_\xi$ is a projector acting in $L^2(\Gamma)$. Since by (4.11) and (6.9)

$$(W_\xi \psi)(\mathbf{q}, \mathbf{p}) = \langle \xi_{\mathbf{q},\mathbf{p}} | \psi \rangle = \int_{\mathbb{R}^3} K_\xi(\mathbf{q}, \mathbf{p}; \mathbf{q}', \mathbf{p}') \langle \xi_{\mathbf{q}',\mathbf{p}'} | \psi \rangle\, d\mathbf{q}'\, d\mathbf{p}', \tag{6.13}$$

the subspace $\mathbb{P}_\xi L^2(\Gamma)$ of $L^2(\Gamma)$ onto which $\mathbb{P}_\xi$ projects includes $W_\xi L^2(\mathbb{R}^3)$. Conversely, if some $\psi' \in L^2(\Gamma)$ is orthogonal to $W_\xi L^2(\mathbb{R}^3)$, then $\psi' \perp W_\xi \xi_{\mathbf{q},\mathbf{p}}$ for all $(\mathbf{q}, \mathbf{p}) \in \Gamma$, and therefore by (6.9) and (6.12)

$$\int_\Gamma K_\xi^*(\mathbf{q}', \mathbf{p}'; \mathbf{q}, \mathbf{p})\, \psi'(\mathbf{q}', \mathbf{p}')\, d\mathbf{q}'\, d\mathbf{p}' = 0. \tag{6.14}$$

Hence, by (6.12) $\psi' \perp \mathbb{P}_\xi L^2(\Gamma)$, so that

$$\mathbb{P}_\xi L^2(\Gamma) \equiv W_\xi L^2(\mathbb{R}^3) = L^2(\Gamma_\xi). \tag{6.15}$$

Consequently, by (6.9), (6.11) and (6.12), for any $\psi \in L^2(\Gamma)$

$$(\mathbb{P}_\xi \psi)(\mathbf{q}, \mathbf{p}) = \langle W_\xi \xi_{\mathbf{q},\mathbf{p}} | \psi \rangle_\Gamma. \tag{6.16}$$

The operators $\mathbb{P}_\xi(B)$ in (6.6) were supposed to act in the subspace $L^2(\Gamma_\xi)$ of $L^2(\Gamma)$. However, the Bochner integral in (6.5) is well defined on all $L^2(\Gamma)$. If we allow $\mathbb{P}_\xi(B)$ to act accordingly on $L^2(\Gamma)$ and use (6.16), (6.9) and (5.8), we get

$$(\mathbb{P}_\xi(B)\psi)(\mathbf{q}, \mathbf{p}) = \int_B K_\xi(\mathbf{q}, \mathbf{p}; \mathbf{q}', \mathbf{p}')(\mathbb{P}_\xi \psi)(\mathbf{q}', \mathbf{p}')\, d\mathbf{q}'\, d\mathbf{p}'. \tag{6.17}$$

Therefore, by (6.11) $\mathbb{P}_\xi(\Gamma)$ coincides with $\mathbb{P}_\xi^2$, so that

$$\mathbb{P}_\xi(\Gamma) = \mathbb{P}_\xi^2 = \mathbb{P}_\xi = \mathbb{P}_\xi^*. \tag{6.18}$$

Furthermore, since $U(g)$ leaves $L^2(\Gamma_\xi)$ invariant, it commutes with $\mathbb{P}_\xi$, and therefore (6.6) remains valid as a relation between operators acting on all $L^2(\Gamma)$.

In view of (2.5a, b), $\mathbb{P}_\xi(B_1 \times \mathbb{R}^3)$ and $\mathbb{P}_\xi(\mathbb{R}^3 \times B_2)$ also give rise to systems of covariance,

$$U(g)\,\mathbb{P}_\xi(B_1 \times \mathbb{R}^3)\,U^{-1}(g) = \mathbb{P}_\xi((gB_1) \times \mathbb{R}^3), \quad g \in \mathcal{G}', \tag{6.19}$$

$$U(g)\,\mathbb{P}_\xi(\mathbb{R}^3 \times B_2)\,U^{-1}(g) = \mathbb{P}_\xi(\mathbb{R}^3 \times (gB_2)), \quad g \in \mathcal{G}', \tag{6.20}$$

on the stochastic configuration and momentum spaces, respectively, that occur in (5.3). In fact, according to (5.1) and (6.5), for $\psi \in L^2(\Gamma_\xi)$

$$\langle \psi \,|\, \mathbb{P}_\xi(B_1 \times \mathbb{R}^3)\psi\rangle_\Gamma = \int_{B_1} \mathrm{d}\mathbf{q} \int_{\mathbb{R}^3} \mathrm{d}\mathbf{x}\, \chi_\mathbf{q}^\xi(\mathbf{x})|(W_\xi^{-1}\psi)(\mathbf{x})|^2, \tag{6.21}$$

$$\langle \psi \,|\, \mathbb{P}_\xi(\mathbb{R}^3 \times B_2)\psi\rangle_\Gamma = \int_{B_2} \mathrm{d}\mathbf{p} \int_{\mathbb{R}^3} \mathrm{d}\mathbf{k}\, \overline{\chi}_\mathbf{p}^\xi(\mathbf{k})|(W_\xi^{-1}\psi)^\sim(\mathbf{k})|^2. \tag{6.22}$$

Finally, to obtain systems of covariance that involve the Galilei group $\mathcal{G}$ rather than just its isochronous subgroup $\mathcal{G}'$, we proceed as in (3.9)–(3.11) and set in the Heisenberg picture

$$\mathbb{P}_\xi(B, t) = \exp\left(\frac{i\mathbf{P}^2}{2m\hbar}\, t\right) \mathbb{P}_\xi(B) \exp\left(-\frac{i\mathbf{P}^2}{2m\hbar}\, t\right), \tag{6.23a}$$

$$F_\xi(\mathring{B}) = \int_{-\infty}^{+\infty} \mathbb{P}_\xi(B_t, t)\, \mathrm{d}t, \tag{6.23b}$$

$$B_t = \{(\mathbf{q}, \mathbf{p}) \,|\, (t, \mathbf{q}, \mathbf{p}) \in \mathring{B}\} \subset \Gamma. \tag{6.23c}$$

Obviously, $F_\xi(\mathring{B})$ is a POV measure on $\mathbb{R}^1 \times \Gamma$, and

$$\bar{U}(g)F_\xi(B)\bar{U}^{-1}(g) = F_\xi(gB), \quad g \in \mathcal{G}. \tag{6.24}$$

This result prepares the ground for later developments in the relativistic realm, where nonrelativistic phase space has to be replaced by its relativistic counterpart $\mathcal{M}_m^+$ in (2.3.4), which is homeomorphic to $\mathbb{R}^1 \times \Gamma$.

To nurture an intuitive grasp of the quantities and relations introduced in this section, let us consider the special but very important class of stochastic phase spaces $\Gamma_\xi(l) = \Gamma^{(l)}$, which are based on the *optimal* resolution generators

$$\xi^{(l)}(\mathbf{x}) = (8\pi^3\hbar^2 l^2)^{-3/4} \exp\left(-\frac{\mathbf{x}^2}{4l^2}\right). \tag{6.25}$$

These resolution generators are optimal in the sense that they give rise to confidence functions $\chi_\mathbf{q}^{(l)}$ and $\overline{\chi}_\mathbf{p}^{(l)}$ which display the smallest spread compatible with the uncertainty relations (5.10):

$$\Delta x^j = l, \ \Delta k^j = \hbar/2l. \tag{6.26}$$

In fact, it follows from well-known results on minimum uncertainty wave functions (Klauder and Sudarshan, 1968, p. 109) that, for given l, (6.25) is apart from an irrelevant multiplicative constant the *unique* resolution generator which satisfies (6.26). In Chapter 4, in the context of quantum spacetime, these optimal resolution generators will turn out to be the nonrelativistic approximations of proper wave functions of ground exciton states of Born's quantum metric operator. Alternatively, in the nonrelativistic harmonic oscillator quark model, they represent the ground state of a quark-antiquark pair constituting a meson (Close, 1979). Thus, in a wide variety of settings the resolution generators (6.25) play a distinctive role, and we shall often use them for specific illustrations of general ideas and arguments.

The confidence functions in the optimal case are

$$\chi_\mathbf{q}^{(l)}(\mathbf{x}) = (2\pi l^2)^{-3/2} \exp\left[-\frac{1}{2}\left(\frac{\mathbf{x}-\mathbf{q}}{l}\right)^2\right], \tag{6.27a}$$

$$\overline{\chi}_\mathbf{p}^{(l)}(\mathbf{k}) = \left(\frac{2l^2}{\pi\hbar^2}\right)^{3/2} \exp\left[-2l^2\left(\frac{\mathbf{k}-\mathbf{p}}{\hbar}\right)^2\right], \tag{6.27b}$$

and the r.m.s. radius r_0 associated with the particles with proper wave function[12] $\xi^{(l)}$ is therefore

$$r_0 = \left[\int_{\mathbb{R}^3} \mathbf{x}^2 \chi_\mathbf{0}^{(l)}(\mathbf{x})\, d\mathbf{x}\right]^{1/2} = \sqrt{3}\, l. \tag{6.28}$$

According to (4.6b), at $(\mathbf{q},\mathbf{p}) \in \Gamma$

$$\xi_{\mathbf{q},\mathbf{p}}^{(l)}(\mathbf{x}) = (8\pi^3\hbar^2 l^2)^{-3/4} \exp\left[-\left(\frac{\mathbf{x}-\mathbf{q}}{2l}\right)^2 + \frac{i}{\hbar}\mathbf{p}\cdot(\mathbf{x}-\mathbf{q})\right], \tag{6.29}$$

and therefore in the momentum representation

$$\tilde{\xi}_{\mathbf{q};\mathbf{p}}^{(l)}(\mathbf{k}) = \left(\frac{l^2}{2\pi^3\hbar^4}\right)^{3/4} \exp\left[\left(-\frac{l^2}{\hbar^2}\right)(\mathbf{k}-\mathbf{p})^2 - \frac{i}{\hbar}\mathbf{k}\cdot\mathbf{q}\right]. \tag{6.30}$$

Since these functions are essentially Gaussian, the integration implicit in (6.9) is of the standard type (Geronimus and Tseyllin, 1965):

$$K^{(l)}(\mathbf{q}', \mathbf{p}'; \mathbf{q}, \mathbf{p}) = \int_{\mathbb{R}^3} \xi^{(l)*}_{\mathbf{q}',\mathbf{p}'}(\mathbf{x})\, \xi^{(l)}_{\mathbf{q},\mathbf{p}}(\mathbf{x})\, d\mathbf{x}$$

$$= (2\pi\hbar)^{-3} \exp\left\{ -\frac{1}{2}\left[\left(\frac{\mathbf{q}'-\mathbf{q}}{2l}\right)^2 + l^2 \left(\frac{\mathbf{p}'-\mathbf{p}}{\hbar}\right)^2 - \right.\right.$$

$$\left.\left. -\frac{i}{\hbar}(\mathbf{q}'-\mathbf{q})\cdot(\mathbf{p}'+\mathbf{p}) \right] \right\}. \tag{6.31}$$

The elements of $L^2(\Gamma^{(l)})$ are

$$\psi_l(\mathbf{q}, \mathbf{p}) = (8\pi^3\hbar^2 l^2)^{-3/4} \int_{\mathbb{R}^3} \exp\left[-\left(\frac{\mathbf{x}-\mathbf{q}}{2l}\right)^2 - \right.$$

$$\left. -\frac{i}{\hbar}(\mathbf{x}-\mathbf{q})\cdot\mathbf{p} \right] \psi(\mathbf{x})\, d\mathbf{x}, \tag{6.32}$$

and we see from (6.32) and (6.30) that

$$\lim_{l\to+0} \left(\frac{\pi\hbar^2}{2l^2}\right)^{3/4} \psi_l(\mathbf{q}, \mathbf{p}) = \psi(\mathbf{q}), \tag{6.33a}$$

$$\lim_{l\to+\infty} (2\pi l^2)^{3/4} \psi_l(\mathbf{q}, \mathbf{p}) = \exp\left(\frac{i}{\hbar}\,\mathbf{p}\cdot\mathbf{q}\right) \tilde{\psi}(\mathbf{p}). \tag{6.33b}$$

Thus, upon renormalization, the configuration space wave function $\psi(\mathbf{q})$ can be recovered from $\psi(\mathbf{q}, \mathbf{p})$ in the sharp-point limit $l\to+0$, and its momentum space representative can be essentially obtained when $l\to+\infty$. Furthermore, since

$$\lim_{l\to+0} \chi^{(l)}_{\mathbf{q}}(\mathbf{x}) = \delta^3(\mathbf{x}-\mathbf{q}), \tag{6.34a}$$

$$\lim_{l\to+\infty} \bar{\chi}^{(l)}_{\mathbf{p}}(\mathbf{k}) = \delta^3(\mathbf{k}-\mathbf{p}), \tag{6.34b}$$

we obtain from (6.21) and (6.22) that

$$\lim_{l\to+0} \langle\psi_l\,|\,P_{\xi^{(l)}}(B_1\times\mathbb{R}^3)\psi_l\rangle = \int_{B_1} |\psi(\mathbf{x})|^2\, d\mathbf{x} = \langle\psi|E^{\mathbf{X}}(B_1)\psi\rangle, \tag{6.35a}$$

$$\lim_{l\to+\infty} \langle\psi_l\,|\,P_{\xi^{(l)}}(\mathbb{R}^3\times B_2)\psi_l\rangle = \int_{B_2} |\tilde{\psi}(\mathbf{k})|^2\, d\mathbf{k} = \langle\psi|E^{\mathbf{P}}(B_2)\psi\rangle. \tag{6.35b}$$

Thus, in a sense, the stochastic phase-space representations of quantum mechanics on $L^2(\Gamma^{(l)})$ are sandwiched between the configuration and momentum representations, which emerge as degenerate cases in the limits $l \to +0$ and $l \to +\infty$, respectively. In fact, if we replace q^j and p^j by the complex variables

$$z_j = \frac{1}{2}\left(\frac{q^j}{l} - i\,\frac{2lp^j}{\hbar}\right), \qquad z_j^* = \frac{1}{2}\left(\frac{q^j}{l} + i\,\frac{2lp^j}{\hbar}\right), \tag{6.36}$$

and set by definition

$$\exp\left(\frac{i}{\hbar}\frac{\mathbf{q}\cdot\mathbf{p}}{2}\right)\psi_l(\mathbf{q},\mathbf{p}) = (2\pi\hbar)^{3/2}\exp\left(-\frac{1}{2}\,\mathbf{z}^*\mathbf{z}\right)f(\mathbf{z}), \tag{6.37}$$

then the mapping $\psi_l \mapsto f_l$ is a unitary transformation (Prugovečki, 1976b) of $L^2(\Gamma^{(l)})$ onto the Bargmann (1961, 1967) space of entire functions $f(\mathbf{z})$ with inner product:

$$\langle f|g\rangle_B = \pi^{-3}\int_{\mathbb{R}^6} f^*(\mathbf{z})g(\mathbf{z})\exp(-\mathbf{z}^*\mathbf{z})\,d(\operatorname{Re} z)\,d(\operatorname{Im} z). \tag{6.38}$$

1.7. Resolution Generators and Gauge Freedom in Stochastic Phase Spaces

In Section 1.4 we introduced reducible unitary ray representations $U(g)$ on $L^2(\Gamma)$, and we then pinpointed the spin-zero irreducible subspaces $L^2(\Gamma_\xi)$. These subspaces were defined as images of isometric mappings (4.6) from the Hilbert space $L^2(\mathbb{R}^3)$ of wave functions in the configuration representation to the Hilbert space $L^2(\Gamma)$. Under this mapping the function

$$\tilde{\xi}(\mathbf{q},\mathbf{p}) = (W_\xi\xi)(\mathbf{q},\mathbf{p}) = \int_{\mathbb{R}^3}\exp\left[\frac{i}{\hbar}\,\mathbf{p}\cdot(\mathbf{q}-\mathbf{x})\right]\xi^*(\mathbf{q}-\mathbf{x})\xi(\mathbf{x})\,d\mathbf{x} \tag{7.1}$$

becomes the stochastic phase space representative of the configuration space resolution generator $\xi(\mathbf{x})$. However in view of (6.15), we could have constructed $L^2(\Gamma_\xi)$ without appeal to the configuration representation by introducing, in accordance with (4.6b), (4.7) and (6.9),

$$K_{\tilde{\xi}}(\mathbf{q},\mathbf{p};\mathbf{q}',\mathbf{p}') = \langle\tilde{\xi}_{\mathbf{q},\mathbf{p}}|\tilde{\xi}_{\mathbf{q}',\mathbf{p}'}\rangle_\Gamma, \tag{7.2a}$$

$$\tilde{\xi}_{\mathbf{q},\mathbf{p}} = U\left(0,\mathbf{q},\frac{\mathbf{p}}{m},I\right)\tilde{\xi}, \tag{7.2b}$$

and then setting by definition $L^2(\Gamma_\xi)$ equal to $\mathbb{P}_{\bar{\xi}} L^2(\Gamma)$, where, as in (6.11),

$$(\mathbb{P}_{\bar{\xi}}\psi)(\mathbf{q},\mathbf{p}) = \int_\Gamma K_{\bar{\xi}}(\mathbf{q},\mathbf{p};\mathbf{q}',\mathbf{p}')\,\psi(\mathbf{q}',\mathbf{p}')\,d\mathbf{q}'\,d\mathbf{p}'. \tag{7.3}$$

Indeed, from the definition (4.5) of $U(g)$ combined with (7.2), we readily derive that

$$K_{\bar{\xi}}(\mathbf{q},\mathbf{p};\mathbf{q}',\mathbf{p}') = K^*_{\bar{\xi}}(\mathbf{q}',\mathbf{p}';\mathbf{q},\mathbf{p})$$

$$= \int_\Gamma K_{\bar{\xi}}(\mathbf{q},\mathbf{p};\mathbf{q}'',\mathbf{p}'')K_{\bar{\xi}}(\mathbf{q}'',\mathbf{p}'';\mathbf{q}',\mathbf{p}')\,d\mathbf{q}''\,d\mathbf{p}'', \tag{7.4}$$

so that $\mathbb{P}_{\bar{\xi}}$ is a projector in $L^2(\Gamma)$. The construction of $L^2(\Gamma_{\bar{\xi}})$ can be therefore carried out exclusively within the confines of $L^2(\Gamma)$ if $\bar{\xi}(\mathbf{q},\mathbf{p})$ is the unique resolution generator for each $\mathbb{P}_{\bar{\xi}} L^2(\Gamma)$.

To establish that this is the case (Prugovečki, 1978f), note first that by (4.5)

$$U\left(0,\mathbf{q}_1,\frac{\mathbf{p}_1}{m},I\right)U\left(0,\mathbf{q}_2,\frac{\mathbf{p}_2}{m},I\right)$$

$$= \exp\left(\frac{i}{\hbar}\mathbf{p}_1\cdot\mathbf{q}_2\right)U\left(0,\mathbf{q}_1+\mathbf{q}_2,\frac{\mathbf{p}_1+\mathbf{p}_2}{m},I\right), \tag{7.5a}$$

and consequently

$$U^*\left(0,\mathbf{q},\frac{\mathbf{p}}{m},I\right) = U^{-1}\left(0,\mathbf{q},\frac{\mathbf{p}}{m},I\right)$$

$$= \exp\left(\frac{i}{\hbar}\mathbf{q}\cdot\mathbf{p}\right)U\left(0,-\mathbf{q},\frac{-\mathbf{p}}{m},I\right). \tag{7.5b}$$

Furthermore, (4.5) and (7.3) imply that if $\psi = \mathbb{P}_{\bar{\xi}}\psi$ then $U(g)\psi = \mathbb{P}_{\bar{\xi}}(U(g)\psi)$, so that $U(g)$ leaves $L^2(\Gamma_{\bar{\xi}})$ invariant. Since $\bar{\xi} \in L^2(\Gamma_{\bar{\xi}})$, by (7.2b) $\bar{\xi}_{\mathbf{q},\mathbf{p}} \in L^2(\Gamma_{\bar{\xi}})$, and consequently by (7.3) and (7.4)

$$\bar{\xi}_{\mathbf{q},\mathbf{p}}(\mathbf{q}',\mathbf{p}') = K_{\bar{\xi}}(\mathbf{q}',\mathbf{p}';\mathbf{q},\mathbf{p}). \tag{7.6}$$

In particular, due to (7.2),

$$\bar{\xi}(\mathbf{q},\mathbf{p}) = \bar{\xi}_{\mathbf{0},\mathbf{0}}(\mathbf{q},\mathbf{p}) = K_{\bar{\xi}}(\mathbf{q},\mathbf{p};\mathbf{0},\mathbf{0}), \tag{7.7}$$

and, consequently, by (7.2) and (7.5b)

$$\bar{\xi}(\mathbf{q}, \mathbf{p}) = \left\langle U\left(0, \mathbf{q}, \frac{\mathbf{p}}{m}, I\right) \bar{\xi} \mid \bar{\xi} \right\rangle_\Gamma$$

$$= \left\langle \bar{\xi} \mid U^* \left(0, \mathbf{q}, \frac{\mathbf{p}}{m}, I\right) \bar{\xi} \right\rangle_\Gamma = \exp\left(\frac{i}{\hbar} \mathbf{q} \cdot \mathbf{p}\right) \bar{\xi}^*(-\mathbf{q}, -\mathbf{p}). \tag{7.8}$$

According to (7.3) and (7.6), the elements of $\mathbb{P}_{\bar{\xi}} L^2(\Gamma)$ are characterized by the identity

$$\psi(\mathbf{q}, \mathbf{p}) = (\mathbb{P}_{\bar{\xi}} \psi)(\mathbf{q}, \mathbf{p}) = \langle \bar{\xi}_{\mathbf{q}, \mathbf{p}} \mid \psi \rangle_\Gamma, \qquad (\mathbf{q}, \mathbf{p}) \in \Gamma. \tag{7.9}$$

If we therefore assume that there is in $\mathbb{P}_{\bar{\xi}} L^2(\Gamma)$ a second resolution generator $\bar{\xi}'$, so that $\mathbb{P}_{\bar{\xi}'} = \mathbb{P}_{\bar{\xi}}$, then (7.9) with $\bar{\xi}$ replaced by $\bar{\xi}'$ would be true for all $\psi \in \mathbb{P}_{\bar{\xi}'} L^2(\Gamma)$, and therefore in particular for $\psi = \bar{\xi}$, so that

$$\bar{\xi}(\mathbf{q}, \mathbf{p}) = \langle \bar{\xi}'_{\mathbf{q}, \mathbf{p}} \mid \bar{\xi} \rangle_\Gamma = \left\langle U\left(0, \mathbf{q}, \frac{\mathbf{p}}{m}, I\right) \bar{\xi}' \mid \bar{\xi} \right\rangle_\Gamma$$

$$= \left\langle \bar{\xi}' \mid U^*\left(0, \mathbf{q}, \frac{\mathbf{p}}{m}, I\right) \bar{\xi} \right\rangle_\Gamma = \exp\left(\frac{i}{\hbar} \mathbf{q} \cdot \mathbf{p}\right) \langle \bar{\xi}' \mid \bar{\xi}_{-\mathbf{q}, -\mathbf{p}} \rangle_\Gamma. \tag{7.10}$$

Hence, using (7.9) with $\psi = \bar{\xi}'$, we get

$$\bar{\xi}(\mathbf{q}, \mathbf{p}) = \exp\left(\frac{i}{\hbar} \mathbf{q} \cdot \mathbf{p}\right) \bar{\xi}'^*(-\mathbf{q}, -\mathbf{p}), \tag{7.11}$$

and a comparison with (7.8) shows that $\bar{\xi}(\mathbf{q}, \mathbf{p}) \equiv \bar{\xi}'(\mathbf{q}, \mathbf{p})$. Consequently, the resolution generator of any of the irreducible subspaces $L^2(\Gamma_{\bar{\xi}})$ is uniquely pinpointed by (7.2b) and (7.9).

Once this fundamental fact has been established, we could regard (7.1) as an equation to be solved for $\xi(\mathbf{x})$ when $\bar{\xi}(\mathbf{q}, \mathbf{p})$ is given, and thus establish the link with the configuration representation. Hence, although as a concession to the universal familiarity with the configuration representation we have chosen it as the starting point in our derivation of the irreducible subspaces $L^2(\Gamma_\xi)$ of $U(g)$, we could have reversed the procedure by starting with $L^2(\Gamma_{\bar{\xi}})$ and then computing $\xi_{\mathbf{q}, \mathbf{p}}(\mathbf{x})$ from $\xi(\mathbf{x})$ by means of (4.6b). This would then have led to the transformation $W_{\bar{\xi}} = W_\xi^{-1}$ in (4.10) from $L^2(\Gamma_\xi)$ to the configuration representation. Hence, as long as we work with the given representation (4.5) of $\mathscr{G}$ (which, as we shall see shortly, corresponds to a specific choice of gauge), we do not have to distinguish between $L^2(\Gamma_\xi)$ and $L^2(\Gamma_{\bar{\xi}})$, or between K_ξ and $K_{\bar{\xi}}$.

Several key properties of the reproducing kernels $K_{\bar{\xi}}$ follow directly from the defining formula (7.2) and the properties of $U(g)$. Thus, by (2.7), with U replacing $\hat{U}$,

$$U(0, R\mathbf{a}, R\mathbf{v}, I) = U(0, \mathbf{0}, \mathbf{0}, R)\, U(0, \mathbf{a}, \mathbf{v}, I)\, U(0, \mathbf{0}, \mathbf{0}, R^{-1}), \tag{7.12}$$

and therefore, by (7.2) and (4.2),

$$K_{\bar{\xi}}(R\mathbf{q}, R\mathbf{p}; R\mathbf{q}', R\mathbf{p}') = K_{\bar{\xi}}(\mathbf{q}, \mathbf{p}; \mathbf{q}', \mathbf{p}'). \tag{7.13}$$

In particular, by (7.7) we get that for $R \in \mathrm{SO}(3)$

$$\bar{\xi}(\mathbf{q}, \mathbf{p}) = \bar{\xi}(R\mathbf{q}, R\mathbf{p}) = \bar{\xi}(-\mathbf{q}, -\mathbf{p}), \tag{7.14}$$

where the second equality follows from the fact that the first one implies that $\bar{\xi}$ depends exclusively on $\mathbf{q}^2$, $\mathbf{p}^2$ and $\mathbf{q} \cdot \mathbf{p}$. Similarly, from (7.2), (7.5) and (7.7), we get that

$$K_{\bar{\xi}}(\mathbf{q}, \mathbf{p}; \mathbf{q}', \mathbf{p}') = \exp\left[\frac{i}{\hbar}\, \mathbf{p}' \cdot (\mathbf{q} - \mathbf{q}')\right] \bar{\xi}(\mathbf{q} - \mathbf{q}', \mathbf{p} - \mathbf{p}'), \tag{7.15}$$

so that the method of constructing $L^2(\Gamma_{\bar{\xi}})$ from a $\bar{\xi}(\mathbf{q}, \mathbf{p})$ with properties (7.14) and norm

$$\|\bar{\xi}\|_\Gamma = \langle \bar{\xi} | \bar{\xi} \rangle_\Gamma^{1/2} = (2\pi\hbar)^{-3/2} \tag{7.16}$$

is now manifestly given by (7.6), (7.9) and (7.15).

The fact that in the configuration representation $|\psi(\mathbf{x})|^2$, and not $\psi(\mathbf{x})$ itself, is directly related to position measurements gives rise to the possibility of gauge transformations

$$U_{\hat{\omega}} : \psi(\mathbf{x}) \longmapsto \exp\left[\frac{i}{\hbar}\, \hat{\omega}(\mathbf{x})\right] \psi(\mathbf{x}) \tag{7.17}$$

for any choice of real (Borel-measurable) function $\hat{\omega}(\mathbf{x})$. Under these unitary mappings the realization (2.2) and (2.12) of the CCRs is replaced by

$$X^j_{\hat{\omega}} = U_{\hat{\omega}} X^j U_{\hat{\omega}}^{-1} = X^j, \qquad P^j_{\hat{\omega}} = U_{\hat{\omega}} P^j U_{\hat{\omega}}^{-1}, \tag{7.18a}$$

$$(P^j_{\hat{\omega}}\, \psi)(\mathbf{x}) = -\left(i\hbar\, \frac{\partial}{\partial x^j} + \frac{\partial \hat{\omega}}{\partial x^j}\right) \psi(\mathbf{x}), \tag{7.18b}$$

as $\check{U}(g)$ in (3.18) is replaced by

$$\check{U}_{\hat{\omega}}(g) = U_{\hat{\omega}}\, \check{U}(g)\, U_{\hat{\omega}}^{-1}, \tag{7.18c}$$

Thus, the PV measure (2.10) stays unchanged, whereas (2.14) goes over into the PV measure

$$E^{\mathbf{P}\hat{\omega}}(B) = U_{\hat{\omega}} E^{\mathbf{P}}(B) U_{\hat{\omega}}^{-1} \tag{7.19}$$

which in general is different from $E^{\mathbf{P}}(B)$.

The gauge transformation (7.17) is represented in each $L^2(\Gamma_\xi)$ by $W_\xi^{-1} U_{\hat{\omega}} W_\xi$, which is a restriction to $L^2(\Gamma_\xi)$ of the globally on $L^2(\Gamma)$ defined unitary operator

$$\bar{U}_{\hat{\omega}} = \exp\left[\frac{i}{\hbar}\hat{\omega}(\mathbf{Q})\right] \tag{7.20}$$

with Q^j given in (4.1). However, in addition to this kind of gauge transformation, which is related to sharp localizability, we can consider gauge transformations

$$U_\omega : \psi(\mathbf{q}, \mathbf{p}) \longmapsto \exp\left[\frac{i}{\hbar}\omega(\mathbf{q}, \mathbf{p})\right]\psi(\mathbf{q}, \mathbf{p}) \tag{7.21}$$

correspondingly motivated by the fact that for localizability in stochastic phase spaces it is $|\psi(\mathbf{q}, \mathbf{p})|^2$, and not $\psi(\mathbf{q}, \mathbf{p})$ itself, that is related to measurements of the stochastic values $(\mathbf{q}, \chi_{\mathbf{q}}^\xi)$ and $(\mathbf{p}, \bar{\chi}_{\mathbf{p}}^\xi)$. Under U_ω in (7.21) the realization (4.1) of the CCRs is replaced by

$$Q_\omega^j = U_\omega Q^j U_\omega^{-1} = i\hbar\frac{\partial}{\partial p^j} + q^j + \frac{\partial\omega}{\partial p^j}, \tag{7.22a}$$

$$P_\omega^j = U_\omega P^j U_\omega^{-1} = -i\hbar\frac{\partial}{\partial q^j} - \frac{\partial\omega}{\partial q^j}, \tag{7.22b}$$

as $U(g)$ in (4.5) is replaced by

$$U_\omega(g) = U_\omega U(g) U_\omega^{-1}, \quad g \in \mathcal{G}. \tag{7.23}$$

Moreover, instead of $\mathbb{P}_\xi(B)$ in (6.5), we have the POV measure

$$\mathbb{P}_\xi^{(\omega)}(B) = U_\omega \mathbb{P}_\xi(B) U_\omega^{-1}, \quad B \subset \Gamma, \tag{7.24}$$

which together with $U_\omega(g)$ gives rise to new Galilean systems of covariance on stochastic phase spaces Γ_ξ. Naturally, with the change to any of these new gauges the resolution generators $\bar{\xi}$ as well as the irreducible subspaces $L^2(\Gamma_{\bar{\xi}})$ change to

$$\bar{\xi}^\omega = U_\omega \bar{\xi}, \qquad L^2(\Gamma_{\bar{\xi}^\omega}) = U_\omega L^2(\Gamma_{\bar{\xi}}). \tag{7.25}$$

Hence, whereas such basic relationships as (7.2b), (7.3), (7.4) and (7.5) remain completely unaltered in their appearance, some other relationships in this section, such as (7.2a), (7.6), (7.8) and (7.9), acquire gauge-dependent factors (such as $\exp[(i/\hbar)\omega(\mathbf{q}, \mathbf{p})]$) on their right-hand sides. Furthermore, if $\omega(\mathbf{q}, \mathbf{p})$ is not rotationally invariant, then clearly

$$\bar{\xi}^\omega(\mathbf{q}, \mathbf{p}) = \exp\left[\frac{i}{\hbar}\omega(\mathbf{q}, \mathbf{p})\right]\bar{\xi}(\mathbf{q}, \mathbf{p}) \tag{7.26}$$

will not satisfy (7.14), nor will (7.13) hold true for $K_{\bar{\xi}^\omega}$.

We note that the realizations (7.22) of the CCR's are of the general form (4.4), with

$$f^j(\mathbf{q}, \mathbf{p}) = q^j + \frac{\partial \omega}{\partial p^j}, \qquad g^j(\mathbf{q}, \mathbf{p}) = -\frac{\partial \omega}{\partial q^j}. \tag{7.27}$$

Conversely, if (4.4) satisfies (4.3) then f^j and g^j must satisfy (7.26) for some choice of ω, which is fixed at least locally.[13] Amongst all feasible choices, $\bar{\omega} = -\mathbf{q} \cdot \mathbf{p}/2$ distinguishes itself by yielding a realization of the CCR's,

$$Q^j_{\bar{\omega}} = i\hbar \frac{\partial}{\partial p^j} + \frac{q^j}{2}, \qquad P^j_{\bar{\omega}} = -i\hbar \frac{\partial}{\partial q^j} + \frac{p^j}{2}, \tag{7.28}$$

which displays manifest reciprocal symmetry under the interchange of position and momentum variables. This observation will turn out to be of significance when, in Chapter 4, we incorporate Born's (1938, 1949) reciprocity theory into the present framework. Indeed, in Born's (1949) view, the symmetric role that position and momentum play in some fundamental physical laws (Hamilton equations, uncertainty principle, etc.) is evidence of a deep underlying symmetry in Nature, which should serve as a guideline in formulating theories for elementary particles. Furthermore, the realization (7.28) of the CCR's is essentially the one adopted by Prigogine (1980, p. 253) in discussing the intimate connection between classical and quantum mechanics at the superoperator level – a connection first pointed out by Born (1955c) and systematically studied by in the superoperator context Ali and Prugovečki (1977b) and Prugovečki (1978b), which will be discussed in Chapter 3.

In summary, for the unitary ray representations in $L^2(\Gamma)$ of the Galilei group, the choice of gauge is paramount to a specific choice of representations, and thereby to a realization of the CCRs. Once this choice has been made, the irreducible subspaces $L^2(\Gamma_{\bar{\xi}})$ are pinpointed by unique resolution generators $\bar{\xi}(\mathbf{q}, \mathbf{p})$ which, by the kinematical operations of space translations by $\mathbf{q}$ and boosting by $\mathbf{p}/m$, give rise to continuous resolutions of the identity $\{\xi_{\mathbf{q},\mathbf{p}}\}$ in (7.2b) determining the projectors $\mathbb{P}_{\bar{\xi}}$ in (7.3) onto $L^2(\Gamma_{\bar{\xi}})$ by means of (7.2a). Thus, the contention that for $\psi \in L^2(\Gamma_{\bar{\xi}})$

$$\psi(\mathbf{q}, \mathbf{p}) = \int_\Gamma \xi_{\mathbf{q},\mathbf{p}}(\mathbf{q}', \mathbf{p}')\, \psi(\mathbf{q}', \mathbf{p}')\, d\mathbf{q}'\, d\mathbf{p}', \tag{7.29}$$

or, in terms of configuration space representation wave function ψ',

$$\psi(\mathbf{q}, \mathbf{p}) = \langle \xi_{\mathbf{q},\mathbf{p}} | \psi' \rangle, \qquad \psi' = W_\xi^{-1} \psi, \tag{7.30}$$

represents the probability amplitude for the observation of stochastic position $\underset{\sim}{\mathbf{q}}$ and stochastic momentum $\underset{\sim}{\mathbf{p}}$ with an extended particle of proper wave function

$\bar{\xi} \in L^2(\Gamma)$ (or $\xi = W_{\xi}^{-1}\bar{\xi} \in L^2(\mathbb{R}^3)$) certainly displays internal consistency. This self-consistency is underlined by the existence of related POV measures

$$(\mathbb{P}_{\bar{\xi}}(B)\psi)(\mathbf{q},\mathbf{p}) = \int_B K_{\bar{\xi}}(\mathbf{q},\mathbf{p};\mathbf{q}',\mathbf{p}')\,\psi(\mathbf{q}',\mathbf{p}')\,d\mathbf{q}'\,d\mathbf{p}', \tag{7.31}$$

giving rise, in accordance with (6.6) and (6.24), to Galilean systems of covariance. Since by (6.33) and (6.35) some of these probability amplitudes occupy an intermediate place between the configuration space probability amplitudes

$$\psi(\mathbf{x}) = (\mathbf{x}|\psi\rangle = \int_{\mathbb{R}^3} \delta^3(\mathbf{x}'-\mathbf{x})\,\psi(\mathbf{x}')\,d\mathbf{x}' \tag{7.32}$$

and momentum space probability amplitudes

$$\tilde{\psi}(\mathbf{k}) = (\mathbf{k}|\psi\rangle = \int_{\mathbb{R}^3} \Phi_{\mathbf{k}}^*(\mathbf{x}')\,\psi(\mathbf{x}')\,d\mathbf{x}', \tag{7.33}$$

$$\Phi_{\mathbf{k}}(\mathbf{x}) = (2\pi\hbar)^{-3/2}\exp\left(\frac{i}{\hbar}\mathbf{k}\cdot\mathbf{x}\right), \tag{7.34}$$

it appears appropriate to similarly rewrite (7.30) as

$$\psi(\mathbf{q},\mathbf{p}) = {}_{\xi}(\mathbf{q},\mathbf{p}\,|\,\psi\rangle = \int_{\mathbb{R}^3} \xi_{\mathbf{q},\mathbf{p}}^*(\mathbf{x}')\,\psi(\mathbf{x}')\,d\mathbf{x}'. \tag{7.35}$$

It should be noted, however, that the right-hand sides of (7.32) and (7.33) assume well-defined mathematical meaning only when $\psi(\mathbf{x})$ is restricted to suitable classes of functions, that can be chosen dense in $L^2(\mathbb{R}^3)$ but not identical to $L^2(\mathbb{R}^3)$ – such as the Schwartz space $\mathscr{S}(\mathbb{R}^3)$ (Bogolubov *et al.*, 1975, p. 11) in either case, or $L^1(\mathbb{R}^3) \cap L^2(\mathbb{R}^3)$ in the second case. On the other hand, the integral in (7.35) is well-defined globally on $L^2(\mathbb{R}^3)$. It can be therefore claimed that the transition to stochastic phase-space representations of a state vector yields a built-in regularization that results in wave functions (7.35) which are always continuous, and in many cases, such as for the optimal choice (6.25) for ξ, also infinitely many times differentiable.

Let us now summarize the main conclusions that can be drawn in the spin-zero case from the results of Sections 1.4–1.7 – and which in fact remain true also in the case of nonzero spin (Ali and Prugovečki, 1983).

For free quantum particles of spin zero there is a family of unitary equivalent ray representations of the Galilei group $\mathscr{G}$ on $L^2(\Gamma)$, any two of these unitary representations $U_{\omega_1}(g)$ and $U_{\omega_2}(g)$ being related by a gauge transformation on phase space,

$$U_{\omega_1}(g) = (U_{\omega_1}U_{\omega_2}^{-1})U_{\omega_2}(g)(U_{\omega_1}U_{\omega_2}^{-1})^{-1}, \quad g \in \mathscr{G}, \tag{7.36}$$

with U_{ω_i}, $i = 1, 2$, defined as in (7.21). The irreducible subrepresentations of $U_\omega(g)$ act on closed subspaces $L^2(\Gamma_{\bar{\xi}\omega})$ of $L^2(\Gamma)$ determined by (unique in a chosen gauge) generators $\bar{\xi}^\omega \in L^2(\Gamma_{\bar{\xi}\omega})$ of continuous resolutions

$$\bar{\xi}^\omega_{\mathbf{q},\mathbf{p}} = U_\omega\left(0, \mathbf{q}, \frac{\mathbf{p}}{m}, I\right)\bar{\xi}^\omega \tag{7.37}$$

of the projector $\mathbb{P}_{\bar{\xi}^\omega}(\Gamma)$ of $L^2(\Gamma)$ onto $L^2(\Gamma_{\bar{\xi}\omega})$, where in general

$$\mathbb{P}_{\bar{\xi}^\omega}(B) = \int_B |\bar{\xi}^\omega_{\mathbf{q},\mathbf{p}}\rangle \, d\mathbf{q}\, d\mathbf{p}\, \langle\bar{\xi}^\omega_{\mathbf{q},\mathbf{p}}| \tag{7.38}$$

is a POV measure on the Borel sets B of Γ. Each of these POV measures constitutes a system of covariance

$$U_\omega(g)\, \mathbb{P}_{\bar{\xi}^\omega}(B) U_\omega^{-1}(g) = \mathbb{P}_{\bar{\xi}^\omega}(gB), \quad g \in \mathscr{G}', \tag{7.39}$$

for the restriction of $U_\omega(g)$ to the isochronous Galilei group $\mathscr{G}'$. Furthermore, if in accordance with (6.23) we set

$$F_{\bar{\xi}^\omega}(\mathring{B}) = \int_{-\infty}^{+\infty} \mathbb{P}_{\bar{\xi}^\omega}(B_t, t)\, dt \tag{7.40}$$

then $F_{\bar{\xi}^\omega}(\mathring{B})$ is a POV measure in the Borel sets $\mathring{B}$ of $\mathbb{R}^1 \times \Gamma$, which constitutes a stochastic phase-space system of covariance for all of $U_\omega(g)$:

$$U_\omega(g) F_{\bar{\xi}^\omega}(\mathring{B}) U_\omega^{-1}(g) = F_{\bar{\xi}^\omega}(g\mathring{B}), \quad g \in \mathscr{G}. \tag{7.41}$$

Any two irreducible subrepresentations $U_\omega^{(i)}(g)$ of given $U_\omega(g)$ that correspond to resolution generators $\bar{\xi}_i^\omega$, $i = 1, 2$, i.e.

$$U_\omega^{(i)}(g)\psi = U_\omega(g)\, \mathbb{P}_{\bar{\xi}_i^\omega}(\Gamma)\psi, \quad \psi \in L^2(\Gamma_{\bar{\xi}_i^\omega}), \tag{7.42}$$

are unitarily equivalent:

$$U_\omega^{(2)}(g) = W_{\bar{\xi}_2^\omega} W_{\bar{\xi}_1^\omega}^{-1} U^{(1)}(g) W_{\bar{\xi}_1^\omega} W_{\bar{\xi}_2^\omega}^{-1}, \quad g \in \mathscr{G}. \tag{7.43}$$

However, this equivalence via $W_{\bar{\xi}_2^\omega} W_{\bar{\xi}_1^\omega}^{-1}$ does not extent to the corresponding POV measures in (7.40) and (7.41). Consequently, the associated possibility measures

$$P_\psi^{\bar{\xi}^\omega}(B) = \langle\psi | \mathbb{P}_{\bar{\xi}^\omega}(B)\psi\rangle_\Gamma \tag{7.44}$$

are not equal. The physical interpretation of this fact is that for $\bar{\xi}_1^\omega \neq \bar{\xi}_2^\omega$ the corresponding values of (7.44) are probabilities over distinct (gauge-independent) stochastic phase spaces $\Gamma_{\bar{\xi}_1}^{\omega} \neq \Gamma_{\bar{\xi}_2}^{\omega}$, which belong to measurements performed on the same system with distinct extended quantum test particles of proper wave functions $\bar{\xi}_1^{\omega}$ and $\bar{\xi}_2^{\omega}$.

1.8. Stochastic Probability Currents and Their Sharp-Point Limits

A further piece of essential corroborating evidence in favour of the interpretation of $\psi(\mathbf{q}, \mathbf{p})$ in (7.35) as a probability amplitude for measurements of stochastic position $\underset{\sim}{\mathbf{q}}$ and stochastic momentum $\underset{\sim}{\mathbf{p}}$ with extended test particles of proper wave function ξ is the existence for real $\xi(\mathbf{x})$ of a conserved and Galilei covariant current $\mathbf{j}^{\xi}(\mathbf{q})$ which in the sharp-point limit

$$\chi_{\mathbf{q}}^{\xi}(\mathbf{x}) = (2\pi\hbar)^3 \, |\xi(\mathbf{x} - \mathbf{q})|^2 \longrightarrow \delta^3(\mathbf{x} - \mathbf{q}) \tag{8.1}$$

converges to its conventional counterpart

$$\mathbf{j}(\mathbf{q}) = \frac{\hbar}{2im} \, \psi^*(\mathbf{q}) \overleftrightarrow{\nabla} \psi(\mathbf{q}). \tag{8.2}$$

To better understand some of the physical implications of this mathematical fact, let us recall that in classical statistical mechanics the state of an ensemble of single particles is described by the distribution function $\rho^{\mathrm{cl}}(\mathbf{q}, \mathbf{p}; t)$, and therefore the configuration space density is:

$$\rho^{\mathrm{cl}}(\mathbf{q}, t) = \int_{\mathbb{R}^3} \rho^{\mathrm{cl}}(\mathbf{q}, \mathbf{p}; t) \, d\mathbf{p}. \tag{8.3}$$

The probability current is then defined in a natural manner as the average velocity, i.e.

$$\mathbf{j}^{\mathrm{cl}}(\mathbf{q}, t) = \int_{\mathbb{R}^3} \frac{\mathbf{p}}{m} \, \rho^{\mathrm{cl}}(\mathbf{q}, \mathbf{p}; t) \, d\mathbf{p}, \tag{8.4}$$

and its conservation follows from Liouville's theorem (Pathria, 1972, p. 35). On the other hand, despite the fact that we can formally write (8.2) as

$$\mathbf{j}(\mathbf{q}) = \tfrac{1}{2} \left[\psi^*(\mathbf{q}) (\mathbf{P}\psi)(\mathbf{q}) + (\mathbf{P}\psi)^*(\mathbf{q}) \psi(\mathbf{q})\right], \tag{8.5}$$

a similar natural interpretation of $\mathbf{j}(\mathbf{q})$ does not exist due to the uncertainty principle, which prohibits the association of any 3-velocity value with a sharp observation of position. Hence, strictly speaking, (8.2) or (8.5) are introduced in an *ad hoc* manner, namely on grounds that one gets the continuity equation

$$\partial_t \rho(\mathbf{x}, t) + \nabla \cdot j(\mathbf{x}, t) = 0, \quad \rho(\mathbf{x}, t) = |\psi(\mathbf{x}, t)|^2, \tag{8.6}$$

as a consequence of the Schrödinger equation

$$\partial_t \psi(\mathbf{x}, t) = \left[-\frac{\hbar^2}{2m} \nabla^2 + V(\mathbf{x}) \right] \psi(\mathbf{x}, t). \tag{8.7}$$

It is therefore remarkable that in stochastic phase space, where we can indeed define by analogy with (8.3) and (8.4)

$$\rho^\xi(\mathbf{q}, t) = \int_{\mathbb{R}^3} |\psi(\mathbf{q}, \mathbf{p}; t)|^2 \, d\mathbf{p}, \quad \psi \in L^2(\Gamma_\xi), \tag{8.8}$$

$$\mathbf{j}^\xi(\mathbf{q}, t) = \int_{\mathbb{R}^3} \frac{\mathbf{p}}{m} |\psi(\mathbf{q}, \mathbf{p}; t)|^2 \, d\mathbf{p}, \tag{8.9}$$

we get not only the continuity equation

$$\partial_t \rho^\xi(\mathbf{q}, t) + \nabla \cdot \mathbf{j}^\xi(\mathbf{q}, t) = 0, \tag{8.10}$$

but also the result that

$$\rho^\xi(\mathbf{q}, t) \longrightarrow \rho(\mathbf{q}, t), \qquad \mathbf{j}^\xi(\mathbf{q}, t) \longrightarrow \mathbf{j}(\mathbf{q}, t) \tag{8.11}$$

in the sharp-point limit (8.1). Indeed, this fact is not only further evidence as to the consistency of the aforementioned interpretation of $\psi(\mathbf{q}, \mathbf{p}; t)$ as a probability amplitude, but it also shows that via the stochastic phase-space interpretation we can relate aspects of the conventional formalism of quantum mechanics which appear basically unrelated if we restrict ourselves to the customary formulation.

In order to prove (Prugovečki, 1978a; Ali *et al.*, 1981) (8.10) and (8.11), let us work in the gauge adopted in Section 1.4, in which a representation $U(g), g \in \mathcal{G}$, is given in the Schrödinger picture by (4.5a). We see from (4.1) and (4.5b) that in this gauge

$$i\hbar \partial_t \psi(\mathbf{q}, \mathbf{p}; t) = -\frac{\hbar^2}{2m} \nabla_{\mathbf{q}}^2 \psi(\mathbf{q}, \mathbf{p}; t), \tag{8.12}$$

so that the conservation of the current

$$\mathbf{j}^\xi(\mathbf{q}, \mathbf{p}; t) = \frac{\hbar}{2im} \psi^*(\mathbf{q}, \mathbf{p}; t) \overleftrightarrow{\nabla}_{\mathbf{q}} \psi(\mathbf{q}, \mathbf{p}; t) \tag{8.13}$$

at the stochastic point $(\mathbf{q}, \chi_{\mathbf{q}}^\xi) \times (\mathbf{p}, \bar{\chi}_{\mathbf{p}}^\xi)$ follows in the same manner as for (8.2):

$$\partial_t |\psi(\mathbf{q}, \mathbf{p}; t)|^2 + \nabla_{\mathbf{q}} \cdot \mathbf{j}^\xi(\mathbf{q}, \mathbf{p}; t) = 0. \tag{8.14}$$

Therefore, the continuity equation (8.10) is obtained by integration in $\mathbf{p} \in \mathbb{R}^3$ if we set

$$\mathbf{j}^\xi(\mathbf{q}, t) = \int_{\mathbb{R}^3} \mathbf{j}^\xi(\mathbf{q}, \mathbf{p}; t)\, d\mathbf{p}. \tag{8.15}$$

To show that (8.15) coincides with (8.9), let us note that by (7.35)

$$\psi(\mathbf{q}, \mathbf{p}; t) = \int_{\mathbb{R}^3} \tilde{\xi}^*_{\mathbf{q},\mathbf{p}}(\mathbf{k})\, \tilde{\psi}(\mathbf{k}, t)\, d\mathbf{k}$$

$$= \int_{\mathbb{R}^3} \tilde{\xi}^*(\mathbf{k} - \mathbf{p})\, \tilde{\psi}_{\mathbf{q}}(\mathbf{k}, t)\, d\mathbf{k}, \tag{8.16}$$

$$\tilde{\psi}_{\mathbf{q}}(\mathbf{k}, t) = \exp\left(\frac{i}{\hbar} \mathbf{k} \cdot \mathbf{q}\right) \tilde{\psi}(\mathbf{k}, t). \tag{8.17}$$

Hence, by a change to the variables $\mathbf{p}' = \mathbf{k} - \mathbf{p}$, the right-hand side of (8.15) assumes the form

$$\int_{\mathbb{R}^3} \frac{\mathbf{p}}{m} |\psi(\mathbf{q}, \mathbf{p}; t)|^2\, d\mathbf{p} + \langle \psi_{\mathbf{q}} | \mathbf{M}_\xi \psi_{\mathbf{q}} \rangle + \langle \mathbf{M}_\xi \psi_{\mathbf{q}} | \psi_{\mathbf{q}} \rangle, \tag{8.18}$$

where $\mathbf{M}_\xi$ is an integral operator with kernel

$$\mathbf{M}_\xi(\mathbf{k}, \mathbf{k}') = \frac{1}{2m} \int_{\mathbb{R}^3} \mathbf{p}\, \tilde{\xi}^*(\mathbf{p})\, \tilde{\xi}(\mathbf{k}' - \mathbf{k} + \mathbf{p})\, d\mathbf{p} \tag{8.19}$$

in the momentum representation. Hence (8.9) and (8.15) are equal iff $\mathbf{M}_\xi$ is an antisymmetric operator. This will be the case if $\xi(\mathbf{x}) = \xi^*(\mathbf{x})$, since due to the rotational invariance (4.16) of $\xi(\mathbf{x})$ we have

$$\tilde{\xi}(\mathbf{k}) = (2\pi\hbar)^{-3/2} \int_{\mathbb{R}^3} \cos\left(\frac{\mathbf{k} \cdot \mathbf{x}}{\hbar}\right) \xi(\mathbf{x})\, d\mathbf{x} = \tilde{\xi}^*(\mathbf{k}). \tag{8.20}$$

By (8.19) the rotational invariance and reality of $\tilde{\xi}(\mathbf{k})$ imply, respectively, that

$$\mathbf{M}_\xi(\mathbf{k}, \mathbf{k}') = -\mathbf{M}_\xi(\mathbf{k}', \mathbf{k}) = -\mathbf{M}^*_\xi(\mathbf{k}', \mathbf{k}). \tag{8.21}$$

The reality and rotational invariance of the momentum space proper wave function $\tilde{\xi}(\mathbf{k})$ will turn out to be very welcome properties when the transition to the relativistic regions is made in Chapter 2. Both these features are present in eigenfunctions of isotropic harmonic oscillators, which will emerge in Section 4.5

as the proper wave functions of exciton states of Born's metric operator. The fact that both these features of ξ emerge independently as essential ingredients for a physically consistent formulation of nonrelativistic quantum mechanics on Γ_ξ is therefore a clearcut manifestation of mutual concurrence of independently derived physical principles and ideas.

In the presence of an external potential, neither the stochastic mean momentum $\mathbf{p}$, nor its sharp counterpart $\mathbf{k}$, is conserved and, therefore, neither (8.12) nor (8.14) is any longer true. Nevertheless, the continuity equation (8.10) remains true for

$$\psi(\mathbf{q}, \mathbf{p}; t) = {}_\xi\langle \mathbf{q}, \mathbf{p} \mid \psi_t\rangle = \int_{\mathbb{R}^3} \exp\left(-\frac{i}{\hbar}\,\mathbf{p}\cdot\mathbf{x}\right)\,\xi(\mathbf{x}-\mathbf{q})\,\psi(\mathbf{x}, t)\,d\mathbf{x}, \tag{8.22}$$

where $\psi(\mathbf{x}, t)$ is a solution of the Schrödinger equation (8.7).

To see that, we insert (8.22) into (8.8), and after differentiating with respect to t, we use (8.6) to get

$$\partial_t \rho^\xi(\mathbf{q}, t) = (2\pi\hbar)^3\,\frac{i\hbar}{2m}\int d\mathbf{x}\,|\xi(\mathbf{x}-\mathbf{q})|^2 \times \\ \times\,[\psi^*(\mathbf{x}, t)\,\nabla^2\psi(\mathbf{x}, t) - \psi(\mathbf{x}, t)\,\nabla^2\psi^*(\mathbf{x}, t)]\,. \tag{8.23}$$

By inserting (8.22) into (8.9), and using properties of Fourier transforms, we also obtain

$$\begin{aligned} \nabla\mathbf{j}^\xi(\mathbf{q}, t) &\\ = (2\pi\hbar)^3\,\frac{\hbar}{im}\int_{\mathbb{R}^3} & d\mathbf{x}\,\xi(\mathbf{x}-\mathbf{q})\,\nabla\psi^*(\mathbf{x}, t)\,\times \\ \times\,[\psi(\mathbf{x}, t)\nabla_\mathbf{x}\xi^*(\mathbf{x}-\mathbf{q}) &+ \xi^*(\mathbf{x}-\mathbf{q})\nabla\psi(\mathbf{x}, t)] + \\ +\,(2\pi\hbar)^3\,\frac{\hbar}{im}\int_{\mathbb{R}^3} & d\mathbf{x}\,\xi(\mathbf{x}-\mathbf{q})\,\psi^*(\mathbf{x}, t)\,\times \\ \times\,[\nabla_\mathbf{x}\xi^*(\mathbf{x}-\mathbf{q})\cdot\nabla\psi(\mathbf{x}, t) &+ \xi^*(\mathbf{x}-\mathbf{q})\nabla^2\psi(\mathbf{x}, t)]\,. \end{aligned} \tag{8.24}$$

When $\xi(\mathbf{x})$ is real, we easily establish that the right-hand sides of (8.23) and (8.24) add up to zero by performing integrations by parts until all partial derivatives of ξ are removed. Thus, the validity of the continuity equation remains true even in the presence of an external potential acting in accordance with (8.7) on a point particle.

In this context, it should be noted that in Section 1.4 we showed that $W_\xi \hat{U}(g) W_\xi^{-1}$ is the restriction of $U(g)$ in (4.5) to $L^2(\Gamma_\xi)$, so that

$$U(g)\,\mathbb{P}_\xi = W_\xi \hat{U}(g) W_\xi^{-1}\,\mathbb{P}_\xi, \tag{8.25}$$

and consequently for Q^j in (4.1) and X^j in (2.2)

$$Q^j\,\mathbb{P}_\xi = W_\xi X^j W_\xi^{-1}\,\mathbb{P}_\xi. \tag{8.26}$$

By the general theory of functions of compatible observables (Prugovečki, 1981a, pp. 269–284) these relations extend to arbitrary functions, and in particular to

$$V(\mathbf{Q})\,\mathbb{P}_\xi = W_\xi V(\mathbf{X}) W_\xi^{-1}\,\mathbb{P}_\xi. \tag{8.27}$$

Hence, the Schrödinger equation (8.7) can be rewritten in $L^2(\Gamma_\xi)$ as follows

$$i\hbar\,\partial_t \psi(\mathbf{q}, \mathbf{p}; t) = \left(-\frac{\hbar^2}{2m}\,\nabla_\mathbf{q}^2 + V(\mathbf{Q})\right)\,\psi(\mathbf{q}, \mathbf{p}; t). \tag{8.28}$$

On the other hand, instead of $V(\mathbf{Q})$ we could also consider interaction Hamiltonians that are functions $V(\mathbf{Q}_{\mathrm{st}})$ of the stochastic position operators (Ali and Prugovečki, 1977a)

$$(Q_{\mathrm{st}}^j \psi)(\mathbf{q}, \mathbf{p}) = q^j \psi(\mathbf{q}, \mathbf{p}), \quad \psi \in L^2(\Gamma). \tag{8.29}$$

Since these operators, however, do not leave any of the subspaces $L^2(\Gamma_\xi)$ invariant, after applying $V(\mathbf{Q}_{\mathrm{st}})$ to any elements of $L^2(\Gamma_\xi)$ we have to project back onto $L^2(\Gamma_\xi)$ by setting

$$H_I^\xi = \mathbb{P}_\xi V(\mathbf{Q}_{\mathrm{st}})\,\mathbb{P}_\xi, \tag{8.30}$$

so that H_I^ξ acts on $\psi \in L^2(\Gamma_\xi)$ as follows:

$$(H_I^\xi \psi)(\mathbf{q}, \mathbf{p}) = \int_\Gamma K_\xi(\mathbf{q}, \mathbf{p}; \mathbf{q}', \mathbf{p}')\, V(\mathbf{q}')\,\psi(\mathbf{q}', \mathbf{p}')\, d\mathbf{q}'\, d\mathbf{p}'. \tag{8.31}$$

Using (4.6), (4.10) and (6.7)–(6.9) we easily compute that

$$(W_\xi^{-1} H_I^\xi W_\xi \psi)(\mathbf{x}) = V^\xi(\mathbf{x})\,\psi(\mathbf{x}), \tag{8.32}$$

$$V^\xi(\mathbf{x}) = (2\pi\hbar)^3 \int_{\mathbb{R}^3} |\xi(\mathbf{x} - \mathbf{q})|^2\, V(\mathbf{q})\, d\mathbf{q} = \int_{\mathbb{R}^3} \chi_\mathbf{q}^\xi(\mathbf{x})\, V(\mathbf{q})\, d\mathbf{q}. \tag{8.33}$$

Since $V^\xi(\mathbf{x})$ is the potential 'seen' by an extended particle with proper wave function ξ, the interaction Hamiltonian (8.30) will be used in Section 1.10 while considering the propagation of such particles in external fields. The essential point for the present is that the universal validity of the continuity equation (8.10) remains unimpaired even when (8.28) is replaced by

$$i\hbar\,\partial_t \psi(\mathbf{q},\mathbf{p};t) = \left[-\frac{\hbar^2}{2m}\nabla_{\mathbf{q}}^2 + V(\mathbf{Q}_{\text{st}})\right]\psi(\mathbf{q},\mathbf{p};t), \tag{8.34}$$

since by (8.32) in the configuration representation the above equation is equivalent to (8.7) with $V(\mathbf{x})$ replaced by $V^{\xi}(\mathbf{x})$.

Finally, let us establish that in the sharp-point limit (8.1) we recover the conventional density and current, i.e. that (8.11) is true. Comparing (8.8) and (5.1a) we see that

$$\rho^{\xi}(\mathbf{q},t) = \int_{\mathbb{R}^3} \chi_{\mathbf{q}}^{\xi}(\mathbf{x})\,|\psi(\mathbf{x},t)|^2\,d\mathbf{x} \longrightarrow \rho(\mathbf{q},t) \tag{8.35}$$

provided, of course, that $|\psi(\mathbf{x},t)|^2$ is continuous in $\mathbf{x}$. Inserting (8.22) into (8.9), we get

$$\mathbf{j}^{\xi}(\mathbf{q},t) = \int_{\mathbb{R}^3} d\mathbf{p} \int_{\mathbb{R}^6} d\mathbf{x}\,d\mathbf{x}'\,\frac{\mathbf{p}}{m}\exp\left[\frac{i}{\hbar}\,\mathbf{p}\cdot(\mathbf{x}'-\mathbf{x})\right] \times$$

$$\times\,\xi(\mathbf{x}'-\mathbf{q})\,\xi^*(\mathbf{x}-\mathbf{q})\,\psi(\mathbf{x},t)\,\psi^*(\mathbf{x}',t). \tag{8.36}$$

Upon performing the integration in $\mathbf{p} \in \mathbb{R}^3$ by Fourier transform theory, and then noting that

$$\xi(\mathbf{x}-\mathbf{q})\nabla_{\mathbf{x}}\xi^*(\mathbf{x}-\mathbf{q}) = \tfrac{1}{2}\nabla_{\mathbf{x}}\xi^2(\mathbf{x}-\mathbf{q}) \tag{8.37}$$

if $\xi(\mathbf{x})$ is real, we obtain after an integration by parts

$$\mathbf{j}^{\xi}(\mathbf{q},t) = (2\pi\hbar)^3\,\frac{\hbar}{2im}\int_{\mathbb{R}^3} \xi^2(\mathbf{x}-\mathbf{q})\,[\psi^*(\mathbf{x},t)\overleftrightarrow{\nabla}\psi(\mathbf{x},t)]\,d\mathbf{x} \longrightarrow \mathbf{j}(\mathbf{q},t) \tag{8.38}$$

provided, of course, that $\psi(\mathbf{x},t)$ possesses continuous first partial derivatives.

In conclusion, we can state that the probability density $\rho^{\xi}(\mathbf{q},t)$ and probability current $\mathbf{j}^{\xi}(\mathbf{q},t)$ at the stochastic configuration space point $(\mathbf{q},\chi_{\mathbf{q}}^{\xi})$ are defined for an arbitrary Schrödinger picture wave function $\psi(\mathbf{x},t)$. If this wave function is such that the (conventional) probability density $\rho(\mathbf{q},t)$ and probability current $\mathbf{j}(\mathbf{q},t)$ at sharp configuration space points $\mathbf{q}$ are defined and depend continuously upon $\mathbf{q}$, then in the sharp-point limit, ρ^{ξ} and $\mathbf{j}^{\xi}$ converge to ρ and $\mathbf{j}$ at each $\mathbf{q} \in \mathbb{R}^3$.

1.9. Propagators in Nonrelativistic Stochastic Phase Spaces

The concept of particle localizability in the nonrelativistic stochastic phase spaces Γ_ξ which we introduced in Section 1.5 is supposed to provide primarily the first stepping stone towards a consistent solution of the localizability problem in relativistic quantum mechanics, which will be eventually derived in Chapter 2. However, even in the nonrelativistic realm this concept can lead to the resolution of some fundamental difficulties, such as those pertaining to the physical meaning and mathematical existence of Feynman path integrals.

The basic idea of path integrals for nonrelativistic quantum particles orginated with a paper by Dirac (1933), who observed that the time evolution

$$\psi_{t''} = T \exp\left[-\frac{i}{\hbar}\int_{t'}^{t''} H(t)\,\mathrm{d}t\right]\psi_{t'} \tag{9.1}$$

of the state vector in the Schrödinger picture for

$$H(t) = H_0 + H_I(t), \quad H_0 = \frac{\mathbf{P}^2}{2m}, \; H_I(t) = V(\mathbf{X}, t) \tag{9.2}$$

is governed by a propagator

$$\mathscr{K}(\mathbf{x}'', t''; \mathbf{x}, t) = \langle \mathbf{x}'' | \hat{\mathscr{U}}(t'', t') | \mathbf{x}' \rangle, \tag{9.3a}$$

$$\hat{\mathscr{U}}(t'', t') = T \exp\left[-\frac{i}{\hbar}\int_{t'}^{t''} H(t)\,\mathrm{d}t\right], \tag{9.3b}$$

that for small time intervals $[t', t'']$, or small $\hbar$ (i.e. in the WKB approximation), can be related to a solution S of the Hamilton–Jacobi equation for the corresponding classical problem. Feynman (1948) later noted that if one identifies S with an averaged form of the classical action functional

$$S(j, j-1) = \int_{t_{j-1}}^{t_j} \left[\frac{\dot{\mathbf{x}}^2(t)}{2m} - V(\mathbf{x}(t), t)\right] \mathrm{d}t \tag{9.4}$$

along broken line paths with joints at $\mathbf{x}(t_j)$, $j = 0, \ldots, N$, then (9.3) can be rewritten in the form of a so-called path integral:

$$\mathscr{K}(\mathbf{x}'', t''; \mathbf{x}', t') = \int \exp\left(\frac{i}{\hbar} S\right) \mathscr{D}[\mathbf{x}(t)]. \tag{9.5}$$

The derivation of (9.5), as well as its meaning, can be arrived at (Nelson, 1964; Schulman, 1981) by using Trotter's (1959) product formula for an H_I that is time-independent, or, more generally, by noting that under mild technical assumptions (Yosida, 1974, Ch. XIV, Sec. 4) on the domain of a time-dependent $H_I(t)$

$$\hat{\mathscr{U}}(t'', t') = \operatorname*{s-lim}_{N\to\infty} \prod_{j=1}^{1} \exp\left[-\frac{i}{\hbar} H_I(t_j)(t_j - t_{j-1})\right] \hat{\mathscr{U}}_0(t_j, t_{j-1}), \quad (9.6a)$$

$$\hat{\mathscr{U}}_0(t'', t') = \exp\left[-\frac{i}{\hbar} H_0(t'' - t')\right], \quad (9.6b)$$

where $t' = t_0 < t_1 < t_2 < \ldots < t_N = t''$ is any subdivision of $[t', t'']$ and

$$\epsilon = \max_{j=1,\ldots,N} |t_j - t_{j-1}| \longrightarrow 0, \quad N \longrightarrow \infty. \quad (9.6c)$$

Indeed, $\hat{\mathscr{U}}_0(t'', t')$ can be related to a free propagator

$$K(\mathbf{x}'', t''; \mathbf{x}', t') = \left[\frac{m}{2\pi i\hbar(t'' - t')}\right]^{3/2} \exp\left[\frac{im}{2\hbar(t'' - t')}(\mathbf{x}'' - \mathbf{x}')^2\right], \quad (9.7)$$

since for the free Schrödinger operator H_0 in (9.2) we have[14]

$$(\hat{\mathscr{U}}_0(t'', t')\psi_{t'})(\mathbf{x}') = \int_{\mathbb{R}^3} K(\mathbf{x}'', t''; \mathbf{x}', t')\psi_t(\mathbf{x}')\,d\mathbf{x}'. \quad (9.8)$$

Furthermore, by inserting the respective identity resolutions

$$\mathbb{1} = \int_{\mathbb{R}^3} |\mathbf{x}_j\rangle\, d\mathbf{x}_j\, \langle\mathbf{x}_j| = \int_{\mathbb{R}^3} |\mathbf{x}_j'\rangle\, d\mathbf{x}_j'\, \langle\mathbf{x}_j'| \quad (9.9)$$

between each pair of consecutive factors in (9.6a), and writing $\mathbf{x}' = \mathbf{x}_0$, $\mathbf{x}'' = \mathbf{x}_N$, we formally obtain:

$$\langle\mathbf{x}''|\, \hat{\mathscr{U}}(t'', t')\, |\mathbf{x}'\rangle = \lim_{N\to\infty} \int d\mathbf{x}_N' \int \prod_{j=N-1}^{1} d\mathbf{x}_j\, d\mathbf{x}_j' \times$$

$$\times \prod_{j=N}^{1} \left\langle \mathbf{x}_j \right| \exp\left[-\frac{i}{\hbar} V(\mathbf{X}, t_j)(t_j - t_{j-1})\right] \left| \mathbf{x}_j' \right\rangle \times$$

$$\times\ K(\mathbf{x}_j', t_j; \mathbf{x}_{j-1}, t_{j-1}). \quad (9.10)$$

Upon noting that, again in purely formal terms,

$$\left\langle \mathbf{x} \right| \exp\left[-\frac{i}{\hbar} V(\mathbf{X}, t)\,\Delta t\right] \left| \mathbf{x}' \right\rangle = \exp\left[-\frac{i}{\hbar} V(\mathbf{x}, t)\,\Delta t\right] \delta^3(\mathbf{x}' - \mathbf{x}), \quad (9.11)$$

and by choosing the subdivision points equidistant, so that

$$\epsilon = t_j - t_{j-1} = (t'' - t')/N, \tag{9.12}$$

we see that in account of (9.7) the formal relation (9.10) yields the familiar expression (Schulman, 1981, p. 7)

$$\mathscr{K}(\mathbf{x}'', t''; \mathbf{x}', t') = \lim_{N\to\infty} \int_{\mathbb{R}^{3(N-1)}} d\mathbf{x}_1 \ldots d\mathbf{x}_{N-1} \left(\frac{m}{2\pi i\hbar\epsilon}\right)^{3N/2} \times$$

$$\times \exp\left\{\frac{i\epsilon}{\hbar}\sum_{j=1}^{N}\left[\frac{m}{2}\left(\frac{\mathbf{x}_j - \mathbf{x}_{j-1}}{\epsilon}\right)^2 - V(\mathbf{x}_j, t)\right]\right\}, \tag{9.13}$$

for the path integral in (9.5).

Albeit among physicists (9.13) provides the accepted mathematical interpretation (Feynman and Hibbs, 1965) of the path integral on the right-hand side of (9.5), it nevertheless represents as it stands a mathematically meaningless entity. Indeed, the integral over $\mathbb{R}^{3(N-1)}$ in (9.13) does not exist in the Riemann, Lebesgue, or any other acceptable mathematical sense since the integrand is an oscillatory function whose absolute value is constant. For that reason, ever since the publication of a basic paper by Gel′fand and Yaglom (1960), many attempts [15] have been made to impart to the 'path integral' in (9.5) a rigorous mathematical meaning – albeit that eventually entailed total departures from the 'definition' (9.13). In view of the fact that if we multiply both sides of (9.13) with $f^*(\mathbf{x}'')g(\mathbf{x}')$, $f, g \in L^2(\mathbb{R}^3)$, and then integrate in $\mathbf{x}', \mathbf{x}'' \in \mathbb{R}^3$ *prior* to performing the integrations in $\mathbf{x}_1, \ldots, \mathbf{x}_{N-1}$, we obtain by (9.6)–(9.8) a perfectly acceptable formula for

$$\langle f \mid \hat{\mathscr{U}}(t'', t')g\rangle = \int_{\mathbb{R}^3} d\mathbf{x}''\, f^*(\mathbf{x}'') \int_{\mathbb{R}^3} d\mathbf{x}'\, \mathscr{K}(\mathbf{x}'', t''; \mathbf{x}', t')g(\mathbf{x}') \tag{9.14}$$

based on Hilbert space methods, it is not surprising that by some suitable rephrasings of the mathematical meaning of (9.5) we can arrive at various mathematically acceptable [16] formulations of Feynman path integrals.

The abandoning of (9.13) as the mathematical interpretation of (9.5) also entails, however, the abandoning of its very attractive heuristic physical interpretation as provided by Feynman (1949) in terms of sums over broken (i.e. polygonal) paths. According to this interpretation (Feynman and Hibbs, 1965, pp. 20, 48) the complex function

$$\mathscr{K}(\mathbf{x}'', t''; B_{N-1}, t_{N-1}; \ldots; B_1, t_1; \mathbf{x}', t')$$

$$= \int_{B_{N-1}} d\mathbf{x}_{N-1} \mathscr{K}(\mathbf{x}'', t''; \mathbf{x}_{N-1}, t_{N-1}) \int_{B_{N-1}} d\mathbf{x}_{N-2} \cdots$$

$$\ldots \int_{B_1} d\mathbf{x}_1 \mathscr{K}(\mathbf{x}_2, t_2; \mathbf{x}_1, t_1)\mathscr{K}(\mathbf{x}_1, t_1; \mathbf{x}', t') \tag{9.15}$$

is the probability amplitude that a particle, upon starting from $\mathbf{x}'$ at time t', would propagate to $\mathbf{x}''$ by time t'', having passed in the meantime through the spatial regions (i.e. 'slits') B_j at the times t_j, $j = 1, \ldots, N-1$. Thus, in picturesque terms the integral in (9.13) can be viewed as a limit of sums over broken paths of probability amplitudes corresponding to a kind of random motion whose classical analogue is the Brownian motion.[17]

Despite its great intuitive appeal, this physical interpretation is plagued, as it stands, not only by the aforementioned mathematical inconsistency, but also by a crucial physical inconsistency, first observed by Blokhintsev (1968, p. 127) for the case $N = 1$, and then extended to arbitrary N by Skagerstam (1976): the mean energy required for any actual observations of the joints in such an imagined broken paths turns out to be infinite. However, both types of inconsistencies are removed as soon as we reconsider the entire problem in the context of stochastic phase spaces (Prugovečki, 1981b; Gagnon, 1983).

The fact that path integrals can be formulated for phase space as well as for configuration space has been first observed by Feynman (1951). Subsequently, various mathematical advantages of phase-space formulations have been pointed out by Schweber (1962), Klauder (1963), Faddeev (1969), De Witt-Morette *et al.* (1977, 1979), and many others.[18] In addition to all those advantages, the stochastic phase-space approach with its physical interpretation of any $\psi(\mathbf{q}, \mathbf{p})$ in $L^2(\Gamma_\xi)$ as a probability amplitude is capable of providing not only mathematical but also physical insights if we replace the configuration space propagator (9.3a) with the stochastic phase-space propagator

$$\mathscr{K}_\xi(\zeta'', t''; \zeta', t') = \langle \xi_{\mathbf{q}'', \mathbf{p}''} | \hat{\mathscr{U}}(t'', t') \xi_{\mathbf{q}', \mathbf{p}'} \rangle \tag{9.16a}$$

$$\zeta' = (\mathbf{q}', \mathbf{p}'), \quad \zeta'' = (\mathbf{q}'', \mathbf{p}'') \in \Gamma. \tag{9.16b}$$

Indeed, by the general interpretation of $|\psi(\mathbf{q}, \mathbf{p})|^2$ as a probability density,

$$\rho(\zeta', t'; \zeta'', t'') = (2\pi\hbar)^3 \; |\mathscr{K}_\xi(\zeta'', t''; \zeta', t')|^2 \tag{9.17a}$$

is the probability density that an extended particle prepared at t' at stochastic position $\underset{\sim}{\mathbf{q}}'$ with stochastic momentum $\underset{\sim}{\mathbf{p}}'$, will be found at t'' at stochastic position $\underset{\sim}{\mathbf{q}}''$ and possessing stochastic momentum $\underset{\sim}{\mathbf{p}}''$. It is easily seen that the mean energy

$$\left\langle \frac{\mathbf{P}^2}{2m} \right\rangle_{t''} = \int_\Gamma \mathscr{K}_\xi^*(\zeta'', t''; \zeta', t') \left(-\frac{\hbar^2}{2m} \nabla_{\mathbf{q}''}^2 \right) \mathscr{K}_\xi(\zeta'', t''; \zeta', t') \, d\zeta'' \tag{9.17b}$$

associated with the measurement of (9.17a) is generally finite.

The stochastic phase-space propagators (9.10) possess the two basic properties that characterize all propagators, namely:

$$\mathscr{K}_\xi(\zeta'', t''; \zeta', t') = \mathscr{K}_\xi^*(\zeta', t'; \zeta'', t''), \tag{9.18a}$$

$$\mathscr{K}_\xi(\zeta'', t''; \zeta', t') = \int_\Gamma \mathscr{K}(\zeta'', t''; \zeta, t) \mathscr{K}(\zeta, t; \zeta', t') \, d\zeta. \tag{9.18b}$$

These two properties reflect the corresponding properties of evolution operators,

$$\hat{\mathscr{U}}^*(t'', t') = \hat{\mathscr{U}}^{-1}(t'', t') = \hat{\mathscr{U}}(t', t''), \tag{9.19a}$$

$$\hat{\mathscr{U}}(t'', t') = \hat{\mathscr{U}}(t'', t)\,\hat{\mathscr{U}}(t, t'). \tag{9.19b}$$

Indeed, (9.18a) is an immediate consequence of (9.16a) and (9.19a), whereas (9.18b) follows from (9.19b) upon using (4.11) between $\hat{\mathscr{U}}(t'', t)$ and $\hat{\mathscr{U}}(t, t')$, and then resorting to the definition (9.16).

It should be noted that (9.18b) is a well-defined integral (in the Lebesgue sense), whereas the same cannot be claimed in the corresponding formal relations for the Feynman propagator (9.3a):

$$\mathscr{K}(\mathbf{x}'', t''; \mathbf{x}', t') = \mathscr{K}^*(\mathbf{x}', t'; \mathbf{x}'', t'')$$

$$= \int_{\mathbb{R}^3} \mathscr{K}(\mathbf{x}'', t''; \mathbf{x}, t)\mathscr{K}(\mathbf{x}, t; \mathbf{x}', t')\, d\mathbf{x}. \tag{9.20}$$

Indeed, the integral in (9.20) does not exist (even in the Lebesgue sense) for the free propagator (9.7), since its absolute value is constant. Of course, in practice this difficulty is bypassed by applying (Feynman and Hibbs, 1965, p. 43) standard formulae for Gaussians outside the realm of their strict validity. This procedure can be formally justified by the Gel'fand–Yaglom (1960) device of adding to the mass m a small imaginary part $i\delta$, and eventually letting $\delta \longrightarrow +0$. However, even so, divergences remain as $t \longrightarrow t'$ or $t \longrightarrow t''$.

One way of understanding the reason why all the aforementioned divergences do not affect the practical utility of Feynman path integrals is to view these integrals as the sharp-point limit of the phase-space path integrals introduced in the next section, which display no similar anomalies. To see how the Feynman theory can be recovered in a sharp-point limit, let us consider stochastic phase-space free propagators

$$K_\xi(\zeta'', t''; \zeta', t') = \left\langle \xi_{\mathbf{q}'', \mathbf{p}''} \mid \exp\left[-\frac{i}{\hbar} H_0(t'' - t')\right] \xi_{\mathbf{q}', \mathbf{p}'} \right\rangle \tag{9.21}$$

for the special case of the optimal resolution generators $\xi^{(l)}$ in (6.25). Expressing these propagators in the momentum representation, so that

$$K^{(l)}(\zeta'', t''; \zeta', t') = \int_{\mathbb{R}^3} \exp\left[\frac{i\mathbf{k}^2}{2m\hbar}(t' - t'')\right] \xi^{(l)*}_{\mathbf{q}'', \mathbf{p}''}(\mathbf{k})\xi^{(l)}_{\mathbf{q}', \mathbf{p}'}(\mathbf{k})\, d\mathbf{k} \tag{9.22}$$

and using (6.30), we find that the resulting integration of Gaussians can be as easily performed as in deriving (6.31), with the following result:

$$K^{(l)}(\mathbf{q}'', \mathbf{p}'', t''; \mathbf{q}', \mathbf{p}', t') = (2\pi\hbar)^{-3} \left[\frac{l^2}{l^2(t''-t')}\right]^{3/2} \times$$

$$\times \exp\left[-\frac{(\mathbf{q}''-\mathbf{q}')^2}{8l^2(t''-t')} - \frac{l^4}{2l^2(t''-t')}\left(\frac{\mathbf{p}''-\mathbf{p}'}{\hbar}\right)^2\right] \times$$

$$\times \exp\left\{\frac{il^2}{2\hbar l^2(t''-t')}\left[(\mathbf{q}''-\mathbf{q}')\cdot(\mathbf{p}''+\mathbf{p}') - \frac{t''-t'}{2m}(\mathbf{p}''^2+\mathbf{p}'^2)\right]\right., \tag{9.23a}$$

$$l^2(t) = l^2 + i\frac{\hbar}{4m}t, \quad t \in \mathbb{R}^1. \tag{9.23b}$$

Comparing (9.23) with (9.7), it becomes very clear that upon appropriate renormalization $K^{(l)}$ will converge to K in (9.7) as long as $t' \neq t''$, namely that

$$\lim_{l\to+0}\left(\frac{\pi\hbar^2}{2l^2}\right)^{3/2} K^{(l)}(\mathbf{q}'', \mathbf{p}'', t''; \mathbf{q}', \mathbf{p}', t') = \begin{cases} K(\mathbf{q}'', t''; \mathbf{q}', t'), & t' \neq t''. \\ \delta^3(\mathbf{q}''-\mathbf{q}'), & t' = t''. \end{cases} \tag{9.24}$$

Thus, indeed, the free propagation of an extended particle with proper wave function $\xi^{(l)}$ does approximate, upon renormalization, that of a point particle arbitrarily well for sufficiently small proper radius, but the latter displays the expected type δ-singularity as $t'' \longrightarrow t'$. The ensuing divergences are mathematically accounted for by the renormalization factor in (9.24), which becomes infinite as $l \longrightarrow +0$. However, their effect in the context of integrals such as in (9.20) is partly countermanded by the absence of **p**-integrations, which would give rise when $l \longrightarrow +0$ to additional divergences in phase-space integrals such as (9.18b), whose integrands become **p**-independent in the sharp-point limit. Hence, it now becomes clear that the earlier mentioned trick of adding $i\delta$ to $\hbar$ or $-i\delta$ to m, and then letting $\delta \longrightarrow +0$, is on account of (9.23b) tantamount to treating point particle propagation as a limit of the propagation of extended particles whose stochastic momentum degrees of freedom have been suppressed by discarding from (9.23a) all **p**-dependent terms.

1.10. Quantum Stochastic Processes and Phase-Space Path Integrals

The striking formal analogies between the properties of the Feynman probability amplitudes (9.15) and stochastic processes – in particular, Markov chains – have been pointed out by Feynman (1948), Montroll (1952), Gel'fand and Yaglom (1960), and many others. It should be recalled that most generally a stochastic process in $t \in \mathbb{R}^1$ can be defined mathematically by supplying a sample space $\mathcal{X}_t$

for each value t of the time parameter (and therefore, implicitly, a Boolean σ-algebra $\mathcal{A}_t$ of events constituting subsets of $\mathcal{X}_t$), and then specifying a family

$$P_{t_1, \ldots, t_N}(B_1 \times \ldots \times B_N), B_1 \in \mathcal{A}_{t_1}, \ldots, B_N \in \mathcal{A}_{t_N} \tag{10.1}$$

of positive-definite cylindrical measures that satisfy the basic consistency conditions

$$P_{t_1, \ldots, t_{j-1}, t_j, t_{j+1}, \ldots, t_N}(B_1 \times \ldots \times B_{j-1} \times \mathcal{X}_{t_j} \times B_{j+1} \times \ldots \times B_N)$$

$$= P_{t_1, \ldots, t_{j-1}, t_{j+1}, \ldots, t_N}(B_1 \times \ldots \times B_{j-1} \times B_{j+1} \times \ldots \times B_N),$$

$$t_1 < \ldots < t_{j-1} < t_j < t_{j-1} < \ldots < t_N, \qquad j = 1, \ldots, N, \tag{10.2}$$

for any $N = 2, 3, \ldots$, whereas for $N = 1$ we have $P_{t_1}(\mathcal{X}_{t_1}) = 1$. In the common case that $\mathcal{X}_t = \mathbb{R}^n$ for all $t \in \mathbb{R}^1$, by Kolmogorov's theorem (Billingsley, 1979) the family of cylindrical measures (10.1) extends to a probability measure over the set

$$(\mathbb{R}^n)^{\mathbb{R}} = \underset{t \in \mathbb{R}^1}{\mathsf{X}} \mathcal{X}_t \tag{10.3}$$

of all functions $\mathbf{x}(t)$, $t \in \mathbb{R}^1$, assuming values in $\mathbb{R}^n$. A function of this type can be thought of as a kind of general path in $\mathbb{R}^n$. For special types of stochastic processes, such as Brownian motion, this measure can be extended so as to yield, with probability 1, the set of continuous paths. However, this procedure does not assign a unit probability, as we would expect one physical grounds, to the special class of smooth paths resulting from Newton's second law, for which the total probability turns out to equal zero (Billingsley, 1979, p. 450).

This last remark indicates that the concept of a stochastic process based on (10.1) – to which we shall refer from now as a *classical* stochastic process – does not necessarily provide a framework that accurately reflects all aspects of physical reality even within the context of classical mechanics. The need for a wider framework becomes, however, imperative in the quantum context since albeit probability amplitudes such as (9.15) satisfy consistency conditions that parallel (10.2), they are definitely not real-valued functions of their arguments. It therefore appears desirable to introduce the notion of a *quantum stochasic process*[19] from (x', t') to (x'', t''), $x' \in \mathcal{X}_{t'}$, $x'' \in \mathcal{X}_{t''}$, defined as a family

$$K(x'', t''; B_{N-1}, t_{N-1}, \ldots, B_1, t_1; x', t'),$$

$$B_1 \in \mathcal{A}_{t_1}, \ldots, B_{N-1} \in \mathcal{A}_{t_{N-1}}, \tag{10.4}$$

of complex-valued cylindrical measures that satisfy consistency conditions analogous to (10.2):

$$K(x'', t''; B_{N-1}, t_{N-1}, \ldots, B_{j+1}, t_{j+1}, \mathscr{X}_{t_j}, t_j, B_{j-1}, t_{j-1}, \ldots, B_1, t_1; x', t')$$
$$= K(x'', t''; B_{N-1}, t_{N-1}, \ldots, B_{j+1}, t_{j+1}, B_{j-1}, t_{j-1}, \ldots, B_1, t_1; x', t'). \tag{10.5}$$

Although at first sight the modification leading from classical to quantum stochastic processes might appear rather slight, it is actually profound and gives rise to fundamentally new mathematical and physical consequences.

Physically, we observe that since K in (10.4) is complex valued, it is not K itself, but rather

$$\rho(x', t'; B_1, t_1, \ldots, B_{N-1}, t_{N-1}; x'', t'')$$
$$= \text{const } |K(x'', t''; B_{N-1}, t_{N-1}, \ldots, B_1, t_1; x', t')|^2 \tag{10.6}$$

that receives the interpretation of (conditional) *probability density for a quantum chain* from (x', t') to (x'', t''): it is the probability density that after the value $x' \in \mathscr{X}_{t'}$ has been prepared at time t' and $N-1$ subsequent *filtration* operations have been executed at $t_1 < \ldots < t_{N-1}$, the value $x'' \in \mathscr{X}_{t''}$ will be determined *if* a measurement is performed at time t''. It should be emphasized that in a filtration operation, as in a preparatory measurement (Prugovečki, 1967, 1973), no actual values $x_j \in B_{t_j}$ are determined, and in fact the existence of a system undergoing the stochastic process is not even ascertained. Rather, *if* a system is detected at t'', then it can be asserted in retrospect that due to the presence of a filter at t_j, its value x_j had to have been within the range B_j in order that it could have passed through the filter. Single-slit diffraction experiments and double-slit interference experiments provide typical illustrations of filtration procedures.

On the mathematical side, the fact that K is no longer real and positive-definite limits the range of validity of Kolmogorov's theorem: that theorem can be recovered (Gagnon, 1983) if the cylindrical measures in (10.4) are of bounded total variation for all choices of $t_1 < \ldots < t_{N-1}$, $N = 2, 3, \ldots$, but not in general. For example, the stochastic phase-space probability amplitudes

$$K_\xi(\zeta'', t''; B_{N-1}, t_{N-1}, \ldots, B_1, t_1; \zeta', t')$$
$$= \int_{B_{N-1}} d\zeta_{N-1}\, K_\xi(\zeta'', t''; \zeta_{N-1}, t_{N-1}) \int_{B_{N-2}} d\zeta_{N-2} \ldots \times$$
$$\times \int_{B_1} d\zeta_1\, K_\xi(\zeta_2, t_2; \zeta_1, t_1) K_\xi(\zeta_1, t_1; \zeta', t'),$$
$$B_1, \ldots, B_{N-1} \subset \Gamma \equiv \mathscr{X}_t, \tag{10.7}$$

that can be derived from (9.22) are not uniformly bounded, as can be checked by explicit computation for the case of $K^{(l)}$ in (9.23) (Gagnon, 1983). In case of the Feynman probability amplitudes (9.15) the situation is, as it can be expected, much worse, since we have no finite total variation even for fixed choices of $t_1, \ldots, t_{N-1}$.

The nonapplicability of Kolmogorov's theorem does not preclude, however, the possibility of defining functional integrals with respect to the cylindrical measures in (10.4), albeit a general theory of integration with respect to cylindrical measures does not exist (Schwartz, 1973, p. 180). Thus, if $F[x(t)]$ is a functional over 'paths' $x(t)$ with values in $\mathscr{X}_t$ from $x' = x(t')$ to $x'' = x(t'')$, then in accordance with (9.10) and (9.13), we set by definition

$$\int_{x',t'}^{x'',t''} F[x(t)]\, \mathrm{d}K[x(t)] = \lim_{\substack{N\to\infty \\ \epsilon\to+0}} \int_{\mathscr{X}_{t_1}\times\ldots\times\mathscr{X}_{t_{N-1}}} \times$$

$$\times F(x_1, \ldots, x_{N-1})\, \mathrm{d}K_{t_1,\ldots,t_{N-1}}(x_1, \ldots, x_{N-1}), \tag{10.8}$$

where $F(x_1, \ldots, x_{N-1})$ is the value of $F[x(t)]$ along a path for which $x_j = x(t_j)$, $j = 1, \ldots, N-1$, whereas $K_{t_1,\ldots,t_{N-1}}$ is a measure that is assumed to exist on the measurable space

$$(\mathscr{X}_{t_1}\times\ldots\times\mathscr{X}_{t_{N-1}}, \mathcal{A}_{t_1}\times\ldots\times\mathcal{A}_{t_{N-1}}) \tag{10.9}$$

and to represent an extension of (10.4):

$$K_{t_1,\ldots,t_{N-1}}(B_1\times\ldots\times B_{N-1})$$
$$= K(x'', t''; B_{N-1}, t_{N-1}, \ldots, B_1, t_1; x', t'). \tag{10.10}$$

The above definition of a functional integral is a generalization of that suggested by Gel'fand and Yaglom (1960, p. 49). Naturally, in the absence of Kolmogorov's theorem, we must verify independently the existence and uniqueness of the limit in (10.8), which should be independent of the choice of paths (from a given category of paths) and of subdivisions of $[t', t'']$. In quantum mechanics, however, these questions pose no serious difficulty since their resolution emerges automatically from the Hilbert space background of quantum mechanics. Of course, this statement does not apply to the Feynman path integral *per se*, since in arriving at the definition (9.13) via (9.10) we could take advantage of that background only in a very heuristic manner. Therefore, it is not surprising that in some cases, such as in the presence of an external vector potential, the counterpart of (9.13) leads to ambiguous answers, and the choice of $F(x_1, \ldots, x_{N-1})$ in (10.8) cannot be used if agreement with the propagation directed by the Schrödinger equation is desired (Schulman, 1981, p. 27).

In the stochastic phase-space approach based on the probability amplitudes (10.7), the path integral (10.8) assumes the form

$$\int_{\zeta', t'}^{\zeta'', t''} F[\zeta(t)]\, \mathrm{d}_\xi \zeta(t)$$

$$= \lim_{N\to\infty} \int_{\Gamma^{N-1}} F(\zeta_1, \ldots, \zeta_{N-1}) \prod_{j=1}^{N-1} K_\xi(\zeta_j, t_j; \zeta_{j-1}, t_{j-1})\, \mathrm{d}\zeta_j. \quad (10.11)$$

Its existence and uniqueness is assured for classes of functionals over broken paths that emerge when the derivation of the Feynman path integral expression for some propagator whose existence has been already established by Hilbert space methods (as is the case with (9.13)) is recast from the configuration representation to a stochastic phase-space representation. For example, upon inserting the identity operator between each pair of consecutive factors in (9.6a), we can use the resolution of the identity in (4.11) instead of the one in (9.9). We thus get the Γ_ξ counterpart of (9.10),

$$\mathcal{K}_\xi(\zeta'', t''; \zeta', t') = \lim_{N\to\infty} \int \mathrm{d}\zeta'_N \int \prod_{j=N=1}^{1} \mathrm{d}\zeta_j\, \mathrm{d}\zeta'_j \times$$

$$\times \prod_{j=N}^{1} \langle \xi_{\zeta_j} \mid \exp\left[-\frac{i}{\hbar} H_I(t_j)(t_j - t_{j-1})\right] \xi_{\zeta'_j}\rangle K_\xi(\zeta'_j; t_j; \zeta_{j-1}, t_{j-1}) \quad (10.12)$$

which holds exactly and without ambiguities or divergencies, but in which there is no counterpart to (9.11), except in an approximative sense for the average value

$$\bar{V}(j, j-1) = \frac{\int \langle \xi_{\zeta_j} \mid V(\mathbf{X}, t)\, \xi_{\zeta'_j}\rangle K_\xi(\zeta'_j, t_j; \zeta_{j-1}, t_{j-1})\, \mathrm{d}\zeta'_j}{K_\xi(\zeta_j; t_j; \zeta_{j-1}, t_{j-1})} \quad (10.13)$$

between two successive joints in a given broken path. In terms of this average interaction, (10.12) assumes the form (10.11):

$$\mathcal{K}_\xi(\zeta'', t''; \zeta', t') = \lim_{N\to\infty} \int_{\Gamma^{N-1}} \exp\left[-\frac{i}{\hbar} \sum_{j=1}^{N} (t_j - t_{j-1}) \bar{V}(j, j-1)\right] \times$$

$$\times \prod_{j=1}^{N-1} K_\xi(\zeta_j, t_j; \zeta_{j-1}, t_{j-1})\, \mathrm{d}\zeta_j. \quad (10.14)$$

The presence of an action term $\exp[(i/\hbar)S]$ with the customary phase space form of the action

$$S(j, j-1) = \int_{t_{j-1}}^{t_j} \left\{ \dot{\mathbf{q}}(t) \cdot \mathbf{p}(t) - \left[\frac{\mathbf{p}^2(t)}{2m} + V(\mathbf{q}(t), \mathbf{p}(t), t) \right] \right\} \mathrm{d}t \qquad (10.15)$$

is not evident in (10.14). If, however, we work in the stochastic phase space $\Gamma^{(l)}$ with the optimal resolution generator $\xi^{(l)}$ in (6.25), then the free propagator between two successive joints of a given broken path can be approximated as follows:

$$K^{(l)}(\mathbf{q}_j, \mathbf{p}_j, t_j; \mathbf{q}_{j-1}, \mathbf{p}_{j-1}, t_{j-1}) = \mathscr{D}_l(\mathbf{q}_j - \mathbf{q}_{j-1}, \mathbf{p}_j - \mathbf{p}_{j-1}, t_j - t_{j-1}) \times$$
$$\times \exp\left[\frac{i}{\hbar} \left(1 - \frac{i\hbar(t_j - t_{j-1})}{4ml^2} \right) (\mathbf{q}_j - \mathbf{q}_{j-1}) \cdot \mathbf{p}_j - \right.$$
$$\left. - (t_j - t_{j-1}) \overline{T}(j, j-1) + O(|t_j - t_{j-1}|^2) \right]. \qquad (10.16)$$

In (10.16) $\overline{T}(j, j-1)$ is the average kinetic energy between two successive joints of the broken path,

$$\overline{T}(j, j-1) = \frac{1}{2m} \frac{\mathbf{p}_j^2 + \mathbf{p}_{j-1}^2}{2}, \qquad (10.17)$$

and $\mathscr{D}_l$ can be formally viewed as a new element of measure:

$$\mathscr{D}_l(\mathbf{q}, \mathbf{p}, t) = (2\pi\hbar)^{-3} \left(1 - \frac{i\hbar t}{4ml^2} \right)^{3/2} \times$$
$$\times \exp\left\{ -\frac{1}{2} \left(1 - \frac{i\hbar t}{4ml^2} \right) \left[\left(\frac{\mathbf{q}}{2l} \right)^2 + l^2 \left(\frac{\mathbf{p}}{\hbar} \right)^2 + \frac{i}{\hbar} \mathbf{q} \cdot \mathbf{p} \right] \right\}. \qquad (10.18)$$

If we now insert (10.16) into (10.14), and denote by $S^{(l)}$ the averaged action (10.15) along broken paths in $\Gamma^{(l)}$, for which, therefore,

$$S^{(l)}(j, j-1) = (\mathbf{q}_j - \mathbf{q}_{j-1}) \cdot \mathbf{p}_j - [\overline{T}(j, j-1) + \overline{V}(j, j-1)](t_j - t_{j-1}), \qquad (10.19)$$

then upon neglecting terms of order $(\Delta t)^2$, $\Delta t \cdot \Delta q^j$ and $\Delta t \cdot \Delta p^j$, we obtain

$$\mathscr{K}^{(l)}(\zeta'', t''; \zeta', t') \stackrel{\text{def}}{=} \langle \xi^{(l)}_{\mathbf{q}'', \mathbf{p}''} \mid \hat{\mathscr{U}}(t'', t') \xi^{(l)}_{\mathbf{q}', \mathbf{p}'} \rangle$$
$$= \int_{\zeta', t'}^{\zeta'', t''} \exp\left(\frac{i}{\hbar} S^{(l)} \right) \mathscr{D}_l[\zeta(t)]. \qquad (10.20)$$

As emphasized by Schulman (1981, p. 249) in the closely related context of path integrals in the coherent state representation,[20] the ignoring of terms of order $\Delta t \cdot \Delta q^j$ and $\Delta t \cdot \Delta p^j$ is mathematically very questionable. Furthermore, $\mathscr{D}_l(\zeta_j, \zeta_{j-1})$ does not satisfy consistency conditions like (9.18), that would give rise to a quantum stochastic process. Hence, the path integral in (10.20) has only heuristic value on par with the one in (9.5), and its correct mathematical interpretation is still contained in (10.14). Nevertheless, we shall see in the next section that the classical action (10.15) does play a mathematically rigorous role when a semiclassical approximation of $\mathscr{K}_\xi$ is derived.

The averaged potential in (10.13) is not the one that would be 'felt' by an extended particle, but rather by a point particle. Consequently, the propagation described by (10.14) is essentially that of a point particle on which, however, measurements of stochastic position and momentum $\zeta'' = (\mathbf{q}'', \mathbf{p}'')$ are performed at t''. For the propagation of an extended particle itself under the influence of an external potential $V(\mathbf{q}, t)$, we have to replace in (10.13) the function $V(\mathbf{X}, t)$ of the sharp position operators X^j in (2.2) by the function $V(\mathbf{Q}_{\text{st}}, t)$ of the stochastic position operators Q^j_{st} in (8.29), so that, as in (8.30), the interaction Hamiltonian would be

$$H_I^\xi(t) = \mathbb{P}_\xi V(\mathbf{Q}_{\text{st}}, t)\mathbb{P}_\xi. \tag{10.21}$$

Re-expressing the resulting integral for this averaged potential $\bar{V}_\xi$ in terms of inner products in $L^2(\Gamma_{\bar{\xi}})$ by means of (7.6), we obtain a physically suggestive expression for $\bar{V}_\xi$:

$$\bar{V}_\xi(j, j-1) = \frac{\int K_\xi(\zeta_j, t_j; \zeta_j'; t_j) V(\mathbf{q}_j', t_j) K_\xi(\zeta_j'; t_j; \zeta_{j-1}, t_{j-1})\, \mathrm{d}\zeta_j'}{K_\xi(\zeta_j, t_j; \zeta_{j-1}, t_{j-1})}. \tag{10.22}$$

Naturally, as can be gathered from (8.33), in the sharp-point limit both $\bar{V}$ and $\bar{V}_\xi$ formally approach upon renormalization the same entity, namely $V(\mathbf{q}_j, t)$.

1.11. Semiclassical Propagators in Stochastic Phase Spaces

The heuristic physical image that emerges from the path-integral expressions (9.5) or (10.20) for the nonrelativistic propagator of a quantum particle is that of random propagation which for infinitesimal time intervals $(t, t + \mathrm{d}t)$ proceeds with a probability amplitude which is the superposition of $\exp[(i/\hbar)S]\, \delta\mathbf{x}(t)$ for all path segments over that time interval, where S is the action along each such path segment. Since the classical path corresponds to minimum action S_{cl}, and therefore it is the solution of the variational equation $\delta S = 0$, the path segments in its immediate neighborhood will provide the major contribution, whereas the phases of the remaining ones will approximately cancel out in view of the rapid oscillations of

$\exp[(i/\hbar)\delta S]$ for large $\hbar^{-1}\,\delta S$ (Feynman and Hibbs, 1965, p. 29). This leads to the WKB semiclassical approximation (Schulman, 1981, p. 94).

$$\mathscr{K}_{\mathrm{cl}}(\mathbf{x}'',t'';\mathbf{x}',t') = \left[\det\left(\frac{i}{2\pi\hbar}\,\frac{\partial^2 S}{\partial\mathbf{x}'\,\partial\mathbf{x}''}\right)\right]^{1/2} \exp\left[\frac{i}{\hbar}\,S_{\mathrm{cl}}(t'';\mathbf{x}',t')\right] \tag{11.1}$$

of the Feynman propagator in (9.5), but rigorous evaluations of the margins of error introduced when $\mathscr{K}$ is replaced by $\mathscr{K}_{\mathrm{cl}}$ are hampered by the difficulty of estimating the actual contribution from all those rapidly oscillating phases for paths not close to the classical one.

These difficulties do not occur, however, when the propagator

$$\mathscr{K}^{(l)}(\zeta'',t'';\zeta',t') = \left\langle \xi^{(l)}_{\zeta''} \mid \exp\left[-\frac{i}{\hbar}H(t''-t')\right]\xi^{(l)}_{\zeta'}\right\rangle \tag{11.2}$$

in $\Gamma^{(l)}$ is approximated by a semiclassical propagator of the form

$$\mathscr{K}^{(l)}_{\mathrm{cl}}(\zeta'',t'';\zeta',t') = \exp\left[\frac{i}{\hbar}\,S_{\mathrm{cl}}(t'';\zeta',t')\right]\langle \xi^{\mathrm{cl}}_{\zeta'',t''} \mid \xi^{\mathrm{cl}}_{\zeta',t'}\rangle, \tag{11.3}$$

where $S_{\mathrm{cl}}(t'';\zeta',t')$ is the action

$$S_{\mathrm{cl}}(t'';\zeta',t') = \int_{t'}^{t''}\left[\frac{\mathbf{p}^2_{\mathrm{cl}}(t)}{2m} - V(\mathbf{q}_{\mathrm{cl}}(t))\right]\mathrm{d}t \tag{11.4}$$

along the classical path starting from $\zeta' = (\mathbf{q}',\mathbf{p}')$ at t' and terminating at t'' at the point $\zeta''_{\mathrm{cl}} = (\mathbf{q}_{\mathrm{cl}}(t''), \mathbf{p}_{\mathrm{cl}}(t''))$ determined by Hamilton's equations of motions:

$$\dot{\mathbf{q}}_{\mathrm{cl}}(t) = \mathbf{p}_{\mathrm{cl}}(t)/m, \qquad \dot{\mathbf{p}}_{\mathrm{cl}}(t) = -\nabla\cdot V(\mathbf{q}_{\mathrm{cl}}(t)). \tag{11.5}$$

Indeed, according to a theorem by Hagedorn (1980), there are elements $\xi^{\mathrm{cl}}_{\mathbf{q},\mathbf{p},t}$ in $L^2(\mathbb{R}^3)$ of the form

$$\xi^{\mathrm{cl}}_{\mathbf{q},\mathbf{p},t}(\mathbf{x}) = \left(\frac{8\pi^3\hbar^2}{l^2}\right)^{-3/4} (\det A(t))^{-1/2} \times$$

$$\times \exp\left\{-(\mathbf{x}-\mathbf{q})\cdot B(t)\left[A^{-1}(t)\,\frac{\mathbf{x}-\mathbf{q}}{4} + \frac{i}{\hbar}B^{-1}(t)\mathbf{p}\right]\right\} \tag{11.6}$$

for which at any t'' there is a constant $C(t''-t')$ such that

$$\left\|\exp\left[-\frac{i}{\hbar}H(t-t')\right]\xi^{(l)}_{\mathbf{q},\mathbf{p}} - \exp\left[\frac{i}{\hbar}S_{\mathrm{cl}}(t;\mathbf{q},\mathbf{p},t')\right]\xi^{\mathrm{cl}}_{\mathbf{q}(t),\mathbf{p}(t),t}\right\| <$$

$$< C(t''-t')\hbar^{\beta/2} \tag{11.7}$$

for all $t \in [t', t'']$; here $\beta > 0$ is a fixed number such that[21] $V \in \mathscr{C}^{\beta+2}(\mathbb{R}^3)$, and $A(t)$ and $B(t)$ are 3×3 matrix-valued functions:

$$A(t) = l^2 \frac{\partial \mathbf{q}_{\text{cl}}(t)}{\partial \mathbf{q}_{\text{cl}}(0)} + \frac{i\hbar}{2} \frac{\partial \mathbf{q}_{\text{cl}}(t)}{\partial \mathbf{p}_{\text{cl}}(0)}, \tag{11.8a}$$

$$B(t) = \frac{\partial \mathbf{p}_{\text{cl}}(t)}{\partial \mathbf{p}_{\text{cl}}(0)} - \frac{2i}{\hbar} l^2 \frac{\partial \mathbf{p}_{\text{cl}}(t)}{\partial \mathbf{q}_{\text{cl}}(0)}. \tag{11.8b}$$

It can be readily established by explicit computation that $\mathscr{K}^{(l)}_{\text{cl}}$ in (11.3) satisfies the consistency conditions of a propagator:

$$\mathscr{K}^{(l)}_{\text{cl}}(\zeta', t'; \zeta'', t'') = \mathscr{K}^{(l)*}_{\text{cl}}(\zeta'', t''; \zeta', t')$$

$$= \int_\Gamma \mathscr{K}^{(l)}_{\text{cl}}(\zeta', t'; \zeta, t) \mathscr{K}^{(l)}_{\text{cl}}(\zeta, t; \zeta'', t'')\, d\zeta. \tag{11.9}$$

According to (11.5), (11.7) and (6.29) $\xi^{\text{cl}}_{\mathbf{q},\mathbf{p},t'}(\mathbf{x})$ coincides with $\xi^{(l)}_{\mathbf{q},\mathbf{p}}(\mathbf{x})$, so that for all $\zeta \in \Gamma$

$$\mathscr{K}^{(l)}(\zeta, t'; \zeta', t') \equiv \mathscr{K}^{(l)}_{\text{cl}}(\zeta, t'; \zeta', t'), \tag{11.10}$$

and by (11.2), (11.3) and (11.7)

$$|\mathscr{K}^{(l)}(\zeta, t; \zeta', t') - \mathscr{K}^{(l)}_{\text{cl}}(\zeta, t; \zeta', t')| < \frac{C(t'' - t')}{(2\pi)^{3/2}} \hbar^{(\beta-3)/2} \tag{11.11}$$

for all $t \in [t', t'']$ and $\zeta \in \Gamma$. We see from (11.3)–(11.6) that $\mathscr{K}^{(l)}_{\text{cl}}(\zeta, t; \zeta', t')$ can be considered to be a semiclassical propagator from the stochastic phase space $\Gamma^{(l)}$ to a phase space Γ'_t of stochastic points whose confidence functions are not those in (6.27), but are nevertheless Gaussians of spreads in position and momentum which correspond to the eigenvalues of the following respective 3×3 matrices:

$$(\operatorname{Re} BA^{-1})^{-1/2} = \frac{1}{l}(AA^*)^{1/2}, \qquad \frac{\hbar}{2}(\operatorname{Re} AB^{-1})^{-1/2} = \frac{\hbar}{2l}(BB^*)^{1/2}. \tag{11.12}$$

Interestingly, the estimate (11.7) remains valid if $\xi^{(l)}$ is replaced by any other eigenfunction of a quantum haromonic oscillator, and the corresponding $\xi^{\text{cl}}_{\mathbf{q},\mathbf{p},t}$ is computed in a manner analogous to (11.6) (Hagedorn, 1981). Consequently, the above theory of semiclassical approximations of quantum propagators $\mathscr{K}_\xi$ remains valid (Gagnon, 1983) for proper wave functions corresponding to nonrelativistic counterparts of higher exciton states of Born's quantum metric operator, which will be discussed at length in Chapter 4.

1.12. The Physical Nature and the Measurement of Proper Wave Functions and of Stochastic Position and Momentum

The fundamental mathematical distinction between quantum point particles and stochastically extended ones is that the latter possess a proper wave function ξ, whereas the former do not – or rather, in the nonrelativistic momentum representation[22] their proper wave function can be viewed as equal (over all momentum space) to the constant corresponding to the limiting process in (8.1) at $\mathbf{q} = \mathbf{0}$. Of course, it would be incorrect to visualize an extended quantum particle as a blob of matter, i.e. to conceptualize its extension in classical terms. Rather, that extension should be conceived in purely stochastic terms that reflect operational measurement procedures, and for which the term 'probabilistic cloud' so very common in quantum mechanics provides the most adequate descriptive connotation.

The 'probabilistic cloud' that embodies the physical attributes contained in the mathematical object which we call a 'proper wave function' might originate from the extended particle with that proper wave function being a bound state of two or more constituent particles, which themselves might be either pointlike or extended. For example, in the conventional nonrelativistic model of the hydrogen atom, the entire atom is viewed as a composite system consisting of two pointlike constituents, namely the proton and electron. If $\mathbf{r}'$ and $\mathbf{r}''$ denote the position vectors for the proton and the electron, respectively, then the hydrogen atom wave function in the configuration representation is (Messiah, 1962)

$$\psi_H(\mathbf{r}', \mathbf{r}'') = \psi_{\text{c.m.}}(\mathbf{R})\,\psi_{nlm}(\mathbf{r}), \tag{12.1a}$$

$$\mathbf{R} = \frac{m'\mathbf{r}' + m''\mathbf{r}''}{m' + m''}, \qquad \mathbf{r} = \mathbf{r}' - \mathbf{r}'', \tag{12.1b}$$

where $\psi_{\text{c.m.}}$ is its center-of-mass part, and ψ_{nlm} is the internal motion part labelled, as customary, by the principal, azimuthal and magnetic quantum numbers n, l and m, respectively. In general $|\psi_{\text{c.m.}}(\mathbf{R})|^2$ will be the probability density in the laboratory frame of the center-of-mass being at $\mathbf{R}$. However, let us imagine that the center-of-mass frame is actually marked experimentally, e.g. by setting a 'classical' (namely, infinitely massive – e.g. the proton itself to a good approximation) point particle at its origin O. Then, from the point of view of the observer in this center-of-mass frame, $|\psi_{nlm}(\mathbf{r})|^2$ provides the probabilistic reduced mass distribution around the center-of-mass O, and $|\tilde{\psi}_{nlm}(\mathbf{k})|^2$ the probability density of 3-momentum of this mass distribution. Hence, in the ground state, i.e. for $n = 1$, $l = m = 0$, $\psi_{100}(\mathbf{x}) = \xi(\mathbf{x})$ behaves like the proper wave function of a stochastically extended particle. In the excited states, i.e. for $n > 1$, the role of proper wave function in the configuration representation is played by the element $\xi_{nl}(\mathbf{x})$ of $l^2(2l+1) \otimes L^2(\mathbb{R}^3)$, i.e. by the vector-valued function of $\mathbf{x} \in \mathbb{R}^3$ with the $2l+1$ components ψ_{nlm}, $m = -l, \ldots, +l$, so that l plays the role of intrinsic spin[23] of the extended system. For an observer stationed at $t = 0$ at the origin O' of a translated inertial frame, which is at $\mathbf{q}$ in relation to O, and moving with velocity

$\mathbf{v} = \mathbf{p}/(m' + m'')$ in relation to the center-of-mass frame, the proper wave function shall be, by (2.6),

$$\xi'_{nl}(\mathbf{x}) = \exp\left[\frac{i}{\hbar}\,\mathbf{p}\cdot(\mathbf{x}-\mathbf{q})\right]\xi_{nl}(\mathbf{x}-\mathbf{q}). \tag{12.2}$$

Naturally, in these considerations, not only is the assumption that the proton is a point particle unwarranted, but so is the one that the center-of-mass can be marked with total accuracy. Both those assumptions shall be dispensed with in Chapter 4.

In case of systems consisting of three or more constituents, random position variables related to the stochastic distribution of internal physical attributes (e.g. charge, magnetic moments, etc.) have to be introduced, and the proper wave functions for each of those attributes are to be expressed then in terms of those variables. For example, for a system consisting of Z identical charged pointlike particles moving in a potential well centered at their center-of-mass (as is the case, approximately, for the electrons in an atom, or the protons in a nucleus), the charge *form factor* is defined in the nonrelativistic context to be (Messiah, 1962; Hofstadter, 1957) for elastic scattering equal to

$$F_Z(\mathbf{s}) = \int \sum_{k=1}^{Z} \exp\left(\frac{i}{\hbar}\,\mathbf{s}\cdot\mathbf{r}_k\right)\,|\psi(\mathbf{r}_1,\ldots,\mathbf{r}_Z)|^2\,d\mathbf{r}_1\ldots d\mathbf{r}_Z, \tag{12.3}$$

where ψ is the configuration space wave function of all the constituents (i.e. electrons or protons, respectively) in the external field that keeps them bound together. Then the proper wave function ξ_Z for the charge distribution in the ground state could be phenomenologically required to satisfy the equation

$$\xi_Z^2(\mathbf{x}) = (2\pi\hbar)^{-3}\,Z^{-1}\int_{\mathbb{R}^3}\exp\left(-\frac{i}{\hbar}\,\mathbf{x}\cdot\mathbf{s}\right)F_Z(\mathbf{s})\,d\mathbf{s}, \tag{12.4}$$

since then indeed $Ze\,|\xi_Z(\mathbf{x})|^2$ represents the effective charge density within the system. We note that the properties of rotational symmetry (Section 1.4) and reality (Section 1.8) of ξ_Z in the spinless cases impose stringent requirements on the form factor $F_Z(\mathbf{s})$, which has to share those properties. When that is the case, the form factor determined from scattering experiments (Hofstadter, 1957) in its turn determines by (12.4) the proper wave function $\xi_Z(\mathbf{x})$ up to the sign of the square root at each $\mathbf{x} \in \mathbb{R}^3$.

It is, however, not necessary to require that a quantum particle should be composite in order that it be extended in the stochastic sense. In fact, when the concept of proper wave function was first introduced by Landé (1939) and Born (1939), it had been presented as a primitive concept, rather than as the internal wave function for the relative motion of the constituents in a composite system. Thus, when we apply the stochastic phase-space formalisms to hadrons, we have two choices: (1) we can regard the proper wave function as the outcome of the confinement of a quark–antiquark pair or of three quarks constituting, respectively, a meson and a baryon in any of the currently popular quark models (Close, 1979); (2) we can regard it as a fundamental entity that characterizes the spatio-temporal distribution of the intrinsic characteristics of an elementary particle in relation to an inertial frame, or, more generally, in relation to other particles playing the role of test bodies.

Albeit it is very tempting to pursue only the first alternative since it enjoys much popular acclaim in leading contemporary elementary particle circles, on a foundational level the second alternative has to be given serious consideration on both epistemological and mathematical grounds. Epistemically, as pointed out very forcefully by Heisenberg (1976) in his last published paper, the notion of constituent of an elementary particle might be physically meaningless if it turns out that in reaching hadronic dimensions one has reached the lower limits in the operational divisibility of space – or of spacetime, in the relativistic context. In other words, albeit the Western cultural tradition that dates back to ancient Greek philosophers (in particular, Plato), psychologically compels us to envisage space and time as a continuum that can be divided into smaller parts without end, that conceptualization would be empirically meaningless if this thought-process of division does not have an experimental counterpart by which bits of matter marking spatial intervals can be split in the laboratory into smaller and smaller parts that delineate the ever-decreasing spatial intervals. Thus, Heisenberg (1976) finds all constituent models of hadrons, with their permanent quark confinement feature, to be physically ill-posed, and in the long run liable to lead to a theoretical dead-end reminiscent of that first encountered in the Ptolemaic theory of planetary motions. More recently, some researchers (e.g. Mignani, 1981) point out that the great proliferation by the late seventies of various species[24] of quarks appears to enhance this analogy, which is underlined by the fact that no consistent mechanism for quark confinement has been discovered as yet in QCD (Altarelli, 1982; Huang, 1982) or other nonstatic quark models. In addition, a great many other internal inconsistencies of quark models have been pointed out in recent years by Santilli (1979, 1981), thus substantiating Heisenberg's negative evaluation of the long-range prospects of constituent models for hadrons.

Nevertheless, it might turn out that quarks do exist[25] as true, independent entities, despite the negative outcome (Marinelli and Marpurgo, 1980) in checking claims (LaRue *et al*., 1977, 1979) as to the experimental observation of fractionally charged particles which might represent the long sought-after hadron constituents. Although such a *direct* experimental result would invalidate Heisenberg's criticism of hadronic constituent models *per se*, it would not nullify his deeper epistemological observations. Indeed, in that case we would be faced with either accepting an infinite regression of prequark (Pati *et al*., 1975; Harari, 1979; Schupe, 1979), pre-prequark, pre-pre-prequark, etc., constituents, or postulating that after a finite number of steps we reach fundamental constituents whose extension is a primitive feature rather than due to the presence of subconstituents.

These emerge as the only two alternatives if we assume that basic relativistic and quantum mechanical postulates still govern the behavior of the postulated constituents of hadrons. We base this conclusion on the fact that relativistic quantum mechanics displays well-known inconsistencies (to be reviewed and discussed in Sections 2.1–2.2) if the concept of sharp localizability is retained, but that, as shown in Chapters 2 and 4, these inconsistencies disappear if that concept is removed, and massive particles are assumed to possess a (nontrivial) proper wave function.

A particle whose proper wave function is not constituent-based can be operationally envisaged as a totality of stochastic distributions of physical attributes (mass, charge, spin, etc.) which are revealed as a result of measurement procedures involving other particles in the role of test bodies. We shall discuss in a self-consistent manner measurement procedures related to spatio-temporal relationships in Chapter 4. For the time being, let us take for granted the existence of classical inertial frames of reference which stand in deterministic relationships to each other, and in relation to which arbitrarily accurate measurements of position and momentum can be performed along the lines of Heisenberg's (1930) well-known *gedanken* experiments – such as measurements of position and momentum with an idealized optical or electron microscope combined with a wave-frequency or velocity spectrograph. In view of (5.9)–(5.10) it is perfectly consistent with the uncertainty principle to assume that if we adjust the accuracy of the apparatus (e.g. the chosen wavelength of the beam of light or of electrons, the aperture of the focusing lens of the microscope, etc.) to the size of the observed extended particles, we can measure simultaneously their stochastic position $(\mathbf{q}, \chi^\xi_\mathbf{q})$ and stochastic momentum $(\mathbf{p}, \bar{\chi}^\xi_\mathbf{p})$ as specified in (5.2)–(5.3b). For example, by recording N successive collisions between the observed particle and the test particles (e.g. the photons or electrons that pass through the aforementioned microscope and spectrograph) we can infer at times $t_0, \ldots, t_N$ the positions $\mathbf{q}_0, \ldots, \mathbf{q}_N$ and momenta $\mathbf{p}_1, \ldots, \mathbf{p}_N$ within respective ranges $B_0, \ldots, B_N$. However, whereas $\mathbf{q}_0, \ldots, \mathbf{q}_N$ are determined by the direction of the incident beam of test particles and the direction of the test particles deflected in the microscope (Heisenberg, 1930), the measurement of the momenta of these test particles before and after collision determines only the recoil momenta, rather than $\mathbf{p}_0, \ldots, \mathbf{p}_N$ themselves. On the other hand, the conditional probability that the observed particle should follow the above described zigzag pattern (by (9.21)) equals[26]

$$(2\pi\hbar)^{3N} \int_{B_N} d\mathbf{q}_N \, d\mathbf{p}_N \ldots \int_{B_1} d\mathbf{q}_1 \, d\mathbf{p}_1 \prod_{j=1}^{N} |K_\xi(\mathbf{q}_j, \mathbf{p}_j, t_j; \mathbf{q}_{j-1}, \mathbf{p}_{j-1}, t_{j-1})|^2, \tag{12.5}$$

if it had started at the instant t_0 from the stochastic position $\underset{\sim}{\mathbf{q}}_0$ with stochastic momentum $\underset{\sim}{\mathbf{p}}_0$. If the above measurement is *determinative*,[27] then the determined values $\mathbf{q}_0$ and $\mathbf{p}_0$ at t_0 are obtained by finding the maximum value of (12.5) subject to the constraints derived from the measurements in the incident and deflected test particles – such as the most likely locations and times of collisions, the reçoil momenta, etc. For N large enough, this maximum will be sufficiently peaked by comparison with $\chi_{\mathbf{q}}^{\xi}$ and $\tilde{\chi}_{\mathbf{p}}^{\xi}$ to justify the adoption of $\mathbf{q}_0$ and $\mathbf{p}_0$ as the values determined with *practical* certainty. On the other hand, in a *preparatory* measurement the conditional probability

$$(2\pi\hbar)^{3N} \int_{B_{N-1}} d\mathbf{q}_{N-1}\, d\mathbf{p}_{N-1} \cdots$$

$$\cdots \int_{B_0} d\mathbf{q}_0\, d\mathbf{p}_0 \prod_{j=0}^{N-1} |K_\xi(\mathbf{q}_{j+1}, \mathbf{p}_{j+1}, t_{j+1}; \mathbf{q}_j, \mathbf{p}_j, t_j)|^2 \tag{12.6}$$

has to be maximized in order to compute the values $\mathbf{q}_N$ and $\mathbf{p}_N$ prepared at t_N.

A variant of the above measurement[28] procedure for stochastic position and momentum which reflects actual contemporary experimental praxis would employ macrodetectors – such as Wilson or bubble chambers – instead of microdetectors, such as the photons or electrons in the abovementioned beams. In this case $\mathbf{q}_0, \ldots, \mathbf{q}_N$ would be the recorded points on the particle 'track' (e.g. the centers of cndensation in a Wilson chamber, or of evaporation in a bubble chamber) resulting from the ionization process caused by the observed charged particle as it passes through those chambers. Since there are no simple means to measure the recoil momenta, an external magnetic field might be introduced, and the determined value $\mathbf{p}_0$, or the prepared value $\mathbf{p}_N$ deduced (cf., e.g., Prugovečki, 1981a, pp. 5–7) from the classical trajectory that best fits the observed track. This procedure is fully in accord with the results of Section 1.11 on the semiclassical approximation (based on (11.3) and (11.4)) for the actual propagation of an extended quantum particle.

The nonrelativistic semiclassical image of an extended quantum particle that emerges from the above consideration is that of a spatially probabilistic cloud that might be either primarily amorphous or markedly granular in its detailed structure described by $\chi_{\mathbf{q}}^{\xi}$, and possessing a diffuse stochastic distribution $\tilde{\chi}_{\mathbf{p}}^{\xi}$ of momenta in relation to classical inertial rest frames. Although mathematically consistent, this description does not pose, however, the deeper epistemological question as to whether strictly deterministic, i.e. truly classical frames and measuring apparatuses are available in actuality. The relativistic counterparts of the considerations in this chapter will eventually force us to abandon this naive image and to view a quantum particle as inserted in the midst of a physical reality that can be described exclusively in quantum terms. Nevertheless, the considerations of this chapter will remain of practical utility since they reflect contemporary modes of experimental methodology and are very close to the prevalent manner of viewing and analyzing the raw experimental data.

Notes

1 Albeit the introduction of mathematical concepts which might be unfamiliar to most physicists will be avoided as much as possible, an exact formulation of fundamental physical concepts does require some use of the basics of measure theory and of Hilbert space theory. For the convenience of the reader, the necessary prerequisites have been summarized in Appendices A and B.

2 Since the act of measurement is liable to 'disturb' the system, a more precise statement necessitates the differentiation between *preparatory* and *determinative* measurements (Prugovečki, 1967, 1973a, 1981a). We shall return to this point in Section 1.12. Note also that due to such relations as (1.8), (4.16) and (5.2), in later applications (cf. (5.1)) we can also interpret $\chi_{\mathbf{q}}(\mathbf{x})$ as the probability density that the reading $\mathbf{q}$ is obtained when the actual sharp value is $\mathbf{x}$.

3 Stochastic values (Prugovečki, 1979) were originally called 'fuzzy' (Prugovečki, 1973–75; Ali and Emch, 1974). The concept itself was, however, implicitly contained in earlier papers by Born (1955a, c), and although the first paper (Prugovečki, 1974) explicitly advocating the present approach to quantum mechanics and simultaneous measurability of observables made use of the extensively developed theory of fuzzy sets (Kaufmann, 1975), later research indicated that the connection was somewhat indirect: in probability theory based on fuzzy sets (Zadeh, 1965, 1968) the events are 'fuzzy', whereas in stochastic phase spaces, and in stochastic geometries in general, it is the sample points themselves that can be 'fuzzy' in the sense of being spread-out stochastic values. Mathematically, these two frameworks can be bridged by the concept (Frank, 1971) of probabilistic topological space.

4 Throughout this monograph we shall follow a suggestion first put forth by von Neumann (1931, 1955) and interpret the CCRs in the sense of the Weyl relations, thus avoiding domain questions (Putnam, 1967; Prugovečki, 1981a, pp. 339–342) and at the same time eliminating anomalous representations that are not physically acceptable (Peierls, 1979).

5 Strictly speaking, for functions $\psi(\mathbf{x}, t)$ that are Lebesgue square-integrable but not integrable in $\mathbf{x}$ over $\mathbb{R}^3$, the Fourier–Plancherel transform is given by a limit-in-the-mean (l.i.m.) rather than by the integral in (2.11), which is then divergent (cf. (B.24)–(B.30) in Appendix B). However, for the sake of notational simplicity we shall abide throughout this book to the custom followed in physics literature of *not* putting l.i.m. in front of the integral symbol in this and similar cases.

6 Many authors (Jauch and Piron, 1967; Gudder, 1979) include (3.6) into the definition of POV measures, so that all their POV measures are normalized. However, as we shall see shortly, the concept of quantum localization in spacetime necessitates the use of non-normalized POV measures, whose general use in defining systems of covariance has been advocated only recently (Cattaneo, 1979; Scutaru, 1979; Castrigiano and Heinrichs, 1980; Ali, 1981, 1982; Giovannini, 1981).

7 Note that in most so-called nonlocal theories, sharp localizability is retained, and the nonlocal features usually occur in the structure of the interaction term, or, in case of quantum field theories, in some formal properties of the fields (cf. Chapter 5). On the other hand, pure *S*-matrix theories require no notion of locality whatsoever.

8 As recalled, these origins do lie in the 'science of measurement' – as expressed by the very term 'geo-metry'. However, due to Euclid's authoritative 'Elements', the budding ideas became fossilized into a deductive discipline, whose axioms and primitive concepts (line, point, etc.) acquired the respective status of '*a priori* truths' and 'ideas intrinsic to the human mind'. The quote from Einstein's writings that heads Chapter 4 reflects the spirit of the analysis carried out in that chapter, which is one of search after a mathematical framework best suited to certain measurement–theoretical ideas, rather than of forcing such ideas into an *a prioristically* given rigid mathematical scheme.

9 Recent and apparently independent rederivations (Aharanov *et al.*, 1981; O'Connell and Wigner, 1981; O'Connell and Rajagopal, 1982; Jannussis *et al.*, 1982) of some of the basic

aspects of this formalism still used the configuration representation as a starting point. However, the subsequently given derivation is not only more elegant and physically incisive, but as it will become clear in Chapter 4, it also has the advantage (Prugovečki, 1982c) of leading to a geometrization of quantum mechanics.

[10] The introduction of proper wave functions as a means of describing elementary particles which are not pointlike was proposed by Landé (1939) and Born (1939) almost two decades prior to the actual discovery of Hofstadter (1956, 1957) of the extended nature of the proton and neutron, and of the subsequent realization that all baryons as well as all mesons possess an r.m.s. radius of the order of 1 fm = 10^{-13} cm. For the time being, we shall leave open the question as to whether this extension is necessarily due to a quark substructure, or possibly has an alternative explanation. This point is discussed in Section 1.12, where we also discuss how actual simultaneous measurements of stochastic position and momentum can be performed.

[11] By (4.6) and Fubini's theorem, the domain of $\mathbb{P}_\xi$ contains the dense set $L^2(\Gamma) \cap L^1(\Gamma)$. Since by (6.10) and (6.12) $\mathbb{P}_\xi$ is idempotent and symmetric, it is necessarily bounded: $\|\mathbb{P}_\xi\psi\|^2 = \langle\psi|\mathbb{P}_\xi\psi\rangle \leqslant \|\psi\|\,\|\mathbb{P}_\xi\psi\|$. Hence, by the extension principle (Prugovečki, 1981a, p. 188) – i.e. by implicitly using l.i.m. in (6.11) – the domain of P_ξ can be taken to be $L^2(\Gamma)$.

[12] These wave functions are actually the kind of proper wave functions first proposed by Landé (1939). They play a central role in the models for the simultaneous measurement of stochastic position $\underset{\sim}{q}$ and stochastic momentum $\underset{\sim}{p}$ studied by Busch (1982) and by Yuen (1982).

[13] This follows by the same argument as the one to be used in the relativistic context in Section 4.6.

[14] The relation (9.8) is a rigorous result for any $\psi_{t'} \in L^2(\mathbb{R}^3)$ if we in general interpret the integral with an l.i.m. in front of it (Prugovečki, 1981a, pp. 437, 492).

[15] Gel'fand and Yaglom (1960) proposed that the covergence of the integrals in (9.13) be secured by adding a small imaginary part $-i\delta$ to $\hbar$, and eventually letting $\delta \longrightarrow +0$. Cameron (1960), however, showed that the method does not yield a functional integral – as Gel'fand and Yaglom had hoped. Some of the later attempts are reviewed in recent conference proceedings, such as those edited by Papadopoulos and Devreese (1978), or by Albeverio *et al.* (1979).

[16] Usually, however, at a price which severely restricts their applicability in practice – such as the requirement that $V(\mathbf{x}, t)$ be a bounded potential (Albeverio and Høegh-Krohn, 1976).

[17] This analogy has been exploited to the ultimate by Nelson (1966, 1967), who proposed the idea of a *stochastic mechanics* in which the propagation of a quantum particle obeying the Schrödinger equation would be associated with a diffusion process for a classical particle in an external random field. Sometimes this approach, in which quantum aspects are purportedly reduced to classical ones, is mistakenly referred to as stochastic 'quantum' mechanics. It is, however, very dissimilar both in spirit and in substance from the present stochastic quantum mechanics, which acknowledges the uncertainty principle as a fundamental aspect of physical reality that, as such, is embedded in the very fabric of the chosen stochastic spaces, whereas in the stochastic mechanics approach it is a derived property of the position – osmotic velocity uncertainty (de Falco *et al.*, 1982). Reviews of Nelson's and other stochastic approaches to quantum mechanics that are not based on stochastic spaces can be found in the articles by De Witt and Elsworthy (1981) and by Zaslansky (1981).

[18] See Schulman (1981), pp. 311–314. A special form of stochastic phase-space path integral is implicit in the work reviewed by Schulman (1981, pp. 256–258) in the context of the coherent state representation. Indeed, the coherent state representation of a wave packet emerges from its $\Gamma^{(l)}$-representation when the transformations (6.36)–(6.38) are executed.

[19] The definition of quantum stochastic processes suggested by Davies (1976, p. 70) and Lewis (1981) is rather different from the present one, but a connection can be established if we integrate the probability density (10.6) over $x'' \in \mathscr{X}_{t''}$ and then express the result in terms of traces in the context of the results in Chapter 3. The present definition has the advantage

of representing a straightforward extrapolation of the conventional definition of stochastic processes, which is customarily given in terms of the cylindrical measures in (10.1) (Billingsley, 1979, p. 431).

20 The broken paths in Γ_ξ are, however, not discontinuous, as they would have to be if **p** were proportional to the velocity of a classical particle following a broken path in configuration space. Thus, their segments are not 'classical paths' (Schulman, 1981, p. 305). This is not surprising if we recall that even in Brownian motion the set of smooth paths has Wiener measure zero, so that they do not contribute to a Wiener (path) integral.

21 For the sake of simplicity, we have specialized Theorem 1.1 of Hagedorn (1980, 1981) to the case $\alpha = 1/2$ and $n = 3$.

22 As in the case with ordinary wave functions in quantum mechanics, the proper wave function ξ can be expressed in the configuration representation as $\xi(\mathbf{x})$, in the momentum representation as $\tilde{\xi}(\mathbf{k}) = (U_F\xi)(\mathbf{k})$ (in accordance with (2.11)) – or in any other spectral representation space $L^2_\mu(\mathbb{R}^n)$ of a complete set of observables. Alternatively, it can be expressed in any stochastic phase-space representation space $L^2(\Gamma_{\xi'})$ for extended particles of proper wave functions ξ', including those with $\xi' = \xi$. Naturally, if the particles of proper wave function ξ have spin and/or other internal degrees of freedom, ξ will assume values in some multidimensional Hilbert space rather than just complex values.

23 For the sake of simplicity in exposition and notation, in this chapter we have not considered the case of nonzero spin $j = ½, 1, \ldots$. However, spin can be incorporated in all considerations of this chapter (Ali and Prugovečki, 1983) by considering stochastic phase-space representations of $\mathcal{G}$ which correspond to representations $D^{(j)}(R)$, $j = ½, 1, \ldots$ (Miller, (1972) of SU(2). In fact, the case of nonzero spin shall be considered in Sections 2.3, 4.6 and 5.4 in the relativistic context. Studies of stochastic spin spaces have also been initiated (Prugovečki, 1977b; Schroeck, 1982b).

24 At present, six flavours and three colours for a total of eighteen species – i.e. greater in number than the number of hadrons known at the time of inception of the quark model, and for whose classification the model was originally intended. This leads to 26 *ad hoc* (Salam, 1979) parameters in six-quark models based on SU(2) × U(1) × $SU_c(3)$. In grand unification schemes of electro-weak and strong interactions with a total of 45 quarks and leptons that are based on SU(45) (Georgi and Jarlskog, 1979), the number of elementary gauge bosons reached 2024 – and rather large groups are still used in more recent attempts at 'grand unification' (Frampton *et al.*, 1980; Konuma and Maskawa, 1981) and 'superunification' (Ellis *et al.*, 1980). Barut (1981, pp. 101–102) presents an instructive scenario as to how atomic theory would have evolved if instead of the Bohr model of atoms the type of reasoning used in quark theory had been adopted: contemporary atomic physics would have had to view atoms as consisting of four species of 'quarks' bound together by 'gluon' forces!

25 It should be realized that all indirect evidence (Close, 1979; Huang, 1982; Altarelli, 1982) for quark 'existence' – such as large transversal components of momentum transfers, the jet phenomenon, etc. – might be, first of all, implicitly dependent on the model underlying the interpretation of the raw experimental data (Ktoridis *et al.*, 1980; Santilli, 1981), and second, it might be ascribable to intrinsic proper wave functions which are not constituent-based, but are nevertheless capable of describing granular distributions of matter. On the other hand, if daring speculators as to quarks being actually leptons in a 'hadronic phase' of the physical vacuum (Rajasekaran, 1980, p. 227) turn out to have any valid theoretical basis, then the question of quark existence would have to be viewed in a new light.

26 That is, assuming that between collisions the particle under observation moves freely, as in the similar considerations of Blokhintsev (1968) and Skagerstam (1976, p. 215), rather than in an external field, in which case (9.16) has to be employed instead of (9.21). The probability (12.5) should not be confused with the one in (10.6): the latter corresponds to a sequence of filtration procedures at $t_1, \ldots, t_{N-1}$, whereas the former is based on actual determinations of position resulting from collision processes at $t_1, \ldots, t_{N-1}$.

[27] An information–theoretical discussion of preparatory versus determinative measurements can be found in Prugovečki (1973, 1975). In most cases, however, we can say that in a *preparatory* measurement we are interested in the values of the measured quantities immediately after the interaction of the system with the test particle (or, in general, with the preparatory section of the apparatus); whereas in a *determinative* measurement, the values just prior to the system–apparatus interaction are deduced. See Prugovečki (1967; 1981a, pp. 5–9) for examples.

[28] A systematic study of *gedanken* experiments, as well as of more realistic designs for the simultaneous measurement of stochastic position and momentum of quantum particles, has been carried out by Busch (1982). Some of the consequences of the resulting stochastic theory of quantum measurement with regard to some of the paradoxes in the standard theory have been investigated by Schroeck (1983a). A specific device for such simultaneous measurements, which is based on heterodyne detection, has been described by Yuen and Shapiro (1980) – see also Yuen (1982). The nonrelativistic scattering theory of extended particles in stochastic phase spaces has been studied by Gagnon (1983), who also derived some interesting mathematical properties of related stochastic phase-space path integrals.

Chapter 2

Relativistic Stochastic Quantum Mechanics

"I believe that certain erroneous developments in particle theory . . . are caused by the misconception by some physicists that it is possible to avoid philosophical arguments altogether. Starting with poor philosophy, they pose the wrong questions. It is only a slight exaggeration to say that good physics has at times been spoiled by poor philosophy."

HEISENBERG (1976, p. 32)

"The whole problem of relativistic quantum mechanics, which has been holding the development of theoretical physics for decades, reduces to finding suitable sets of ten operators satisfying the commutation relations [for the Lie algebra of the Poincaré group]."

DIRAC (1978b, p. 6)

We have seen in the last chapter that when we take in nonrelativistic stochastic quantum mechanics the sharp-point limit in either position or momentum we do obtain well-defined limits for such basic measurable quantities as probability densities, currents, transition probabilities, etc., and in this manner recover conventional nonrelativistic quantum theory. The existence of such limits demonstrates that the textbook idealizations that tacitly postulate perfectly exact position or momentum measurements are theoretically compatible with the other basic tenets of nonrelativistic quantum mechanics (canonical commutation relations, Schrödinger and Heisenberg equations of motion, etc.), but it does not establish the fundamental ontological validity of such a postulate. Clearly, we are dealing here with the question as to whether space is infinitely divisible or not, and as to whether there actually are particles that are pointlike in a literal sense rather than as approximations of physical reality. The fact that we can *imagine* in our minds the division of space into smaller and smaller partitions by means of lines, planes and other idealized geometrical objects might constitute sufficient proof of this infinite divisibility within the context of Plato's and subsequent schools of philosophic idealism, but it cannot be taken as relevant in the context of any empirically based philosophy that underlies scientific epistemology. As strongly emphasized by Heisenberg (1976), all indications are that when carried out in operational, empirically based terms, any sequence of increasingly finer subdivisions of space or time might reach the stage where any further subdivision loses all meaning. This might be due to the inability of either subdividing very small 'objects' (such as elementary particles) used to delineate small regions of space, or of distinguishing between the 'whole' and the 'parts' resulting from such a subdivision.

The issues at stake here are of fundamental concern to both relativity and quantum mechanics as succinctly explained by Born (1949b, p. 463):

> Relativity postulates that all laws of nature are invariant with respect to such linear transformations of spacetime [variables] $x^k = (\mathbf{x}, t)$ for which the quadratic form $R = x_k x^k = t^2 - \mathbf{x}^2$ is invariant (the velocity of light is taken to be unity). The underlying physical assumption is that the 4-dimensional distance $r = R^{1/2}$ has an absolute significance and can be measured. This is a natural and plausible assumption as long as one has to do with macroscopic dimensions where measuring rods and clocks can be applied. But is it still plausible in the domain of atomic phenomena?
>
> Doubts have been expressed a long time ago, for example, by Lindemann (1932) in his instructive little book. I think that the assumption of the observability of the 4-dimensional distance of two events inside atomic dimensions is an extrapolation which can only be justified by its consequences; and I am inclined to interpret the difficulties which quantum mechanics encounters in describing elementary particles and their interactions as indicating the failure of that assumption.

Thus, the fundamental issue as to the 'absolute significance' of conventional relativity in the subatomic realm cannot be settled in either a purely deductive manner or by exclusively empirical means. Indeed, any deductive argument would have to be based on axioms, which themselves could be questioned. On the other hand, experimental observations carried out at the high energies technologically available at present indicate that all hadrons (i.e. the overwhelming majority of known particles) display an intrinsic extension of the order of 1 fm = 10^{-13} cm, whereas the leptons are by comparison pointlike. However, these observations cannot offer, all by themselves, a lasting epistemological foundation, since the discovery of new particles and the attainment of higher energies might easily challenge our present image of the world of elementary particles.

As implicitly indicated by the earlier quotation of Born, a much more reliable indicator that something is seriously amiss with the conventional ideas about the structure of spacetime and of the elementary particles is supplied by the numerous inconsistencies that plague conventional relativistic quantum physics. These inconsistencies range from the nonexistence of relativistically covariant position operators and probability currents (which will be discussed in Sections 2.1 and 2.2) to the presence of a host of divergences in so-called 'local' quantum field theory. This last set of inconsistencies has prompted Dirac (1978b, p. 6) to urge that "one must seek a new relativistic quantum mechanics and one's prime concern must be to base it on sound mathematics".

In this chapter we shall indeed formulate a new and mathematically sound relativistic quantum mechanics by the simple expedient of allowing the Poincaré group to take over the role which the Galilei group played in the preceding chapter. We shall show that in this manner we arrive at a relativistic stochastic quantum-mechanical theory, which displays none of the defects of conventional relativistic quantum mechanics (nonexistence of probability densities or currents, instability of one-particle states in external fields, etc.), but which nevertheless merges into its nonrelativistic counterpart in the limit $c \to +\infty$. On the other hand, this theory does not possess a well-defined sharp-point limit for the position observables, thus confirming the fact that there can be no consistent conventional quantum mechanics based on wave functions at sharp spacetime points. We interpret this feature as an indication that the concept of sharp localization is not compatible with the tenets of relativistic quantum theory (Poincaré covariance, canonical commutation relations, etc.), and that, therefore, the conventional mathematical description of spacetime is in need of reformulation in the quantum realm. Such a reformulation will be presented in Chapter 4, where we shall return to the discussion of the basic epistemological aspects of our present ideas about spacetime, and propose alternatives to the present assumptions about its structure.

2.1. The Nonexistence of Relativistic Probability Currents at Sharp Spacetime Points

The sharp localizability problem which one encounters when extrapolating from nonrelativistic to relativistic quantum mechanics consists, on the mathematical level, of finding position operators X^j, $j = 1, 2, 3$, that are self-adjoint, commute, transform as a 3-vector under space rotations, satisfy the CCRs in conjunction with corresponding momentum operators P^j, $j = 1, 2, 3$, and at the same time provide a relativistically covariant meaning to 3-vector position. There are formal proofs (cf., e.g., Jordan and Mukunda, 1963) as to the nonexistence of such operators. Yet scores of attempts[1] have been made to bypass the difficulty by simply dropping one or more of the above requirements on $\mathbf{X}$ – albeit they all appear essential to sharp localization on physical grounds, and are indeed satisfied by the operators (1.2.2) (i.e. the operators in Equation (2.2) of Chapter 1) in the nonrelativistic context, when Poincaré covariance is replaced by Galilei covariance.

The root of the problem has already manifested itself in the spin-zero case, where the Klein–Gordon equation

$$\left(\Box + \frac{m^2c^2}{\hbar^2}\right)\phi(x) = 0, \quad \Box = \frac{\partial^2}{\partial x_\mu \partial x^\mu}, \quad x = (x^0 = ct, \mathbf{x}), \tag{1.1}$$

takes over the role of the free Schrödinger equation (1.3.12). The most general solutions of (1.1) that are considered to be reasonable candidates for wave functions are those of the form

$$\phi(x) = (2\pi\hbar)^{-3/2} \int_{\mathscr{V}_m} \exp\left(-\frac{i}{\hbar}\, x \cdot k\right) \tilde{\phi}(k)\, \mathrm{d}\Omega_m(k), \tag{1.2}$$

$$x \cdot k = x_\mu k^\mu = x^0k^0 - x^1k^1 - x^2k^2 - x^3k^3, \tag{1.3}$$

where $\tilde{\phi}$ is an arbitrary element of the Hilbert space $L^2(\mathcal{V}_m)$ consisting of functions on the mass hyperboloid

$$\mathcal{V}_m = \mathcal{V}_m^+ \cup \mathcal{V}_m^-, \mathcal{V}_m^\pm = \{k \,|\, k^2 = k \cdot k = m^2c^2,\ k^0 \gtrless 0\}, \tag{1.4}$$

that are square integrable with respect to the relativistically invariant measure

$$\mathrm{d}\Omega_m(k) = \delta(k^2 - m^2c^2)\,\mathrm{d}^4k. \tag{1.5}$$

The reason for concentrating on this kind of solution is that (1.2) is *formally* reminiscent of the transition (1.2.11) between the momentum and configuration representation in the nonrelativistic case, and that at the same time $\tilde{\phi}(k)$ can be regarded as a probability amplitude in the momentum representation.

The interpretation of $|\tilde{\phi}(k)|^2$ as a probability density in momentum space stems from the existence in $L^2(\mathcal{V}_m)$ of a spin-zero representation of the (restricted) Poincaré group $\mathcal{P}$. The general element (a, Λ) of $\mathcal{P}$ consists of a spacetime translation by the 4-vector a^μ and a proper Lorentz transformation Λ, and therefore constitutes an extrapolation of the Galilei group $\mathcal{G}$ to the relativistic regime. Indeed, the transition

$$(b, \mathbf{a}, \mathbf{v}, R) \longmapsto (a = (b, \mathbf{a}), \Lambda = \Lambda_v R), \quad v = \gamma_\mathbf{v}(c, \mathbf{v}), \tag{1.6}$$

can be executed due to the fact (Bogolubov *et al.*, 1975, Eq. (6.31)) that every proper Lorentz transformation Λ can be uniquely decomposed into the product of a rotation R and a boost Λ_v to the 3-velocity $\mathbf{v}$, namely the pure Lorentz transformation

$$x^0 \longmapsto x'^0 = \gamma_\mathbf{v}\left(x^0 + \frac{\mathbf{v}\cdot\mathbf{x}}{c}\right), \quad \gamma_\mathbf{v} = \left(1 - \frac{\mathbf{v}^2}{c^2}\right)^{-1/2}, \tag{1.7a}$$

$$\mathbf{x} \longmapsto \mathbf{x}' = \mathbf{x} + \mathbf{v}\frac{\gamma_\mathbf{v}}{c}\left(x^0 + \frac{\gamma_\mathbf{v}}{1+\gamma_\mathbf{v}}\,\frac{\mathbf{v}\cdot\mathbf{x}}{c}\right), \tag{1.7b}$$

and, consequently, we find that (1.2.5) is obtained when we let $c \longrightarrow \infty$ in the Lorentz transformation Λ occurring in the Poincaré transformation:

$$x \longmapsto x' = a + \Lambda x, \ \Lambda = \Lambda_v R, \tag{1.8a}$$

$$k \longmapsto k' = \Lambda k = \Lambda_v(k^0, R\mathbf{k}). \tag{1.8b}$$

Therefore, the mapping $(a, \Lambda) \longmapsto \tilde{U}(a, \Lambda)$, where

$$\tilde{U}(a, \Lambda) : \tilde{\phi}(k) \longmapsto \tilde{\phi}'(k) = \exp\left(\frac{i}{\hbar}\,a\cdot k\right)\tilde{\phi}(\Lambda^{-1}k), \tag{1.9}$$

preserves the inner products

$$\langle\tilde{\phi}_1 | \tilde{\phi}_2\rangle_{\mathscr{V}_m^\pm} = \int_{\mathscr{V}_m^\pm} \tilde{\phi}_1^*(k)\tilde{\phi}_2(k)\, \mathrm{d}\Omega_m(k) \tag{1.10}$$

in both $L^2(\mathscr{V}_m^+)$ and $L^2(\mathscr{V}_m^-)$ due to the manifest invariance of (1.5):

$$\mathrm{d}\Omega_m(k) = \mathrm{d}\Omega_m(k'),\ k' = \Lambda^{-1}k. \tag{1.11}$$

It can be easily ascertained that

$$\tilde{U}(a_1, \Lambda_1)\tilde{U}(a_2, \Lambda_2) = \tilde{U}(a_1 + \Lambda_1 a_2, \Lambda_1\Lambda_2), \tag{1.12a}$$

$$(a_1 + \Lambda_1 a_2, \Lambda_1\Lambda_2) = (a_1, \Lambda_1)(a_2, \Lambda_2), \tag{1.12b}$$

so that (1.9) is a unitary (vector) representation of $\mathscr{P}$, which is irreducible on account of Wiener's theorem (Prugovečki, 1981a, p. 576).

If P_μ, $\mu = 0, \ldots, 3$, denote the infinitesimal generators of spacetime translations, so that

$$\tilde{U}(a, I) = \exp\left(\frac{i}{\hbar}\, a \cdot \tilde{P}\right) \tag{1.13}$$

then it is immediately seen from (1.9) that

$$(\tilde{P}_\mu\tilde{\phi})(k) = k_\mu\tilde{\phi}(k). \tag{1.14}$$

Hence, the joint spectral measure $E^{\tilde{P}}(B)$ of $\{\tilde{P}^\nu\}$ is

$$(E^{\tilde{P}}(B)\tilde{\phi})(k) = \chi_B(k)\tilde{\phi}(k), \tag{1.15}$$

and by (1.9) we have

$$\tilde{U}(a, \Lambda)E^{\tilde{P}}(B)\tilde{U}^{-1}(a, \Lambda) = E^{\tilde{P}}(\Lambda B) \tag{1.16}$$

so that $\{\tilde{U}(a, \Lambda), E^{\tilde{P}}(B)\}$ constitutes a transitive system of imprimitivity, as defined in Section 1.2. We see that by (1.15), for Borel sets B in $\mathscr{V}_m^+$ or $\mathscr{V}_m^-$ we have

$$\langle\tilde{\phi}|E^{\tilde{P}}(B)\tilde{\phi}\rangle_{\mathscr{V}_m^\pm} = \int_{B\cap\mathscr{V}_m^\pm} |\tilde{\phi}(k)|^2\, \mathrm{d}\Omega_m(k), \tag{1.17}$$

so that, if we interpret P^μ to be the (sharp) 4-momentum operators, then indeed $\tilde{\phi}(k)$ is the probability amplitude for relativistic momentum values $k \in \mathscr{V}_m^\pm$ – where, by convention, $\mathscr{V}_m^+$ is associated with particles and $\mathscr{V}_m^-$ with the corresponding antiparticles.

This interpretation of the operators P_μ in (1.14) is not only in keeping with the corresponding interpretation of $\{H_0, \mathbf{P}\}$ *vis-à-vis* (1.13), (1.2.9b) and (1.2.14), respectively, but it also allows us to view the identification

$$\tilde{\psi}(\mathbf{k}) = \tilde{\phi}(k), \qquad k^0 = (\mathbf{k}^2 + m^2c^2)^{1/2}, \tag{1.18}$$

as a transition from the nonrelativistic momentum space probability amplitude $\tilde{\psi}(\mathbf{k})$ to its relativistic counterpart $\tilde{\phi}(k)$. Indeed, if we regard k^j, $j = 1, 2, 3$, as coordinates of points in $\mathscr{V}_m^+$ and $\mathscr{V}_m^-$, respectively, then in this coordinate system (1.5) assumes the form

$$\mathrm{d}\Omega_m(\mathbf{k}) = \frac{\mathrm{d}\mathbf{k}}{2(\mathbf{k}^2 + m^2c^2)^{1/2}}. \tag{1.19}$$

Therefore in the limit $c \longrightarrow \infty$ with $\tilde{\phi}$ kept normalized, (1.17) goes over into the nonrelativistic expression $\langle\psi|E^{\mathbf{P}}(B)\psi\rangle$ that corresponds to (1.2.14).

For all these reasons, the interpretation of $\tilde{\phi}(k)$ as a momentum probability amplitude in the context of special relativity[2] is universally accepted. However, fundamental difficulties have been experienced with the candidate for configuration space amplitude in (1.2) from the very first time Klein (1926) and Gordon (1926) proposed (1.1) as a relativistic equation of motion for spin-zero particles. Indeed, $|\phi(x)|^2$ cannot be interpreted as a probability density since, first of all, when inserted in (1.2.4b) it does not yield the same value for all $\tilde{\phi} \in L^2(\mathscr{V}_m^+)$, so that the ensuing probability would not be normalizable in a covariant manner; second, the nonrelativistic expression (1.2.4a) is not Poincaré covariant since Lorentz boosts (1.7) do not preserve simultaneity, so that the hyperplane

$$\sigma = \{x \mid x^0 = n \cdot x = 0\}, \quad n^0 = 1, n^1 = n^2 = n^3 = 0, \tag{1.20}$$

in a boosted inertial frame is no longer described by $x'^0 = 0$. In fact, the oriented surface element

$$\mathrm{d}\sigma_\mu(x) = \frac{1}{3!}\,\epsilon_{\mu\alpha\beta\gamma}\,\mathrm{d}x^\alpha\,\mathrm{d}x^\beta\,\mathrm{d}x^\gamma \tag{1.21}$$

points in the direction of n^μ, so that

$$\mathrm{d}\sigma_0(x) = \mathrm{d}x^1\,\mathrm{d}x^2\,\mathrm{d}x^3, \quad \mathrm{d}\sigma_j(x) = 0, \tag{1.22}$$

when $n^0 = 1, n^1 = n^2 = n^3 = 0$. Consequently $\mathrm{d}\sigma_\mu$ transforms as a covariant 4-vector (i.e. 1-form), so that we obviously have to associate with each solution $\phi(x)$ of (1.1) a probability current $j^\mu(x)$, in terms of which the probability $P^X(B)$ of detecting the particle within any Borel set B in σ could be written in the covariant form

$$P_\sigma^X(B) = \int_B j^\mu(x)\,\mathrm{d}\sigma_\mu(x), \quad B \subset \sigma. \tag{1.23}$$

On the other hand, if probability is conserved, we should have in the originally chosen inertial frame, where (1.22) holds,

$$P_\sigma^X(\sigma) = \int_\sigma j^0(x)\, \mathrm{d}\mathbf{x} = 1, \qquad \sigma = \{x \mid x^0 = \text{const}\}, \tag{1.24}$$

at all times $t = x^0/c$. A sufficient condition for (1.24) to hold is that the current $j^\mu(x)$ be conserved, i.e.,

$$\partial_\mu j^\mu(x) = 0, \qquad \partial_\mu = \partial/\partial x^\mu. \tag{1.25}$$

This current conservation is easily seen to be also a necessary condition if (1.24) is supposed to hold for arbitrary spacelike hypersurfaces, since then (1.25) follows by Gauss' theorem from

$$\int_\sigma j^\mu(x)\, \mathrm{d}\sigma_\mu(x) = \int_{\sigma'} j^\mu(x)\, \mathrm{d}\sigma_\mu(x) \tag{1.26}$$

when σ' is arrived at by an infinitesimal deformation of σ around a given point $x \in \sigma$.

The originally made choice for $j^\mu(x)$, which eventually became established as the Klein–Gorden charge[3] current $ej^\mu(x)$, was

$$j^\mu(x) = \frac{i\hbar}{2m}\, \phi^*(x) \overleftrightarrow{\partial}^\mu \phi(x). \tag{1.27}$$

Indeed, heuristically (1.27) can be related as $c \longrightarrow \infty$ to the nonrelativistic ρ and j in (1.8.2) and (1.8.6) (Schweber, 1961, Section 3a). Furthermore, in apparent concordance with (1.24), j^μ makes its appearance in the inner products

$$\langle \phi_1 \mid \phi_2 \rangle_{\sigma^\pm} = \pm i\hbar \int_\sigma \phi_1^*(x) \overleftrightarrow{\partial}_\mu \phi_2(x)\, \mathrm{d}\sigma^\mu(x) \tag{1.28}$$

when we set $\phi_1 = \phi_2 = \phi$ in

$$\phi_1^*(x) \overleftrightarrow{\partial}_\mu \phi_2(x) = \phi_1^*(x)\, \partial_\mu \phi_2(x) - \phi_2(x)\, \partial_\mu \phi_1^*(x). \tag{1.29}$$

The above inner products occur naturally in the spaces

$$L^2(\sigma^\pm) = \mathscr{W}_0 L^2(\mathscr{V}_m^\pm) \tag{1.30}$$

when $\mathscr{W}_0$ is the integral transform which maps $\tilde{\phi}(k)$ into $\phi(x)$ according to (1.2). Indeed, using the unitarity property of the Fourier–Plancherel transform (1.2.11), we easily establish that

$$\langle \mathscr{W}_0 \tilde{\phi}_1 \mid \mathscr{W}_0 \tilde{\phi}_2 \rangle_{\sigma^\pm} = \langle \tilde{\phi}_1 \mid \tilde{\phi}_2 \rangle_{\mathscr{V}_m^\pm}. \tag{1.31}$$

Nevertheless, $j^\mu(x)$ in (1.27) *cannot be interpreted as a probability current even if we restrict ourselves only to positive-energy solutions* $\phi(x)$ (i.e. solutions for which $\tilde{\phi} \in L^2(\mathscr{V}_m^+)$ in (1.2)) since $j^0(x)$ is not positive definite. In fact, we can explicitly construct (Blokhintsev, 1973, p. 81) functions $\phi \in L^2(\sigma^+)$ for which $j^0(x') > 0$ and $j^0(x'') < 0$ at distinct $x', x'' \in \sigma$.

For this reason the Klein–Gordon equation fell into disrepute for about seven years after it had been first proposed (Schweber, 1961, p. 55), and the focus of interest shifted[4] to the Dirac (1928) equation

$$\left(i\gamma^\mu \partial_\mu - \frac{mc}{\hbar}\right) \psi(x) = 0, \tag{1.32}$$

where $\psi(x)$ is a (4-component) bispinor. It originally appeared that the 4-current

$$j^\mu(x) = \bar{\psi}(x)\gamma^\mu \psi(x), \qquad \bar{\psi}(x) = \psi^*(x)\gamma^0 \tag{1.33}$$

supplied the sought-after probability current, at least in the case of spin-½ particles, since indeed $j^0(x) = \psi^*(x)\psi(x)$ is positive definite. In fact, in the beginning Dirac considered (Kálnay, 1971) the multiplication operators

$$X_{\mathrm{D}}^\mu : \psi(x) \mapsto x^\mu \psi(x) \tag{1.34}$$

for $\mu = 1, 2, 3$ to be the representatives of position. However, the first doubts appeared when Breit (1928) found out that the associated velocity operators had $\pm c$ as the only eigenvalues, then Klein (1929) came upon his celebrated paradox for transmitted currents, and later Schrödinger (1930) discovered *Zitterbewegung*. This last phenomenon manifests itself in the Dirac current (1.33) for wave packets containing both positive and negative energy solutions in the form of rapidly oscillating cross terms between those two types of solutions. To these cross terms no straightforward (Barut and Bracken, 1981) physical interpretation can be given (Bjorken and Drell, 1964, p. 38) since the motion governed by (1.32) is supposed to be motion by inertia.

The incorrectness of the interpretation of $j^\mu(x)$ in (1.33) as a localized probability current has been established beyond any doubt with the discovery by Foldy and Wouthuysen (1950) of a unitary transformation U_{FW} which carries every solution $\psi(x)$ of the Dirac equation (1.32) into a solution $\psi_{\mathrm{FW}}(x)$ of the Schrödinger-like equation

$$i\hbar\partial_0 \psi_{\mathrm{FW}}(x) = \gamma^0 (\mathbf{P}^2 + m^2c^2)^{1/2} \psi_{\mathrm{FW}}(x) \tag{1.35}$$

in which $\mathbf{P}$ has as components the operators P^j, where

$$(P_\mu \psi_{\mathrm{FW}})(x) = i\hbar\partial_\mu \psi_{\mathrm{FW}}(x). \tag{1.36}$$

Since γ^0 can be diagonalized, and it has two eigenvalues equal to +1 and −1, respectively, that have multiplicity 2, each of the four components of ψ_{FW} satisfies one of the equations

$$i\partial_0\,\phi(x) = \pm\left(-\nabla^2 + \frac{m^2c^2}{\hbar^2}\right)^{1/2}\phi(x), \tag{1.37}$$

and therefore coincides with either a positive-energy solution $\phi \in L^2(\sigma^+)$ or a negative-energy solution $\phi \in L^2(\sigma^-)$ of the Klein–Gordon equation (1.1). Hence, the positive-definiteness of $j^0(x)$ in (1.33) is due to the fact that in the Dirac representation the spin and the 4-momentum degrees of freedom are intertwined in such a manner that the multiplication operators $\mathbf{X}_D$ in (1.34), which supposedly would describe the position of a spin-½ particle, do not actually leave invariant the space of positive-energy solutions describing all states of a such particle, and therefore *de facto* are not related to that particle alone (Pryce, 1948; Dixon, 1965). Consequently, one is forced to conclude that when it comes to the localization problem, "the Klein–Gordon equation is neither better nor worse than the Dirac equation" (Wightman, 1972, p. 98).

The situation is certainly not better for higher-spin equations. In fact, it was in equations for spin $\geqslant 1$ that additional difficulties were first discovered by Velo and Zwanziger (1969). These later discovered properties of solutions $\psi(x)$ of Lorentz invariant equations lead to direct violations of relativistic causality[5] if $\psi(x)$ is assumed to be in any way related to a phenomenon localized at a spacetime point x (Wightman, 1971; Velo and Wightman, 1978).

To summarize: neither the Klein–Gordon current (1.27) nor the Dirac current (1.33) is a *probability* current that is physically related to sharply localizable point particles since the timelike component $j^0(x)$ of the first is not positive definite, whereas for the second $j^0(x) \geqslant 0$ but the variables x^j, $j = 1, 2, 3$, do not represent position coordinates due to the fact that the operators X^j_D in (1.34) are not position operators.

2.2. The Unfeasibility of Sharp Localization in Relativistic Quantum Mechanics

The fact that neither the Klein–Gordon nor the Dirac currents turned out to be probability currents for sharply localizable relativistic quantum point particles has not stopped the search after a *formal* solution of the (sharp) localization problem in relativistic quantum mechanics. In fact, once it is decided that not *all* the conditions on position operators X^j which have been spelled out at the beginning of the last section are essential ingredients of an acceptable solution of the sharp localization problem for point particles in relativistic quantum mechanics, the quest for a 'solution' is reduced to settling the issue by decreeing which properties should be retained and which should be discarded or modified. Clearly, the ensuing subjectivity can be removed only by a reconsideration of the foundations of both quantum mechanics and relativity theory, which would substitute new and yet carefully appraised physical criteria in place of the old ones. We shall present one such set of criteria in the remaining sections of this chapter as well as in Chapter 4. However, first we have to exhibit the key drawbacks of those 'solutions' which, in the past, had earned at least temporarily some partial acceptance. We shall concentrate in the spin-zero case and the Klein–Gordon equation, but all the subsequent conclusions remain valid in case of nonzero spin.[6]

Among all candidates for position operators in the spaces (1.30) of solutions of the Klein–Gordon equation, there is first of all the obvious choice

$$(\hat{Q}^\mu \phi)(x) = x^\mu \phi(x), \quad \phi \in L^2(\sigma^+), \tag{2.1}$$

which includes the candidate $\hat{Q}^0$ for a 'time operator'. This choice has the advantage of manifest covariance under the representation

$$\hat{U}(a, \Lambda) = \mathscr{W}_0 \tilde{U}(a, \Lambda) \mathscr{W}_0^{-1} \tag{2.2}$$

with $\tilde{U}(a, \Lambda)$ given by (1.9) and W_0 mapping $\tilde{\phi}(k)$ into $\phi(x)$ by (1.2). Indeed, by (1.9) and (1.11)

$$\hat{U}(a, \Lambda) : \phi(x) \longmapsto \phi'(x) = \phi(\Lambda^{-1}(x - a)), \tag{2.3}$$

and consequently

$$\hat{U}^{-1}(a, \Lambda)\hat{Q}^\mu \hat{U}(a, \Lambda) = a^\mu I + \Lambda^\mu{}_\nu \hat{Q}^\nu. \tag{2.4}$$

Furthermore, the infinitesimal generators $M^{\mu\nu}$ corresponding to the Lorentz transformations $\hat{U}(a, \Lambda)$ can be written in terms of $\hat{Q}^\mu$ and of the momentum operators

$$\hat{P}_\mu = W_0 \tilde{P}_\mu W_0^{-1} = i\hbar \frac{\partial}{\partial x^\mu} \tag{2.5}$$

in $L^2(\sigma^+)$ in a form that generalizes the one for rotations of inertial frames (Bogolubov *et al.*, 1975, Section 6.1):

$$\hat{M}^{\mu\nu} = \hat{Q}^\mu \hat{P}^\nu - \hat{Q}^\nu \hat{P}^\mu. \tag{2.6}$$

These features appear all the more striking when one notes that $\{\hat{Q}^\mu, \hat{P}^\mu\}$ supply a representation of the relativistic canonical commutation relations (RCCRs)

$$[\hat{Q}^\mu, \hat{Q}^\nu] = [\hat{P}^\mu, \hat{P}^\nu] = 0, \qquad [\hat{Q}^\mu, \hat{P}^\nu] = -i\hbar g^{\mu\nu} \tag{2.7}$$

if $g^{\mu\nu}$ are the components (in Gaussian normal coordinates) of the Minkowski metric tensor:

$$g^{00} = -g^{11} = -g^{22} = -g^{33} = 1, \qquad g^{\mu\nu} = 0, \quad \mu \neq \nu. \tag{2.8}$$

All the above features of $\hat{Q}^\mu$ will, in Sections 4.5 and 4.6, be found to be of physical relevance. Yet, $\hat{Q}^j$ cannot be interpreted as position operators since they do not leave $L^2(\sigma^+)$ invariant. In fact, on account of the RCCRs in (2.7), these operators do not commute with the mass operator $M^2 = \hat{P}_\nu \hat{P}^\nu$,

$$[\hat{Q}^\mu, \hat{P}_\nu \hat{P}^\nu] = -2i\hbar \hat{P}^\mu \neq 0, \tag{2.9}$$

in terms of which the Klein-Gordon equation (1.1) assumes the form

$$M^2 \phi = m^2 \phi, \quad M^2 = \hat{P}_\nu \hat{P}^\nu. \tag{2.10}$$

Hence, $\hat{Q}^j$ can refer only to the 'localization' of 'states' which do not relate to a given mass value (and therefore not to an elementary particle in the accepted sense of the word), but rather to the entire mass spectrum, including even imaginary masses (Johnson, 1969, 1971; Broyles, 1970).

The next obvious choice $\tilde{X}^j$ for relativistic position operators is arrived at by first executing the identification (1.18) of relativistic with nonrelativistic wave functions, and then allowing the nonrelativistic position operators (1.2.2) expressed in the momentum representation, i.e. as $\tilde{X}^j = U_{\mathrm{F}} X^j U_{\mathrm{F}}^{-1}$, to act in the usual manner:

$$(\tilde{X}^j \tilde{\phi})(k) = i\hbar \frac{\partial}{\partial k^j} \tilde{\phi}(([\mathbf{k}^2 + m^2 c^2]^{1/2}, \mathbf{k})). \tag{2.11}$$

However, in that case by (1.10) and (1.19)

$$\begin{aligned} \langle \tilde{\phi}_1 | \tilde{X}^j \tilde{\phi}_2 \rangle_{\mathcal{V}_m^+} &= \int_{\mathbb{R}^3} \tilde{\phi}_1^* \left(i\hbar \frac{\partial \tilde{\phi}_2}{\partial k^j} \right) \frac{d\mathbf{k}}{2(\mathbf{k}^2 + m^2 c^2)^{1/2}} \\ &= \int_{\mathbb{R}^3} \left(i\hbar \frac{\partial \tilde{\phi}_1}{\partial k^j} - \frac{i\hbar k^j \tilde{\phi}_1}{\mathbf{k}^2 + m^2 c^2} \right)^* \tilde{\phi}_2 \frac{d\mathbf{k}}{2(\mathbf{k}^2 + m^2 c^2)^{1/2}} \neq \\ &\neq \langle \tilde{X}^j \tilde{\phi}_1 | \tilde{\phi}_2 \rangle_{\mathcal{V}_m^+}. \end{aligned} \tag{2.12}$$

Hence, the operators $\tilde{X}^j$, $j = 1, 2, 3$, are nonsymmetric, and therefore they have been almost generally dismissed as unphysical (Schweber, 1961, Section 3c). However, several authors (Kálnay and Toledo, 1967; Gallardo *et al.*, 1967) have attempted to impart a physical meaning to $\tilde{X}^j$, $j = 1, 2, 3$, by starting from the assumption that nonsymmetric operators can represent[7] legitimate observables, and by questioning the assumption that localization should be sharp, i.e. based on the concept of pointlike elementary particles. Thus, they argue that for every complex value $\lambda'^j + i\lambda''^j$ in the spectrum of these 'observables', λ'^j can be associated with the mean position of an extended elementary particle, and λ''^j with its 'diameter'.

The idea that all massive particles are extended, which is implicit in this proposal, comes very close in spirit to the stochastic phase space solution of the localization problem. However, this proposal does not supply results (or even hints) as to the operational procedures for the measurement of such a position (e.g. as those discussed in Sections 1.12, 4.2 and 4.3), nor does it supply a truly relativistic framework in which the space and time degrees of freedom are treated on an equal footing,[8] and therefore inconsistencies result (Jancewicz, 1975). On the other hand, the symmetric part of $\tilde{X}^j$ does coincide with the Newton–Wigner operators X^j_{NW} which will be discused next, and in turn these operators will be seen in Section 2.5 to be indeed related to the mean stochastic position of extended particles in inertial frames in which the measuring apparatus is at rest.

The relativistic position operators proposed by Newton and Wigner (1949) are automatically obtained from (2.12) when we *impose* symmetry by taking half of the value of the term in the second integral (which spoils the symmetry) and transferring it to the first integral. Hence, the operators X^j_{NW},

$$(X^j_{\mathrm{NW}}\tilde{\phi})(k) = \left(i\hbar\frac{\partial}{\partial k^j} - \frac{1}{2}\frac{i\hbar k^j}{\mathbf{k}^2 + m^2c^2}\right)\tilde{\phi}(([\mathbf{k}^2 + m^2c^2]^{1/2}, \mathbf{k})) \tag{2.13}$$

which result from this procedure are symmetric:

$$\langle\tilde{\phi}_1 | X^j_{\mathrm{NW}}\tilde{\phi}_2\rangle = \langle X^j_{\mathrm{NW}}\tilde{\phi}_1 | \tilde{\phi}_2\rangle. \tag{2.14}$$

In fact, we see that

$$\tilde{X}^j = X^j_{\mathrm{NW}} + i\frac{\hbar}{2}\frac{\tilde{P}_j}{\tilde{\mathbf{P}}^2 + m^2c^2}, \tag{2.15}$$

so that indeed X^j_{NW} emerge as the symmetric parts of $\tilde{X}^j$.

The Newton–Wigner operators in (2.13) are not only symmetric, but also self-adjoint (cf. Appendix B) and commuting. As a matter of fact, they constitute a complete set of operators in $L^2(\mathscr{V}^+_m)$. Indeed, the unitary transformation U_{NW} from $L^2(\mathscr{V}^+_m)$ to their spectral representation space (cf. Appendix B) $L^2(\mathbb{R}^3)$ can be easily computed, and we get

$$U_{\mathrm{NW}}: \tilde{\phi}(k) \longmapsto \phi_{\mathrm{NW}}(y) = (\Phi_y | \tilde{\phi})_{\mathscr{V}^+_m} = \int_{\mathscr{V}^+_m} \Phi^*_y(k)\tilde{\phi}(k)\, \mathrm{d}\Omega_m(k), \tag{2.16}$$

where $\Phi_y(k)$ is an eigenfunction expansion (cf. (B.23) in Appendix B) for $\{X^j_{\mathrm{NW}}\}$,

$$\Phi_y(k) = (2\pi\hbar)^{-3/2}\, k_0^{1/2} \exp\left(\frac{i}{\hbar}\, y \cdot k\right), \tag{2.17}$$

which at $y = 0$ is purported to describe a spin-zero relativistic point particle 'localized' at the origin of the inertial frame. In fact, we easily verify that for $y^0 = 0$

$$(\Phi_y | X^j_{\mathrm{NW}} \tilde{\phi}\rangle_{\mathscr{V}^+_m} = (y^j \Phi_y | \tilde{\phi}\rangle_{\mathscr{V}^+_m} = y^j \phi_{\mathrm{NW}}(y), \tag{2.18}$$

and the unitarity of the Fourier–Plancherel transform yields

$$\langle \tilde{\phi}_1 | \tilde{\phi}_2 \rangle_{\mathscr{V}^+_m} = \int_{\mathbb{R}^3} (U_{\mathrm{NW}} \tilde{\phi}_1)^* (y) \, (U_{\mathrm{NW}} \tilde{\phi}_2)(y) \, \mathrm{d}y, \tag{2.19}$$

as well as the fact that $L^2(\mathbb{R}^3) = U_{\mathrm{NW}} L^2(\mathscr{V}^+_m)$, so that $L^2(\mathbb{R}^3)$ is the spectral representation space of $\{\tilde{X}^j_{\mathrm{NW}}\}$ under U_{NW}. Thus we find that

$$(U_{\mathrm{NW}} X^j_{\mathrm{NW}} \tilde{\phi})(y) = y^j (U_{\mathrm{NW}} \tilde{\phi})(y), \tag{2.20a}$$

$$(U_{\mathrm{NW}} \tilde{P}^j \tilde{\phi})(y) = -i\hbar \frac{\partial}{\partial y^j} (U_{\mathrm{NW}} \tilde{\phi})(y), \tag{2.20b}$$

so that $\{U_{\mathrm{NW}} X^j_{\mathrm{NW}} U^{-1}_{\mathrm{NW}}, U_{\mathrm{NW}} \tilde{P}^j U^{-1}_{\mathrm{NW}}\}$ coincides in this spectral representation space for $\{X^j_{\mathrm{NW}}\}$ with the Schrödinger representation of the CCRs. However, under $U(a, \Lambda)$ in (1.9) $\phi_{\mathrm{NW}}(y)$ does not transform covariantly if Λ is a pure Lorentz transformation:

$$\phi'_{\mathrm{NW}}(y) = (U_{\mathrm{NW}} \tilde{\phi}')(y) \neq \phi_{\mathrm{NW}}(\Lambda^{-1}(y - a)). \tag{2.21}$$

Indeed, although the integral in (2.16) appears to be manifestly covariant, that is not so since (2.17) is not covariant due to the presence of $k_0^{1/2}$.

Thus, the Newton–Wigner operators do not lead to relativistically covariant probability amplitudes $\phi_{\mathrm{NW}}(y)$, nor to covariant probability densities $|\phi_{\mathrm{NW}}(y)|^2$. This can be physically justified (Fleming, 1965; Castrigiano and Mutze, 1982) by arguing that these operators are related only to the laboratory frame in which the detecting apparatus is at rest,[9] and that therefore manifest covariance is not to be expected. However, even if this argument were totally correct, these operators would not yield a relativistically consistent concept of sharp localization since they lead to the violation of relativistic causality (Fleming, 1965; Ruijsenaars, 1981).

This last fundamental observation does not apply only to Newton–Wigner operators but to any sensible notion of sharp localizability (as discussed Section 1.2), and demonstrates that the principles of relativity, uncertainty and sharp localizability are not logically compatible. The concept of relativistic causality which it involves *would be* of the most general conceivable nature *if* sharp localizability could be transferred from classical to quantum relativistic mechanics. Its general formulation is as follows:

The propagation of a quantum (point) particle in (classical) Minkowski space is *relativistically causal* if whenever the particle is to be found with certainty (i.e. with probability equal to 1) in a spatial region B at time $t = x^0/c$, then the same particle would be found with certainty within the causal future cone C_B^+ of B,

$$C_B^+ = \bigcup_{\mathbf{x} \in B} \{x' \mid (x'_\mu - x_\mu)(x'^\mu - x^\mu) \geqslant 0, x'^0 \geqslant x^0\} \tag{2.22}$$

at all later times $t' = x'_0/c$, i.e. the probability of its being at t' in a spatial region B_0 for which $\{x'_0\} \times B_0$ is disjoint from C_B^+ equals zero.

The above notion of relativistic causality is exactly the one which lies at the foundation of particle propagation in classical relativistic mechanics, since it is an immediate consequence of the assumption that no ordinary material body (i.e. tachyons excluded) can exceed the speed of light. Nevertheless, by the following theorem, first proved by Hegerfeldt (1974) for arbitrary quantum particles in the context of sharp localizability discussed in Section 1.2 (and later generalized by Skagerstam (1976) to Jauch–Piron (1967) type of localizability), we cannot construct relativistic quantum theories of particles of positive-definite energy (Jancewicz, 1977) in which those particles are sharply localizable and yet obey relativistic causality in the earlier defined sense (cf. Perez and Wilde, 1977, as well as Hegerfeldt and Ruijsenaars, 1980, for a thorough study and results on multiparticle systems).

HEGERFELDT'S THEOREM. *In a relativistic quantum theory of point particles there are no one-particle states that are ever localized in a finite spatial (or space-time) region and yet obey at later times the principle of relativistic causality.*

If the interpretation of the Newton–Wigner operators as relativistic position observables were correct, then every state whose Newton–Wigner amplitude $\phi_{\mathrm{NW}}(y)$ in (2.16) is zero at some instant y^0/c for $\mathbf{y}$ outside some spatial region B would be sharply localized within B, and by Hegerfeldt's theorem it would violate relativistic causality, albeit, paradoxically (Ruijsenaars, 1981), the velocity operators corresponding to Newton–Wigner operators predict propagation speeds that are always smaller than c. Since the range of the Newton–Wigner transform U_{NW} equals $L^2(\mathbb{R}^3)$, there is an infinity of such states, and therefore a physical inconsistency results. By (1.9)–(1.12) the one-particle theory is relativistic, so that the fault must lie in the interpretation of the Newton–Wigner operators as sharp position operators (cf., however, (5.23)–(5.28)).

In summary, we can state that no relativistic theory of quantum particles allows for a consistent notion of sharp localizability.[10] Although this conclusion has been arrived at by mathematical arguments, it can be obtained also by simple physical reasoning (Prugovečki, 1978f, Section V) if one considers the problem in operational terms and takes into consideration the presence of confidence functions $\chi_{\mathbf{q}}(\mathbf{x})$ and $\chi_{q^0}(x^0)$ that characterize any measurement process. Indeed, from such a perspective pointlike localizability has to be viewed (cf. Section 1.2) as the outcome of a sequence of measurements of steadily increasing precision for which the corresponding $\chi_q(x) = \chi_{q^0}(x^0)\chi_{\mathbf{q}}(\mathbf{x})$ form a δ-sequence. In classical physics, this asymptotic procedure can be carried out in the rest frame of the particle. However, due to the uncertainty principle, no such frame *literally* exists in operational terms in the quantum context. In the nonrelativistic case that does not really matter since the δ-sequence feature is frame-independent due to the fact that Galilei transformations preserve simultaneity. However, this is not the case for general Lorentz transformations, and therefore the operationally based notion of sharp localization becomes self-contradictory if we assume the validity of *both* the relativity and the uncertainty principles.

2.3. Phase-Space Representations of the Poincaré Group

In view of Hegerfeldt's theorem, which establishes the inconsistency of the notion of sharp localizability in relativistic quantum mechanics, it might appear strange that there is an extensive literature, incorporating hosts of papers and a number of textbooks, which seems to tacitly[11] assume that there is a consistent local relativistic quantum theory. A careful perusal of this literature reveals, of course, that the variables x^μ in the Klein–Gordon or Dirac wave functions $\phi(x)$ and $\psi(x)$, respectively, are treated *as if* they were *bona fide* spacetime variables. The ensuing inconsistencies are then simply ignored by concentrating on asymptotic behavior and on S-matrix techniques in which only the 4-momentum variables k^μ play a role in experimentally verifiable statements, whereas x^μ occur only in statements of indirect physical import. Of course, this kind of approach to fundamental difficulties can provide only a cosmetic cure, and the thereby hidden contradictions re-emerge in the later stages of this framework in the form of divergencies that characterize all so-called 'local' quantum field theoretical models, whose short-comings will be discussed at greater length in Chapter 5.

Nevertheless, some of the numerical successes of this approach indicate that perhaps a consistent theory underlies at least some of the formal manipulations, based on lax mathematical standards,[12] which characterize the mainstream approach to relativistic quantum mechanics. In the remaining part of this chapter we shall show that this is indeed the case if stochastic localizability is substituted in place of sharp localizability. However, in the search for physical as well as mathematical consistency, we shall eventually find it necessary to probe into the very notion of spacetime before being able to present in the second part of this monograph a frame-work that is consistent at all levels – mathematical, physical and epistemological.

In the present chapter the strategy will be to extrapolate the main results of the preceding chapter to the relativistic regime by substituting each key nonrelativistic concept by its relativistic counterpart in the hope of arriving at a consistent concept of *stochastic localizability* in relativistic quantum mechanics. Thus, first of all, the role of the proper Galilei group $\mathcal{G}$ is taken over by the proper Poincaré group $\mathcal{P}$, which acts on points x in the Minkowski space $M(1,3)$ and on 4-momenta $p \in \mathcal{V}_m$ as follows:

$$x \longrightarrow x' = a + \Lambda x, \quad x \in M(1,3), \tag{3.1a}$$

$$p \longrightarrow p' = \Lambda p, \quad p \in \mathcal{V}_m. \tag{3.1b}$$

In (3.1) Λ denotes a proper Lorentz transformation. It should be recalled that any such Λ can always be uniquely decomposed in the form (Bogolubov *et al.*, 1975, p. 159)

$$\Lambda = \Lambda_v R, \qquad R \in \mathrm{SO}(3), \tag{3.2}$$

where Λ_v is a Lorentz boost to the 4-velocity $v = (c\gamma_v, \mathbf{v}\gamma_v)$, so that by (1.7) Λ assumes the form:

$$x^0 \mapsto x'^0 = \gamma_v \left(x^0 + \frac{\mathbf{v} \cdot R\mathbf{x}}{c} \right), \qquad \gamma_v = \left(1 - \frac{\mathbf{v}^2}{c^2} \right)^{-1/2} \tag{3.3a}$$

$$\mathbf{x} \mapsto \mathbf{x}' = R\mathbf{x} + \frac{\gamma_v}{c} \left(x^0 + \frac{\gamma_v}{1 + \gamma_v} \frac{\mathbf{v} \cdot R\mathbf{x}}{c} \right) \mathbf{v}. \tag{3.3b}$$

The relativistic phase space (Ehlers, 1971) of a classical particle of rest mass m equals

$$\mathscr{M}_m^+ = M(1, 3) \times \mathscr{V}_m^+ \tag{3.4}$$

in special relativity. If we define a *coherent flow* of classical test particles as a family of worldlines normal to an initial-data hypersurface σ of operational simultaneity (Lifshitz and Khalatnikov, 1963) and delineated by these particles in free fall, then $\mathscr{M}_m^+$ can be related in special relativity to the nonrelativistic phase space Γ by assigning to each $x \in \mathscr{M}_m^+$ that lies on a worldline the (unique) point $q \in \sigma$ where that worldline intersects σ, and to each $p = (p^0, \mathbf{p}) \in \mathscr{V}_m^+$ its 3-vector $\mathbf{p}$. Furthermore, since in special relativity σ is a spacelike hyperplane, there are inertial frames in which σ can be designated to be the set of points q for which $q^0 = \text{const}$. Hence, if we introduce in

$$\Sigma_m = \Sigma_m^+ \cup \Sigma_m^-, \qquad \Sigma_m^\pm = \sigma \times \mathscr{V}_m^\pm, \tag{3.5}$$

the covariant element of measure (with $\epsilon(p^0) = \pm 1$ for $p^0 \gtrless 0$)

$$\mathrm{d}\Sigma_m(q, p) = 2\epsilon(p^0) p^\nu \mathrm{d}\sigma_\nu(q) \delta(p^2 - m^2c^2)\, \mathrm{d}^4 p, \tag{3.6}$$

where $\mathrm{d}\sigma_\nu$ points in the direction n^μ of the normal to σ, then in any frame where $q^0 = \text{const}$ for all $q \in \sigma$ we shall have $n^0 = 1$ and $\mathbf{n} = \mathbf{0}$, so that

$$\mathrm{d}\Sigma_m(q, p) = \mathrm{d}\mathbf{q}\, \mathrm{d}\mathbf{p}, \qquad q^0 = \text{const}. \tag{3.7}$$

The measure element $\mathrm{d}\Sigma_m$ belongs to the unique measure along spacelike hypersurfaces in $\mathscr{M}_m$ that possesses this property (Ehlers, 1971).

These considerations immediately suggest that the role of $L^2(\Gamma)$ should be taken over in the relativistic case by the Hilbert spaces $L^2(\Sigma_m^{\pm})$ of function $\phi(q, p)$ with inner product,

$$\langle \phi_1 | \phi_2 \rangle_{\Sigma_m^{\pm}} = \int_{\Sigma_m^{\pm}} \phi_1^*(q, p) \phi_2(q, p)\, \mathrm{d}\Sigma_m(q, p), \tag{3.8}$$

where Σ_m^+ corresponds to particles, and Σ_m^- to antiparticles. In fact, by (3.7) the inner product (3.8) assumes the form (1.4.2) in the laboratory frame, where q^0 = const for all $q \in \sigma$. Thus, by analogy with (1.4.5a) we can then define (Prugovečki, 1978d, f)

$$U(a, \Lambda) : \phi(q, p) \longmapsto \phi'(q, p) = \phi(\Lambda^{-1}(q - a), \Lambda^{-1} p), \tag{3.9a}$$

where the relativistic analogue of (1.4.5b) is obtained from the counterpart

$$U(a, I) = \exp\left(\frac{i}{\hbar}\, a \cdot P\right), \quad P^\mu = i\hbar \frac{\partial}{\partial q_\mu}, \tag{3.9b}$$

of (1.13) by setting $a^0 = -ct$ and $\mathbf{a} = \mathbf{0}$. Imposing the condition that

$$M^2 = c^{-2} P_\mu P^\mu \tag{3.10}$$

should be the (square) mass operator with eigenvalue m^2, we get from (3.9b) that

$$\left(\Box_q + \frac{m^2 c^2}{\hbar^2}\right) \phi(q, p) = 0 \tag{3.11}$$

at all $p \in \mathscr{V}_m$ for any ϕ that is in the domain of M^2.

Clearly, $U(a, \Lambda)$ constitutes a vector representation of $\mathscr{P}$ which, like its counterpart $U(g)$ in (1.4.5b), is highly reducible. We shall derive its irreducible subrepresentations in the next section. In the remainder of this section we shall outline an alternative[13] derivation of phase-space representations of the Poincaré group (Ali, 1979, 1980), which is based on Mackey's (1963, 1968) including procedure, and which illucidates the relationship between classical and quantum phase-space representations of the Poincaré group in terms of the earlier defined coherent flows.

The sets $\mathscr{T}$ of time-translations and SO(3) of space rotations constitute subgroups of $\mathscr{P}$. Let us denote by $\mathscr{N}$ the set of left cosets in $\mathscr{P}$ with respect to $\mathscr{T} \otimes$ SO(3), i.e.

$$\mathscr{N} = \mathscr{P} / (\mathscr{T} \otimes \mathrm{SO}(3)). \tag{3.12}$$

Since any $(a, \Lambda) \in \mathscr{P}$ can be decomposed in accordance with (3.2) as follows,

$$(a, \Lambda) = ((0, \mathbf{b}), \Lambda_v)((q^0, \mathbf{0}), R), \quad q^0 = \gamma_v^{-1} a^0, \ \mathbf{b} = \mathbf{a} - \frac{a^0}{c} \mathbf{v}, \tag{3.13}$$

we can parametrize $\mathcal{N}$ by means of **a** and **v**, or, equivalently, in terms of

$$\zeta = (\mathbf{q}, \mathbf{p}), \quad \mathbf{q} = \mathbf{a},\ \mathbf{p} = m\gamma_v \mathbf{v}. \tag{3.14}$$

Since $\mathcal{P}$, $\mathcal{T}$ and SO(3) are unimodular groups, it follows that $\mathcal{N}$ admits an invariant measure $\mu_{\mathcal{N}}$ (Barut and Rączka, 1977). The invariant Haar measure on $\mathcal{P}$ is

$$\mathrm{d}\mu_{\mathcal{P}}(a, \Lambda) = \mathrm{d}^4 a\ \mathrm{d}\Omega_m(p)\ \mathrm{d}R, \qquad \Lambda = \Lambda_{p/m} R. \tag{3.15}$$

Taking (3.13) into consideration, and substituting at each fixed $p = mv \in \mathcal{V}_m^+$ the variable q^0 instead of a^0, we have for any $\mu_{\mathcal{P}}$-integrable function $f(a, \Lambda)$

$$\int f\, \mathrm{d}\mu_{\mathcal{P}} = \int \mathrm{d}\mathbf{q}\ \frac{\mathrm{d}\mathbf{p}}{2p^0} \int f\gamma_v\ \mathrm{d}q^0\ \mathrm{d}R. \tag{3.16}$$

Since $\mathrm{d}q^0\ \mathrm{d}R$ is the invariant measure on $\mathcal{T} \otimes \mathrm{SO}(3)$, we get

$$\mathrm{d}\mu_{\mathcal{N}}(\mathbf{q}, \mathbf{p}) = 2mc\gamma_v\ \mathrm{d}\mathbf{q}\ \frac{\mathrm{d}\mathbf{p}}{2p^0} = \mathrm{d}\mathbf{q}\ \mathrm{d}\mathbf{p}. \tag{3.17}$$

Naturally, since all Haar measures, such as (3.15), are unique only up to a positive multiplicative constant, the same is true of (3.17).

The parametrization of $\mathcal{N}$ and the invariance of (3.16) can be easily understood physically in terms of the earlier defined coherent flow of classical particles, for which **q** and **p** can serve as initial data in the laboratory frame at $q^0 = 0$. Indeed, each element of $\mathcal{N}$ can be then identified with the world-line of a test particle in some coherent flow. If we decompose **q** and **p** into components $q_{\parallel}, p_{\parallel}$ parallel and components $\mathbf{q}_{\perp}, \mathbf{p}_{\perp}$ orthogonal to some 3-velocity vector **u**,

$$\mathbf{q} = q_{\parallel}\mathbf{u}\ |\mathbf{u}|^{-1} + \mathbf{q}_{\perp}, \qquad \mathbf{u} \cdot \mathbf{q}_{\perp} = 0, \tag{3.18a}$$

$$\mathbf{p} = p_{\parallel}\mathbf{u}\ |\mathbf{u}|^{-1} + \mathbf{p}_{\perp}, \qquad \mathbf{u} \cdot \mathbf{p}_{\perp} = 0, \tag{3.18b}$$

then by (3.3) Λ_u maps q_0, $q_{\parallel}$ and $\mathbf{q}_{\perp}$ into

$$q'_0 = \gamma_u(q_0 + \beta_u q_{\parallel}), \qquad q'_{\parallel} = \gamma_u(q_{\parallel} + \beta_u q_0), \qquad \mathbf{q}'_{\perp} = \mathbf{q}_{\perp}, \tag{3.19}$$

whereas Λ_u^{-1} maps p'_0, $p'_{\parallel}$ and $\mathbf{p}'_{\perp}$ into

$$p_0 = \gamma_u(p'_0 - \beta_u\,|\mathbf{p}|), \qquad p_{\parallel} = \gamma_u(p'_{\parallel} - \beta_u p'_0), \qquad \mathbf{p}_{\perp} = \mathbf{p}'_{\perp}, \tag{3.20}$$

where $\beta_u = |\mathbf{u}|\ c^{-1}$. Hence, if we measure momentum elements of volume in relation to the rest frames of particles in a given coherent flow, we get in other frames

$$\mathrm{d}\mathbf{p} = \left[\gamma_u \left(1 - \beta_u \frac{p_{\parallel}}{p_0}\right)\right]_{\mathbf{p}'=0} \mathrm{d}p'_{\parallel}\ \mathrm{d}\mathbf{p}'_{\perp} = \gamma_u\ \mathrm{d}\mathbf{p}' \tag{3.21a}$$

since $p'_{\|} = 0$ when $\mathbf{p}' = \mathbf{0}$. For measurements of space volume in the laboratory frame, i.e. at $q^0 = \text{const}$, we get

$$\mathrm{d}\mathbf{q}' = \gamma_u \, \mathrm{d}q_{\|} \, \mathrm{d}\mathbf{q}_{\perp} = \gamma_u \, \mathrm{d}\mathbf{q}. \tag{3.21b}$$

Hence, the invariance of the Γ-space element of measure follows:

$$\mathrm{d}\mathbf{q}' \, \mathrm{d}\mathbf{p}' = (\gamma_u \, \mathrm{d}\mathbf{q}) \, (\gamma_u^{-1} \, \mathrm{d}\mathbf{p}) = \mathrm{d}\mathbf{q} \, \mathrm{d}\mathbf{p}. \tag{3.22}$$

This is a feature of $\mathrm{d}\mathbf{q} \, \mathrm{d}\mathbf{p}$ that remains valid even in general relativity (Misner *et al.*, 1973, Section 22.6).

The invariance of $\mathrm{d}\mu_{\mathscr{N}}$ in (3.17) is a sufficient (but not necessary) condition for the method of induced representations (Mackey, 1968, p. 24) to work. Hence, let

$$L^{j,m}(q_0, R)\xi = \exp\left(\frac{i}{\hbar} mcq^0\right) D^{(j)}(R)\xi \tag{3.23}$$

be an irreducible representation of $\mathscr{T} \otimes \mathrm{SO}(3)$ in the Hilbert space $l^2(2j+1)$ of one-column complex matrices with $2j+1$ rows, where $D^{(j)}(R)$ is a unitary irreducible representation of SO(3) (or, strictly speaking, of SU(2)) for spin values $j = 0, \frac{1}{2}, 1, \frac{3}{2}, \ldots$. Then according to Mackey's (1968) inducing procedure, we can construct a unitary representation of $\mathscr{P}$ on the Hilbert space $\mathscr{H}^{j,m}$ of functions $f(q, \Lambda) = f((q, \Lambda) \cdot g), g \in \mathscr{N} \otimes \mathrm{SU}(2)$, with values in $l^2(2j+1)$ and inner product

$$\langle f_1 | f_2 \rangle_{\mathscr{N}} = \int_{\mathscr{N}} f_1^*(q, \Lambda) \cdot f_2(q, \Lambda) \, \mathrm{d}\mu_{\mathscr{N}} \tag{3.24}$$

where $f_1^* \cdot f_2$ is the inner product of f_1 and f_2 in $l^2(2j+1)$ for each fixed $(q, \Lambda) \in \mathscr{P}$. The elements of this representation are the operators

$$U^{j,m}(a, \Lambda) : f(q', \Lambda') \mapsto f'(q', \Lambda') = f(\Lambda^{-1}(q' - a), \Lambda^{-1}\Lambda') \tag{3.25}$$

acting in $\mathscr{H}^{j,m}$ as unitary operators.

From (3.23), (3.25) and (3.13) we easily compute that

$$f((q^0, \mathbf{q}), \Lambda_{p/m}) = \exp\left(-\frac{i}{\hbar} \frac{m^2c^2}{p^0} q^0\right) f\left(\left(0, \mathbf{q} - \frac{\mathbf{p}}{p^0} q^0\right), \Lambda_{p/m}\right). \tag{3.26}$$

Consequently, upon setting

$$f(q, p) = f(q, \Lambda_v), \quad p = mv, \tag{3.27}$$

and noting that each element of $\mathscr{H}^{j,m}$ is an equivalence class of $\mu_{\mathscr{N}}$-almost-everywhere equal functions $f(q, \Lambda)$, we can identify uniquely each such element with a function $f((0, \mathbf{q}), (p^0, \mathbf{p}))$ from $L^2(\Gamma)$. Then (3.26) can be rewritten in the form

$$f(q,p) = \left(\exp\left(-\frac{i}{\hbar} H_{\text{cl}} t\right) f\right)((0, \mathbf{q}), p), \quad q^0 = ct, \tag{3.28}$$

$$H_{\text{cl}} = \frac{c}{p^0}(m^2c^2 + \mathbf{p}\cdot\mathbf{P}), \qquad \mathbf{P} = -i\hbar\nabla_{\mathbf{q}}, \tag{3.29}$$

so that, as could be expected, the time-evolution (3.28) obeys a classical law of motion:

$$(p_\mu P^\mu - m^2c^2) f(q,p) = 0, \qquad P^0 = c^{-1} H_{\text{cl}}. \tag{3.30}$$

This kind of classical evolution is the relativistic counterpart of the Liouville equation for the nonrelativistic classical distribution function $\rho_t^{\text{cl}}(\mathbf{q}, \mathbf{p})$ which will be discussed in Sections 3.1, 3.4–3.7. In the present relativistic context $\rho_t^{\text{cl}}(\mathbf{q}, \mathbf{p})$ is replaced by

$$\rho^{\text{cl}}(q,p) = f^*(q,p)\cdot f(q,p), \tag{3.31}$$

which for $j = 0$ equals $|f(q, p)|^2$. Indeed, a system of imprimitivity in $\mathcal{H}^{j,m}$ is obtained if we introduce the PV measure

$$(E^{j,m}(B)f)(q, \Lambda) = \chi_B(q, \Lambda) f(q, \Lambda) \tag{3.32}$$

on the Borel subsets of $\mathcal{N}$, since by (3.25)

$$U^{j,m}(a, \Lambda) E^{j,m}(B) U^{j,m}(a, \Lambda)^* = E^{j,m}((a, \Lambda)B). \tag{3.33}$$

Since $\Lambda^{-1}\Lambda_v = \Lambda_{\Lambda^{-1}v} R$ for some $R \in \mathrm{SO}(3)$, we also have

$$U^{j,m}(a, \Lambda) : f(q,p) \mapsto f'(q,p) = D^{(j)}(R) f(\Lambda^{-1}(q-a), \Lambda^{-1}p), \tag{3.34}$$

and the above system of imprimitivity can be re-expressed on $L^2(\Gamma)$ in terms of (3.34) and the PV measure

$$(E^{j,m}(B)f)(q,p) = \chi_B(q,p) f(q,p), \; q^0 = 0, \tag{3.35}$$

which corresponds to the phase-space distribution function (3.31) at $q^0 = 0$:

$$\langle f | E^{j,m}(B) f \rangle_{\mathcal{N}} = \int_{q^0 = 0} \chi_B(q,p) \rho(q,p) \, d\mu_{\mathcal{N}}. \tag{3.36}$$

Comparing (3.9a) and (3.34) at $j = 0$ (so that $f(q, p)$ is a complex-valued function and $D^{(0)}(R) \equiv 1$), we note the identical form of the action of $U(a, \Lambda)$ and $U^{0,m}(a, \Lambda)$ on the respective wave functions ϕ and f. The crucial distinction lies in the time-evolution law obeyed by these wave functions, which in the case of ϕ is given by (3.11), or equivalently by

$$(P_\mu P^\mu - m^2c^2)\phi(q, p) = 0, \tag{3.37}$$

whereas in the case of f it is given by (3.30). Thus, $|f(q, p)|^2$ as a function of q^0 describes statistically the evolution of an ensemble of classical systems in phase space – $\hbar$ having been introduced in (3.28) only for the sake of facile comparison with the quantum case. As we shall see in subsequent sections, $|\phi(q, p)|^2$ serves the same role for a quantum system in an appropriate stochastic phase space.

The time-evolution law (3.30) was implicit in the $(2j + 1)$-dimensional representation (3.23) of $\mathscr{T} \otimes \mathrm{SO}(3)$ from which we have induced the (infinite-dimensional) unitary representation (3.25) of $\mathscr{P}$. Since all other finite-dimensional irreducible and unitary representations of $\mathscr{T} \otimes \mathrm{SO}(3)$ are equivalent to those in (3.23), no inducing procedure can yield the stochastic phase-space representations that appear in (3.9a). By combining the classical and quantum time evolutions we can derive such representations by inducing from SO(3) (Ali, 1979, 1980), but the more direct route described in the next section yields the same results more rapidly, and at the same time affords easy comparison with the Galilean case studied in Section 1.4.

2.4. Resolution Generators and Relativistic Phase-Space Representations

To arrive at the spin-zero irreducible subrepresentations of the phase-space representation $U(a, \Lambda)$ of $\mathscr{P}$ in (3.9) we proceed as we did in case of the representation $U(b, \mathbf{a}, \mathbf{v}, R)$ of $\mathscr{G}$ in (1.4.5). Thus, first we choose an element $\tilde{\eta} \in L^2(\mathscr{V}_m^{\pm})$ and subject it to the kinematical operations of spacetime translations by the 4-vector q and of boosting by the 4-velocity $v = p/m$, so as to arrive in accordance with (1.9) at

$$\tilde{\eta}_{q,p}(k) = (\tilde{U}(q, \Lambda_v)\tilde{\eta})(k) = \exp\left(\frac{i}{\hbar}\, q \cdot k\right) \tilde{\eta}(\Lambda_v^{-1} k). \tag{4.1}$$

The above outcome of these operations represents the exact counterpart of the nonrelativistic outcome $\xi_{\mathbf{q},\mathbf{p}}$ of the same operations (except for time translations) executed on ξ in arriving at (1.4.6b). Of course, in (1.4.6b) the configuration representation was used (although the momentum representation could have been used just as well), whereas in the relativistic case the momentum representation has decided advantages in view of the fact that, as discussed in Section 2.1, the 'configuration' representation (1.2) actually possesses no direct physical significance.

Pursuing the same strategy as in the nonrelativistic case, we now define the relativistic counterpart of the transformation W_ξ in (1.4.6a)

$$\mathscr{W}_\eta : \tilde{\phi}(k) \longmapsto \phi(q, p) = \int_{\mathscr{V}_m^\pm} \tilde{\eta}^*_{q,p}(k)\tilde{\phi}(k)\, \mathrm{d}\Omega_m(k), \tag{4.2}$$

under the tacit agreement that $p \in \mathscr{V}_m^\pm$ when $k \in \mathscr{V}_m^\pm$, respectively. We intend to show that if $\tilde{\eta}$ is rotationally invariant and such that (cf. (1.4.9))

$$\int_{k\in\mathscr{V}_m^\pm} |\tilde{\eta}(k)|^2 \, \mathrm{d}\mathbf{k} = 2mc(2\pi\hbar)^{-3}, \tag{4.3}$$

then $\mathscr{W}_\eta$ defines an isometric mapping of $L^2(\mathscr{V}_m^\pm)$ into $L^2(\Sigma_m^\pm)$ for each fixed value of $q^0 \in \mathbb{R}^1$. Indeed, by (3.7), (3.8) and (4.1)

$$\langle\phi_1|\phi_2\rangle_{\Sigma_m^\pm} = \int_\Gamma \mathrm{d}\mathbf{q}\,\mathrm{d}\mathbf{p} \int_{\mathscr{V}_m^\pm\times\mathscr{V}_m^\pm} \mathrm{d}\Omega_m(k')\,\mathrm{d}\Omega_m(k'')\exp\left[\frac{i}{\hbar}q\cdot(k''-k')\right] \times$$

$$\times\, \tilde{\eta}^*(\Lambda_v^{-1}k')\tilde{\eta}(\Lambda_v^{-1}k'')\tilde{\phi}_1^*(k')\tilde{\phi}_2(k''). \tag{4.4}$$

Taking into account (1.19), we can take advantage of the isometry property of Fourier–Plancherel transforms (i.e. on a formal level, we interchange the orders of integration in $\mathbf{q}$ and $(\mathbf{k}', \mathbf{k}'')$, which results in the appearance of a $\delta^3(\mathbf{k}' - \mathbf{k}'')$ function), so that

$$\langle\phi_1|\phi_2\rangle_{\Sigma_m^\pm} = \pm(2\pi\hbar)^3 \int_{\mathbb{R}^3} \mathrm{d}\mathbf{p} \int_{\mathscr{V}_m^\pm} \mathrm{d}\Omega_m(k)\,(2k^0)^{-1}\,|\tilde{\eta}(\Lambda_v^{-1}k)|^2\,\tilde{\phi}_1^*(k)\tilde{\phi}_2(k). \tag{4.5}$$

The rotational invariance of $\tilde{\eta}(k)$ implies that, as a function of $k \in \mathscr{V}_m^\pm$, $\tilde{\eta}$ depends exclusively on $|\mathbf{k}|$ (and not on the orientation of $\mathbf{k}$) or, equivalently, on k^0, so that

$$\tilde{\eta}(k) = \eta(mck^0), \qquad k = (k^0, \mathbf{k}) \in \mathscr{V}_m^\pm, \tag{4.6}$$

where η is a function of a single variable. Since

$$mck^0 = [\Lambda_v(mc, \mathbf{0})] \cdot \Lambda_v k = p \cdot \Lambda_v k, \; p = mv, \tag{4.7}$$

we immediately obtain from (4.6) that

$$\tilde{\eta}(\Lambda_v^{-1}k) = \eta(mc(\Lambda_v^{-1}k)^0) = \eta(p\cdot k), \quad p = mv. \tag{4.8}$$

Hence, upon interchanging in (4.5) the orders of integration in k and p by using Fubini's theorem, we conclude that

$$\langle \phi_1 | \phi_2 \rangle_{\Sigma_m^\pm} = \int_{\mathscr{V}_m^\pm} \tilde{\phi}_1^*(k)\tilde{\phi}_2(k)\, d\Omega_m(k) = \langle \tilde{\phi}_1 | \tilde{\phi}_2 \rangle_{\mathscr{V}_m^\pm} \tag{4.9}$$

if and only if for all $k \in \mathscr{V}_m^\pm$

$$\int_{p=mv} d\mathbf{p}\, |\eta(p \cdot k)|^2 = \pm 2k^0 (2\pi\hbar)^{-3}. \tag{4.10}$$

To prove that (4.10) is indeed true, we first note that due to the invariance of $p \cdot k$ and $d\Omega_m(p)$

$$\int_{p \in \mathscr{V}_m^\pm} |\eta(p \cdot k)|^2\, d\mathbf{p} = \int_{p' \in \mathscr{V}_m^\pm} |\eta(mcp_0')|^2 \frac{p_0}{p_0'}\, d\mathbf{p}', \qquad p' = \Lambda_{k/m}^{-1} p. \tag{4.11}$$

Upon decomposing $\mathbf{p}'$ into components $p_{||}'$ parallel to $\mathbf{k}$ and $\mathbf{p}_\perp'$ orthogonal to it, i.e.,

$$p_{||}' = |\mathbf{k}|^{-1} \mathbf{k} \cdot \mathbf{p}', \qquad \mathbf{p}_\perp' = \mathbf{p}' - p_{||}' |\mathbf{k}|^{-1} \mathbf{k}, \tag{4.12}$$

we can apply (3.3a) and write

$$p_0 = \gamma_u \left(p_0' + \frac{\mathbf{u} \cdot \mathbf{p}'}{c} \right) = \frac{\pm k^0}{mc} p_0' + \frac{|\mathbf{k}|}{mc} p_{||}', \tag{4.13}$$

where $\mathbf{u}$ is the 3-velocity associated with the 4-velocity $u = k/m$, so that

$$\mathbf{k} = m\gamma_u \mathbf{u}, \qquad \gamma_u = \left(1 - \frac{\mathbf{u}^2}{c^2} \right)^{-1/2} = \pm k^0/mc. \tag{4.14}$$

Consequently, after inserting (4.13) into (4.11), we get

$$\int_{p \in \mathscr{V}_m^\pm} |\eta(p \cdot k)|^2\, d\mathbf{p} = \int_{p \in \mathscr{V}_m^\pm} |\eta(mcp_0')|^2 \left(\pm \frac{k^0}{mc} + \frac{|\mathbf{k}|}{mc} \frac{p_{||}'}{p_0'} \right) d\mathbf{p}'. \tag{4.15}$$

In view of the fact that

$$p_0' = \pm (p_{||}'^2 + \mathbf{p}_\perp'^2 + m^2 c^2)^{1/2}, \qquad d\mathbf{p}' = dp_{||}'\, d\mathbf{p}_\perp', \tag{4.16}$$

the second term in the brackets on the right-hand side of (4.15) is an odd function of $p'_{||}$ and therefore contributes zero. Hence

$$\int_{p \in \mathscr{V}_m^\pm} |\eta(p \cdot k)|^2\, d\mathbf{p} = \pm \frac{k^0}{mc} \int_{p' \in \mathscr{V}_m^\pm} |\tilde{\eta}(p')|^2\, d\mathbf{p}', \tag{4.17}$$

and by (4.3) we conclude that (4.10) holds, and therefore (4.9) is valid for all $\tilde{\phi}_1$, $\tilde{\phi}_2 \in L^2(\mathscr{V}_m^{\pm})$.

We have thus established that the mapping $\mathscr{W}_\eta$ in (4.2) is isometric from $L^2(\mathscr{V}_m^{\pm})$ to $L^2(\Sigma_m^{\pm})$, where $\Sigma_m^{\pm} = \sigma \times \mathscr{V}_m^{\pm}$ and σ is any q^0 = const hyperplane in Minkowski space. Since $\mathscr{W}_\eta$ is obviously also linear, its range $\mathscr{W}_\eta L^2(\mathscr{V}_m^{\pm})$ has to be a closed subspace of $L^2(\Sigma_m^{\pm})$. To show that this subspace carries an irreducible subrepresentation of $U(a, \Lambda)$ in (3.9), we note that for $\tilde{U}(a, \Lambda)$ in (1.9)

$$(\mathscr{W}_\eta \tilde{U}(a, \Lambda)\tilde{\phi})(q, p) = \int_{\mathscr{V}_m^{\pm}} \exp\left[-\frac{i}{\hbar}(q-a)\cdot k\right] \eta^*(p\cdot k)\tilde{\phi}(\Lambda^{-1}k)\, \mathrm{d}\Omega_m(k) \tag{4.18}$$

by (4.2), since on account of (4.8) we can rewrite (4.1) in the form

$$\tilde{\eta}_{q,\,p}(k) = \exp\left(\frac{i}{\hbar} q\cdot k\right)\eta(p\cdot k),\; p, k \in \mathscr{V}_m^{\pm}. \tag{4.19}$$

Setting now $k' = \Lambda^{-1}k$ and taking into consideration the Poincaré invariance of $\mathrm{d}\Omega_m(k)$ as well as of $(q-a)\cdot k$ and $p\cdot k$, we get

$$(\mathscr{W}_\eta \tilde{U}(a, \Lambda)\tilde{\phi})(q, p) = \int_{\mathscr{V}_m^{\pm}} \exp\left[-\frac{i}{\hbar}k'\cdot\Lambda^{-1}(q-a)\right] \times$$

$$\times\, \eta^*(k'\cdot\Lambda^{-1}p)\tilde{\phi}(k')\, \mathrm{d}\Omega_m(k') = (\mathscr{W}_\eta\tilde{\phi})(\Lambda^{-1}(q-a), \Lambda^{-1}p). \tag{4.20}$$

Comparing (4.20) with (3.9), we see that $\mathscr{W}_\eta$ maps the momentum representation $\tilde{U}(a, \Lambda)$ of $\mathscr{P}$ into a subrepresentation of the phase-space representation $U(a, \Lambda)$ of $\mathscr{P}$. Naturally, this subrepresentation is defined on the range $\mathscr{W}_\eta L^2(\mathscr{V}_m^{\pm})$ of $\mathscr{W}_\eta$, so that indeed this range is an invariant subspace of $U(a, \Lambda)$. Since as a mapping from $L^2(\mathscr{V}_m^{\pm})$ to $\mathscr{W}_\eta L^2(\mathscr{V}_m^{\pm})$ the operator $\mathscr{W}_\eta$ is unitary, it has an inverse $\mathscr{W}_\eta^{-1}$ that isometrically maps $\mathscr{W}_\eta L^2(\mathscr{V}_m^{\pm})$ back into $L^2(\mathscr{V}_m^{\pm})$. Thus our earlier conclusions can be summarized in the statement that *$\mathscr{W}_\eta \tilde{U}(a, \Lambda)\mathscr{W}_\eta^{-1}$ is a unitary representation of $\mathscr{P}$ which acts in the closed subspaces $\mathscr{W}_\eta L^2(\mathscr{V}_m^{\pm})$ of $L^2(\Sigma_m^{\pm})$, and it is an irreducible subrepresentation of the phase-space representation $U(a, \Lambda)$ in* (3.9). Of course, the irreducibility property follows from the irreducibility of $\tilde{U}(a, \Lambda)$ in (1.9) and from the unitarity of $\mathscr{W}_\eta$ as a mapping from $L^2(\mathscr{V}_m^{\pm})$ onto $\mathscr{W}_\eta L^2(\mathscr{V}_m^{\pm}) \subset L^2(\Sigma_m^{\pm})$.

We had arrived at the above fundamental conclusion about $\mathscr{W}_\eta \tilde{U}(a, \Lambda)\mathscr{W}_\eta^{-1}$ by requesting that $\tilde{\eta}(k)$ should be rotationally invariant, or, equivalently, that (4.6) be true. It can be easily established (Prugovečki, 1978d, p. 2263) that, conversely, the rotational invariance of $\tilde{\eta}(k)$ is also a necessary (and not just a sufficient) condition for this conclusion to hold true. This can be proven by replacing

$\eta(k' \cdot \Lambda^{-1}p)$ with $\tilde{\eta}(\Lambda_p^{-1}\Lambda^{-1}k')$ in (4.20), and then investigating the consequence of the request that (4.20) be true for all $\tilde{\phi} \in L^2(\mathscr{V}_m^{\pm})$.

Thus the rotational invariance of $\tilde{\eta}(k)$ embodied in (4.6) proves to be as essential to pinpointing the spin-zero subrepresentation of $U(a, \Lambda)$ in the relativistic case as the rotational invariance of $\xi(\mathbf{x})$ (or, equivalently, of $\tilde{\xi}(\mathbf{k})$), embodied in (1.4.16) was to finding the spin-zero subrepresentations of $U(b, \mathbf{a}, \mathbf{v}, R)$ in the nonrelativistic case. In fact, if we compare (1.4.9) with (4.3), we see that we can establish a one-to-one correspondence between the nonrelativistic resolution generators $\tilde{\xi}(\mathbf{k})$ expressed in the momentum representations and the $\tilde{\eta}(k)$ on the mass hyperboloids $\mathscr{V}_m^{\pm}$ by setting (cf. Ali and Prugovečke, 1983b, for nonzero spin)

$$\tilde{\eta}(k) = (2mc)^{1/2}\, \tilde{\xi}(\mathbf{k}), \quad k = (k^0, \mathbf{k}) \in \mathscr{V}_m^{\pm}, \tag{4.21}$$

since then the properties (1.4.9) and (1.4.16) of ξ secure the properties (4.3) and (4.6) of $\tilde{\eta}$, which were essential to (4.20) and to the unitarity of $\mathscr{W}_\eta$. In turn, the unitarity of $\mathscr{W}_\eta$ ensures that the relativistic counterpart of (4.11) is true; namely, that

$$\int_{\substack{q^0 = \text{const} \\ p \in \mathscr{V}_m^{\pm}}} |\tilde{\eta}_{q,p}\rangle\, d\mathbf{q}\, d\mathbf{p}\, \langle \tilde{\eta}_{q,p}| = \tilde{\mathbb{1}}_{\pm}, \tag{4.22}$$

where $\tilde{\mathbb{1}}_{\pm}$ stands for the respective identity operators in $L^2(\mathscr{V}_m^{\pm})$. This is easily seen to be true if we note that (4.2) is equivalent to

$$(\mathscr{W}_\eta \tilde{\phi})(q, p) = \langle \tilde{\eta}_{q,p} | \tilde{\phi} \rangle_{\mathscr{V}_m^{\pm}}, \tag{4.23}$$

so that by (3.7) and (3.8)

$$\langle \mathscr{W}_\eta \tilde{\phi}_1 \mid \mathscr{W}_\eta \tilde{\phi}_2 \rangle_{\Sigma_m^{\pm}} = \int \langle \tilde{\phi}_1 \mid \tilde{\eta}_{q,p} \rangle_{\mathscr{V}_m^{\pm}} \langle \tilde{\eta}_{q,p} \mid \tilde{\phi}_2 \rangle_{\mathscr{V}_m^{\pm}}\, d\mathbf{q}\, d\mathbf{p} \tag{4.24}$$

for all $\tilde{\phi}_1, \tilde{\phi}_2 \in L^2(\mathscr{V}_m^{\pm})$. In other words, $\tilde{\eta}(k)$ in (4.21) is a resolution generator for continuous resolutions of the identity

$$\tilde{\eta}_{q,p} = \tilde{U}(q, \Lambda_{p/m})\tilde{\eta}, \quad (q, p) \in \Sigma_m^{\pm}, \tag{4.25}$$

in $L^2(\mathscr{V}_m^{\pm})$ that are obtained from $\tilde{\eta}$ by the same kinematical operations (in the inertial frames where $q \in \Sigma_m^{\pm} = \sigma \times \mathscr{V}_m^{\pm}$ is given by $q^0 = 0$) as the continuous resolution of the identity $\xi_{\mathbf{q},\mathbf{p}}$ in (1.4.6b), or, equivalently,

$$\tilde{\xi}_{\mathbf{q},\mathbf{p}} = \tilde{U}\left(0, \mathbf{q}, \frac{\mathbf{p}}{m}, I\right)\tilde{\xi}, \quad (\mathbf{q}, \mathbf{p}) \in \Gamma, \tag{4.26}$$

corresponding to $\tilde{U}$ in (1.2.22), are obtained from $\xi(\mathbf{x})$ and $\tilde{\xi}(\mathbf{k})$, respectively. Naturally, the mathematical executions of these kinematical operations are carried out by representatives of the Poincaré group $\mathscr{P}$ in case of $\tilde{\eta}$, and of the Galilei group $\mathscr{G}$ in case of ξ or $\tilde{\xi}$, as befits the assumptions that we make about the nature and properties of the physical laws governing such operations in the real world.

2.5. Localizability in Relativistic Stochastic Phase Spaces

In Sections 2.3 and 2.4 we have discovered that the Hilbert spaces $L^2(\Sigma_m^\pm)$ over hypersurfaces $\Sigma_m^\pm = \sigma \times \mathscr{V}_m^\pm$ in the relativistic phase spaces $\mathscr{M}_m^\pm$ in (3.4) play in the relativistic case a role analogous to the one played by $L^2(\Gamma)$ in the nonrelativistic case if we work in inertial frames where σ are the q^0 = const hyperplanes. In order to underline this analogy, we chose to carry out the considerations of Section 2.4 in those frames where $\mathrm{d}\Sigma(q, p)$ in (3.6) assumes the form $\mathrm{d}\mathbf{q}\,\mathrm{d}\mathbf{p}$. It is, however, easily verified that all those considerations are frame-independent and that every single formula in Section 2.4 can be rewritten in a frame-independent manner. The result is in most cases a covariant expression, i.e. one in which no single frame, or proper subset of inertial frames, plays a distinguished role. For example,

$$\int_{\Sigma_m^\pm} |\tilde{\eta}_\zeta\rangle\, \mathrm{d}\Sigma_m(\zeta)\, \langle\tilde{\eta}_\zeta| = \tilde{\mathbb{1}}_\pm, \quad \zeta = (q, p) \tag{5.1}$$

represents the covariant counterpart of (4.22). The exceptions to this general observation are (4.10), and its satellite expressions (4.11) and (4.17), in which the frames where $\Sigma_m^\pm$ consists of (q, p) with q^0 = const play a distinguished role, so that, for example, the expression for (4.10) valid in any given (classical) inertial frame is

$$\int |\eta(p' \cdot k')|^2 p^0\, \mathrm{d}\Omega_m(p') = \pm(2\pi\hbar)^{-3} k^0, \tag{5.2}$$

where p'_μ and k'_μ are coordinates in that frame, but k^0 and p^0 still refer to the frame where q^0 = const for $(q, p) \in \Sigma_m^\pm$. We shall refer to the special frames in which

$$\Sigma_m^\pm = \sigma \times \mathscr{V}_m^\pm, \quad \sigma = \{q \mid q^0 = \text{const}\}, \tag{5.3}$$

as the *stochastic rest frames* for the chosen hypersurface $\Sigma_m^\pm$ in $\mathscr{M}_m^\pm$. The set of all stochastic rest frames for given $\Sigma_m^\pm$ obviously constitutes an equivalence class, with two frames belonging to the same class if and only if they can be related to each other by the operations of spacetime translation and/or rotations – but definitely belonging to distinct classes if a boost is required.

The term 'stochastic rest frame' has a deeper physical foundation, since such frames will gradually emerge (in the light of later considerations) to be the stochastic rest frames of extended quantum test particles in some chosen coherent flow of quantum particles with proper wave functions $\tilde{\eta}(k)$ in the momentum representation. The first indication of the possibility of such an interpretation for $\tilde{\eta}$ was presented by the parallelism between the role which $\tilde{\eta}$ and $\tilde{U}(a, \Lambda)$ have played opposite $\tilde{\xi}$ and $\tilde{U}(b, \mathbf{a}, \mathbf{v}, R)$ in Section 2.4. In pure physical terms, that role suggests that the interpretation of $\tilde{\xi}(\mathbf{k})$ as the momentum-space proper wave function of a quantum particle stochastically at rest at the origin of an inertial frame is taken over in the relativistic regime by $\tilde{\eta}(k)$ related to $\tilde{\xi}(\mathbf{k})$ by (4.2).

Much additional evidence is required to substantiate this conjecture. Thus, first of all, we have to investigate the extent to which the considerations of Section 1.5 can be duplicated in the present relativistic context. In view of the marginality properties (1.5.1) of nonrelativistic stochastic phase-space distributions, consider therefore

$$\rho^{\eta}(p) = \int_{\mathbb{R}^3} |\phi(q, p)|^2 \, \mathrm{d}\mathbf{q}, \quad \phi(q, p) = (\mathscr{W}_{\eta}\tilde{\phi})(q, p), \tag{5.4}$$

at q^0 = const in a stochastic rest frame for $\Sigma_m^{\pm}$ (so that the above integral is q^0-independent). By (4.2) and (4.19)

$$(\mathscr{W}_{\eta}\tilde{\phi})(q, p) = \int_{\mathscr{V}_m^{\pm}} \exp\left(-\frac{i}{\hbar} q \cdot k\right) \eta^*(p \cdot k) \tilde{\phi}(k) \, \mathrm{d}\Omega_m(k), \tag{5.5}$$

so that we can perform the integration in (5.4) by using the unitarity property of Fourier–Plancherel transforms:

$$\rho^{\eta}(p) = \pm(2\pi\hbar)^3 \int_{\mathscr{V}_m^{\pm}} |\eta(p \cdot k)|^2 \, |\tilde{\phi}(k)|^2 \, (2k^0)^{-1} \, \mathrm{d}\Omega_m(k). \tag{5.6}$$

Introducing $k_{(p)} = \Lambda_{p/m}^{-1} k$ as new variables of integration and executing the identifications (4.6), (4.21) and (1.5.2), as well as (1.19) after setting $\mathrm{d}\Omega_m(k) = \mathrm{d}\Omega_m(k_{(p)})$, we get

$$\rho^{\eta}(p) = \pm mc \int_{\mathbb{R}^3} \overline{\chi}_0^{\xi}(\mathbf{k}_{(p)}) \, |\tilde{\phi}(\Lambda_p k_{(p)})|^2 \, [2(\Lambda_{p/m} k_{(p)})^0]^{-1} \, \frac{\mathrm{d}\mathbf{k}_{(p)}}{k_{(p)}^0}. \tag{5.7}$$

Hence, in the sharp-momentum limit

$$\overline{\chi}_0^{\xi}(\mathbf{k}_{(p)}) \longrightarrow \delta^3(\mathbf{k}_{(p)}) \tag{5.8}$$

we obtain (upon noting that $k = \Lambda_p k_{(p)} = p$ when $k_{(p)} = (\pm mc, \mathbf{0})$) that

$$\rho^{\eta}(p) \longrightarrow (2p^0)^{-1} \, |\tilde{\phi}(p)|^2 \tag{5.9}$$

at any $p \in \mathscr{V}_m^\pm$ at which $|\tilde{\phi}(p)|^2$ is continuous. Thus, in the sharp-momentum limit, the conventional relativistic probabilities in momentum space are recovered, i.e.

$$\int_B \mathrm{d}\mathbf{p} \int_{\mathbb{R}^3} |\phi(q,p)|^2 \,\mathrm{d}\mathbf{q} \longrightarrow \int_B |\tilde{\phi}(p)|^2 \,\mathrm{d}\Omega_m(p). \tag{5.10}$$

If we try, however, to duplicate the same considerations in the case of

$$\rho_\eta(q) = \int_{\mathbb{R}^3} |\phi(q,p)|^2 \,\mathrm{d}\mathbf{p}, \quad p = (p^0, \mathbf{p}) \in \mathscr{V}_m^\pm, \tag{5.11}$$

then we immediately discover that the **p**-integration cannot be explicitly performed in any manner that would give rise to $\chi_{\mathbf{q}}^\xi(x)$ via (1.5.2), (4.6) and (4.21). Of course, this is very much in keeping with the negative outcome of the many searches after a consistent configuration representation for relativistic quantum mechanics, and indirectly confirms the fact that such a representation does not exist, as implied by Hegerfeldt's theorem. On the other hand, as will be shown in Section 2.8, $\rho_\eta(q)$ in (5.11) nevertheless equals the component $j_\eta^0(q)$ of a relativistically covaraint and conserved probability current $j_\eta^\mu(q)$.

This fact indicates that $|\phi(q, p)|^2$ possesses physical significance despite the impossibility of sharply localizing relativistic quantum particles. This significance emerges in part if we use (4.6) and (4.21) to write (5.5) for $p^0 = \pm mc$ in the form

$$(\mathscr{W}_\eta \tilde{\phi})(q,p)\big|_{\mathbf{p}=\mathbf{0}} = (2mc)^{1/2} \int_{k \in \mathscr{V}_m^\pm} \exp\left(-\frac{i}{\hbar}\, q \cdot k\right) \xi^*(\mathbf{k})\tilde{\phi}(k) \frac{\mathrm{d}\mathbf{k}}{\pm 2k^0}. \tag{5.12}$$

Comparing $\phi(q, p)$ and $\psi(\mathbf{q}, \mathbf{p}; t)$ after performing the identification (1.18), so that for $q^0 = ct$

$$\phi(q,p) = (\mathscr{W}_\eta \tilde{\psi})(q,p), \quad \psi(\mathbf{q},\mathbf{p};t) = \left(\mathscr{W}_\xi \exp\left(-\frac{i}{\hbar} H_0 t\right) \psi\right)(\mathbf{q},\mathbf{p}), \tag{5.13}$$

we immediately see that if $|\tilde{\psi}(\mathbf{k})|$ is significantly larger than zero only within a nonrelativistic neighborhood of $\mathbf{k} = \mathbf{0}$, where

$$\pm k^0 = (\mathbf{k}^2 + m^2c^2)^{1/2} \approx mc + \frac{1}{2} \frac{\mathbf{k}^2}{m^2c^2} \tag{5.14}$$

holds true with good accuracy, then

$$|\phi(q,p)|^2 = |\psi(\mathbf{q},\mathbf{p};t)|^2 + O\left(\frac{\mathbf{p}^2}{m^2c^2}\right) \tag{5.15}$$

for sufficiently small $|t| = |q^0/c|$, i.e. as long as relativistic time evolution can be approximated by nonrelativistic time evolution. In fact, since

$$p \cdot k = m^2c^2 - \tfrac{1}{2}(p-k)^2, \qquad p, k \in \mathscr{V}_m^{\pm}, \tag{5.16}$$

so that *de facto* η in (5.5) can be viewed as a function of $(p-k)^2$, this conclusion remains valid as long as the support of $\tilde{\phi}(k)$ is essentially within a nonrelativistic neighborhood of any given value of $p \in \mathscr{V}_m^{\pm}$, as may be verified by going to a stochastic rest frame of the test particle with proper wave function $\eta_{q,p}$, i.e. a frame where $\mathbf{p}' = \mathbf{0}$ for $p' = \Lambda p$ (e.g. for $\Lambda = \Lambda_{p/m}^{-1}$).

The above considerations indicate that $|\phi(q', p')|^2$ can be viewed as an approximate probability density at the stochastic point

$$\underset{\sim}{\zeta}' = (\zeta', \chi^{\eta}_{\zeta'}),\ \chi^{\eta}_{q',p'}(\mathbf{x}, \mathbf{k}) = \chi^{\xi}_{\mathbf{q}'}(x)\bar{\chi}^{\xi}_{\mathbf{p}'}(k) \tag{5.17}$$

in the stochastic rest frame of the test particle, i.e. for $\mathbf{p}' = \mathbf{0}$, with the approximation getting better as the rest mass m of the test particle increases. Converting to the laboratory frame by using (1.7) for $\Lambda_{p/m}$ and noting that

$$\mathrm{d}\mathbf{x}'\,\mathrm{d}\mathbf{k}' = \left(1 - \frac{\mathbf{k}\cdot\mathbf{p}}{k^0p^0}\right)\mathrm{d}\mathbf{x}\,\mathrm{d}\mathbf{k} \tag{5.18}$$

if we change from the simultaneity in the rest frame of the test particle to that in the rest frame of the observer, we get

$$\chi^{\eta}_{q,p}(\mathbf{x}, \mathbf{k}) = \frac{p^0k^0}{p\cdot k}\,\chi^{\xi}_{\mathbf{0}}\left(\mathbf{x} - \mathbf{q} - \frac{\mathbf{p}\cdot(\mathbf{x}-\mathbf{q})}{p^0(mc+p^0)}\,\mathbf{p}\right) \times$$
$$\times\, \bar{\chi}^{\xi}_{\mathbf{0}}\left(\mathbf{k} - \frac{\mathbf{p}}{mc}\left(k^0 - \frac{\mathbf{p}\cdot\mathbf{k}}{p^0+mc}\right)\right). \tag{5.19}$$

Hence, the *relativistic stochastic phase space* for the resolution generator η can be defined in terms of (3.4) and (5.19) as the set

$$\mathscr{M}^{\pm}_{m,\eta} = \{\underset{\sim}{\zeta} = (\zeta, \chi^{\eta}_{\zeta}) \mid \zeta \in \mathscr{M}^{\pm}_m = M(1,3) \times \mathscr{V}^{\pm}_m\}. \tag{5.20}$$

The hypersurfaces in $\mathscr{M}^{\pm}_{m,\eta}$ corresponding to the nonrelativistic Γ_{ξ} in (1.5.5) are now

$$\Sigma^{\pm}_{m,\eta} = \{\underset{\sim}{\zeta} = (\zeta, \chi^{\eta}_{\zeta}) \mid \zeta \in \Sigma^{\pm}_m = \sigma \times \mathscr{V}^{\pm}_m\}, \tag{5.21}$$

where σ is any spacelike hyperplane in Minkowski space $M(1, 3)$. We note, however, that neither $\mathscr{M}^{\pm}_{m,\eta}$ nor $\Sigma^{\pm}_{m,\eta}$ are Cartesian products (cf. Section 1.1) of configuration and momentum stochastic spaces, as was the case with Γ_{ξ}.

Pursuing the analogy with the nonrelativistic case, and in particular (1.5.7), we shall write

$$L^2(\Sigma^{\pm}_{m,\eta}) \stackrel{\text{def}}{=} \mathscr{W}_\eta L^2(\mathscr{V}^{\pm}_m) \subset L^2(\Sigma^{\pm}_m). \tag{5.22}$$

Then, formally speaking, $L^2(\Sigma^{\pm}_{m,\eta})$ can be viewed as the space of wave functions over the hypersurfaces $\Sigma^{\pm}_{m,\eta}$ in the relativistic stochastic phase spaces $\mathscr{M}^{\pm}_{m,\eta}$. It has to be kept in mind, however, that this interpretation holds only in the asymptotic limit $m \longrightarrow \infty$ of infinitely massive quantum test particles, which therefore asymptotically acquire the feature of classical test particles in having their position and 3-velocity simultaneously well defined. It is this idealization – which is intimately related to the one assuming the existence of classical frames – that we shall dispense with in Chapter 4, and thereby overcome the need for any approximations in the interpretation of $\phi(q, p)$ as a probability amplitude at $(q, p) \in \mathscr{M}_{m,\eta}$. In this chapter, however, we shall retain the naive interpretation of $\mathscr{M}_{m,\eta}$ based on (5.19) and (5.20).

It is interesting to establish the role played by the Newton–Wigner operators X^j_{NW}, defined on $L^2(\mathscr{V}^+_m)$ by (2.13), in the relativistic stochastic phase-space framework. Let us therefore introduce, by analogy with (1.8.29), the stochastic time and position operators

$$(Q^\mu_{\text{st}}\phi)(q, p) = q^\mu \phi(q, p), \quad \mu = 0, \ldots, 3, \tag{5.23}$$

defined in any $L^2(\Sigma^+_m)$, $\Sigma^+_m = \sigma \times \mathscr{V}^+_m \subset \mathscr{M}^+_m$. Let us consider now the orthogonal projection operators $\mathbb{P}_\eta(\Sigma^{\pm}_m)$ onto the subspaces in (5.22), and let us introduce the operators

$$Q^\mu_\eta(\sigma) = \mathbb{P}_\eta(\Sigma^+_m) Q^\mu_{\text{st}} \mathbb{P}_\eta(\Sigma^+_m) \tag{5.24}$$

for the special cases of inertial frames where σ is a q^0 = const hypersurface in $M(1, 3)$. We shall prove now that in any such frame we have, when η is real, that

$$Q^j_\eta(\sigma) = \mathscr{W}_\eta X^j_{\text{NW}} \mathscr{W}^{-1}_\eta, \quad j = 1, 2, 3, \tag{5.25}$$

where X^j_{NW} are the Newton–Wigner operators defined in (2.13). Indeed, using (5.5) we get

$$(\mathscr{W}^{-1}_\eta \mathbf{Q}_\eta \mathscr{W}_\eta \tilde{\phi})(k) = \int d\mathbf{q}\, d\mathbf{p} \int \exp\left[\frac{i}{\hbar}\, \mathbf{q} \cdot (\mathbf{k}' - \mathbf{k})\right] \times$$

$$\times\, q^j \eta(p \cdot k) \eta(p \cdot k') \tilde{\phi}(k')\, d\Omega_m(k'). \tag{5.26}$$

After replacing above $\mathbf{q}$ by $-i\hbar\nabla_{\mathbf{k}'}$ applied to the exponential, then integrating by parts, and in the end using (1.18) and (1.19), we arrive at

$$(\mathscr{W}^{-1}_\eta \mathbf{Q}_\eta \mathscr{W}_\eta \tilde{\phi})(k) = i\hbar \nabla_{\mathbf{k}} \tilde{\psi}(\mathbf{k}) - (2\pi\hbar)^3\, \frac{\tilde{\psi}(\mathbf{k})}{4k^0} \int \nabla_{\mathbf{k}} \eta^2(k \cdot p)\, d\mathbf{p}, \tag{5.27}$$

where $k^0 = (\mathbf{k}^2 + m^2c^2)^{1/2}$. Taking (4.10) into account, we finally obtain

$$(\mathscr{W}_\eta^{-1}\mathbf{Q}_\eta\,\mathscr{W}_\eta\tilde{\phi})\,(k) = i\hbar\left[\nabla_{\mathbf{k}} - \tfrac{1}{2}\,\mathbf{k}(\mathbf{k}^2 + m^2c^2)^{-1}\right]\tilde{\psi}(\mathbf{k}), \tag{5.28}$$

which coincides with (2.13).

It is interesting that if we introduce the stochastic momentum operators

$$(P_{\mathrm{st}}^\mu\phi)\,(q,p) = p^\mu\phi(q,p), \quad \mu = 0, \ldots, 3, \tag{5.29}$$

then a similar but somewhat lengthier argument (Ali and Prugovečki, 1981, p. 194) shows that

$$(\mathscr{W}_\eta^{-1}\mathbf{P}_\eta\,\mathscr{W}_\eta\tilde{\phi})\,(k) = C_\eta\mathbf{k}\tilde{\psi}(\mathbf{k}), \tag{5.30}$$

if $\mathbf{P}_\eta$ is the restriction of $\mathbb{P}_\eta(\Sigma_m^+)\mathbf{P}_{\mathrm{st}}\mathbb{P}_\eta(\Sigma_m^+)$ to $L^2(\Sigma_{m,\eta}^+)$. In (5.30) C_η is a positive constant which is independent of $\tilde{\psi}$, so that in view of (2.20), $\mathbf{Q}_\eta$ and $C_\eta^{-1}\mathbf{P}_\eta$ constitute in $L^2(\Sigma_{m,\eta}^+)$ an irreducible representation of the CCRs, despite the fact that $\mathbf{Q}_{\mathrm{st}}$ and $\mathbf{P}_{\mathrm{st}}$ commute. Of course, this is due to the fact that

$$[\mathbf{Q}_{\mathrm{st}}, \mathbb{P}_\eta(\Sigma_m^+)] \neq 0, \qquad [\mathbf{P}_{\mathrm{st}}, \mathbb{P}_\eta(\Sigma_m^+)] \neq 0, \tag{5.31}$$

so that the spectral measures of $\mathbf{Q}_\eta$ and $\mathbf{P}_\eta$ do not coincide with the restrictions to $L^2(\Sigma_{m,\eta}^+)$ of the spectral measures of $\mathbf{Q}_{\mathrm{st}}$ and $\mathbf{P}_{\mathrm{st}}$, and therefore do not commute. This also means that although

$$\langle\tilde{\phi}\mid \mathbf{X}_{\mathrm{NW}}\tilde{\phi}\rangle_{\mathscr{V}_m^+} = \langle\mathscr{W}_\eta\tilde{\phi}\mid \mathbf{Q}_{\mathrm{st}}\,\mathscr{W}_\eta\tilde{\phi}\rangle_{\Sigma_m^+}, \tag{5.32}$$

the above equality between the expectation values of the Newton–Wigner operators and the stochastic position operators does not extend to functions of these operators – except in an approximative sense when the spread of the proper wave function $\xi(\mathbf{x})$ resulting from the identification (4.21) is sufficiently small. Naturally, this reflects the fact that the localization described by the probability amplitudes $\phi(q, p)$ is stochastic in a non-sharp sense. In fact, it should also be emphasized that (5.32) is true only in stochastic rest frames for $\Sigma_m^+ = \sigma \times \mathscr{V}_m^+$, i.e. when σ are q^0 = const hypersurfaces. Indeed, Q_η^μ transform covariantly together with Q_{st}^μ, whereas X_{NW}^j do not! Thus, the physical significance of the Newton–Wigner operators that emerges from this analysis is that of mean stochastic position operators, but only in stochastic rest frames of Σ_m^+, i.e. only in the stochastic rest frames of the microdetectors in the chosen coherent flow of test particles.

2.6. Systems of Covariance in Relativistic Stochastic Phase Spaces

The interpretation of $\phi(q, p) = (\mathscr{W}_\eta \tilde{\phi})(q, p)$ as a probability amplitude at points $(q, p) \in \Sigma^+_{m,\eta}$, where $\Sigma^{\pm}_{m,\eta}$ is any hypersurface (5.21) in the relativistic stochastic phase spaces $\mathscr{M}^{\pm}_{m,\eta}$ in (5.20), implies that by (4.23)

$$P^{\eta}_{\phi}(B) = \langle \tilde{\phi} | \tilde{\mathrm{I\!P}}_\eta(B) \tilde{\phi} \rangle_{\mathscr{V}^{\pm}_m} \tag{6.1}$$

is a probability measure over Borel sets $B \subset \Sigma_{m,\eta}$ if $\tilde{P}_\eta(B)$ denotes the Bochner integral

$$\tilde{\mathrm{I\!P}}_\eta(B) = \int_B |\tilde{\eta}_{q,p}\rangle \, \mathrm{d}\Sigma_m(q, p) \, \langle \tilde{\eta}_{q,p}|. \tag{6.2}$$

Thus, the formal analogy with (1.6.1) and (1.6.2) is almost complete. It would be totally complete if (1.6.2) were written in terms of the POV measure $U_{\mathrm{F}} \hat{\mathrm{I\!P}}_\xi(B) U_{\mathrm{F}}^{-1}$ in the nonrelativistic momentum spectral representation space, to parallel its relativistic counterpart $\tilde{\mathrm{I\!P}}_\eta$ which acts in the relativistic momentum representation spaces $L^2(\mathscr{V}^{\pm}_m)$ for particles and antiparticles of spin zero.

On the other hand, $\tilde{\mathrm{I\!P}}_\eta(B)$ gives rise to a system of covariance related to $\tilde{U}(a, \Lambda)$ in (1.9)

$$\tilde{U}(g) \tilde{\mathrm{I\!P}}_\eta(B) \tilde{U}^{-1}(g) = \tilde{\mathrm{I\!P}}_\eta(gB), \qquad g \in \mathscr{E}(\sigma) \tag{6.3}$$

only if we restrict $g = (a, \Lambda) \in \mathscr{P}$ to the Euclidean subgroup $\mathscr{E}(\sigma)$ of $\mathscr{P}$ that equals the semidirect product of space translations and rotations which leave a given $\Sigma^{\pm}_m = \sigma \times \mathscr{V}^{\pm}_m$ invariant. Since proper Lorentz transformations do not preserve simultaneity, in relativity theory we do not have a counterpart of the isochronous Galilei group $\mathscr{G}'$, and therefore we have no counterpart of (1.6.4).

Let us introduce, however, by analogy with (1.6.23), the POV measure

$$\tilde{F}_\eta(\mathring{B}) = \int_{-\infty}^{+\infty} \tilde{\mathrm{I\!P}}_\eta(B_{q^0}) \, \mathrm{d}q^0, \tag{6.4a}$$

$$B_{q^0} = \{(q, p) \mid (q, p) \in \Sigma^{(q^0)\pm}_{m,\eta} \cap \mathring{B}\}, \tag{6.4b}$$

on the Borel sets $\mathring{B}$ in $\mathscr{M}^{\pm}_{m,\eta}$, where as q^0 ranges over $\mathrm{I\!R}^1$, the hypersurfaces $\Sigma^{(q^0)\pm}_{m,\eta}$ provide a foliation of $\mathscr{M}^{\pm}_{m,\eta}$ corresponding to that of $M(1, 3)$ into hyperplanes σ_{q^0} consisting of points q with $q^0 = \text{const}$. In other words, $\Sigma^{(q^0)\pm}_{m,\eta}$ share the same stochastic rest frames and the corresponding hypersurfaces $\Sigma^{(q^0)\pm}_m = \sigma_{q^0} \times \mathscr{V}^{\pm}_m$ belong in these frames to the hyperplanes $q^0 = \text{const}$.

It is trivial to establish that $\tilde{F}_\eta(\mathring{B})$ is a POV measure over the Borel sets in $\mathscr{M}^{\pm}_{m,\eta}$ of positive-definite operators acting in $L^2(\mathscr{V}^{\pm}_m)$. It is, however, not as evident that this POV measures gives rise in conjunction with $\tilde{U}(a, \Lambda), (a, \Lambda) \in \mathscr{P}$, to a system of covariance, i.e. that

$$\tilde{U}(g)\tilde{F}_\eta(\mathring{B})\tilde{U}^{-1}(g) = \tilde{F}_\eta(g\mathring{B}), \quad g \in \mathscr{P}. \tag{6.5}$$

In fact, the definition (6.4) is not manifestly covariant, so that $\tilde{F}_\eta$ might be dependent on the choice of coordinates, with the coordinates in stochastic rest frames of the hypersurfaces $\Sigma_{m,\eta}^{(q^0)\pm}$ being favored.

To show that (6.5) is nevertheless true, we have to recast (6.4a) in a manifestly covariant form. This can be achieved by recasting the inner products

$$\langle\phi_1|\phi_2\rangle_{\Sigma_{m,\eta}^\pm} = \int_{\Sigma_m^\pm} \phi_1^*(\zeta)\phi_2(\zeta)\,d\Sigma_m(\zeta) \tag{6.6}$$

in the subspaces $L^2(\Sigma_{m,\eta}^\pm)$ of $L^2(\Sigma_m^\pm)$ defined by (5.22) into the new form

$$\langle\phi_1|\phi_2\rangle_{\Sigma_{m,\eta}^\pm} = \frac{i\hbar}{Z_\eta}\int_{\Sigma_m^\pm} \phi_1^*(\zeta)\overleftrightarrow{\partial}_\mu\phi_2(\zeta)\,d\sigma^\mu(q)\,d\Omega_m(p), \tag{6.7a}$$

$$\phi_1^*(\zeta)\overleftrightarrow{\partial}_\mu\phi_2(\zeta) = \phi_1^*(q,p)\frac{\partial}{\partial q^\mu}\phi_2(q,p) - \phi_2(q,p)\frac{\partial}{\partial q^\mu}\phi_1^*(q,p), \tag{6.7b}$$

where $Z_\eta > 0$ is a renormalization constant which shall play a prominent role in many later developments in this monograph.[14]

To establish (6.7a) and compute Z_η, let us work in some stochastic rest frame for $\Sigma_{m,\eta}^\pm$, so that (6.6) assumes the form

$$\langle\phi_1|\phi_2\rangle_{\Sigma_{m,\eta}^\pm} = \int_{\substack{q^0 = \text{const} \\ p \in \mathscr{V}_m^\pm}} \phi_1^*(q,p)\phi_2(q,p)\,d\mathbf{q}\,d\mathbf{p} \tag{6.8}$$

in accordance with (3.7). Using (5.5) with $\tilde{\phi}_i = \mathscr{W}_\eta^{-1}\phi_i$, $i = 1, 2$, we get on account of the unitarity of Fourier–Plancherel transforms on $L^2(\mathbb{R}^3)$ in the variables $\mathbf{q}, \mathbf{k} \in \mathbb{R}^3$ that

$$i\hbar \int_{q^0 = \text{const}} \phi_1^*(\zeta)\overleftrightarrow{\partial}_0\phi_2(\zeta)\,d\mathbf{q}\,d\Omega_m(p)$$

$$= (2\pi\hbar)^3 \int_{\mathscr{V}_m^\pm} d\Omega_m(p) \int_{\mathscr{V}_m^\pm} d\Omega_m(k)\,|\eta(p\cdot k)|^2\,\tilde{\phi}_1^*(k)\tilde{\phi}_2(k). \tag{6.9}$$

Reversing in (6.9) the orders of integration in k and p, and noting that

$$\int_{\mathscr{V}_m^{\pm}} |\eta(p \cdot k)|^2 \, d\Omega_m(p) = \int_{\mathscr{V}_m^{\pm}} |\eta(mcp_0')|^2 \, d\Omega_m(p') \tag{6.10}$$

is actually a k-independent constant, we conclude that

$$i\hbar \int_{q^0 = \text{const}} \phi_1^*(\zeta) \overleftrightarrow{\partial}_0 \phi_2(\zeta) \, d\mathbf{q} \, d\Omega_m(p) = Z_\eta \langle \tilde{\phi}_1 | \tilde{\phi}_2 \rangle_{\mathscr{V}_m^{\pm}} \tag{6.11}$$

if we combine (6.10) with the remaining constants in (6.9) into

$$Z_\eta = (2\pi\hbar)^3 \int_{\mathscr{V}_m^{\pm}} |\eta(mcp^0)|^2 \, d\Omega_m(p)$$

$$= (2\pi\hbar)^3 \, mc \int_{\mathbb{R}^3} |\xi(\mathbf{k})|^2 \frac{d\mathbf{k}}{(\mathbf{k}^2 + m^2c^2)^{1/2}} , \tag{6.12}$$

where the last expression is based on the identifications (4.6) and (4.21). Comparing now (6.11) with (6.7), and taking into account the unitarity of $\mathscr{W}_\eta$ as an operator from $L^2(\mathscr{V}_m^{\pm})$ onto $L^2(\Sigma_{m,\eta}^{\pm})$, we conclude that (6.7a) is indeed true for all ϕ_1, $\phi_2 \in L^2(\Sigma_{m,\eta}^{\pm})$. Note, however, that since the renormalization constant Z_η is η-dependent, (6.7a) is not true globally on $L^2(\Sigma_m^{\pm})$, i.e. (3.8) remains the only globally valid expression for the inner product over all of $L^2(\Sigma_m^{\pm})$.

Combining now (6.2), (6.4) and (6.11) we get

$$\langle \tilde{\phi}_1 | \tilde{F}_\eta(B) \tilde{\phi}_2 \rangle_{\mathscr{V}_m^{\pm}} = \frac{i\hbar}{Z_\eta} \int_{-\infty}^{+\infty} dq^0 \int_{B_{q^0}} \phi_1^*(\zeta) \overleftrightarrow{\partial}_0 \phi_2(\zeta) \, d\mathbf{q} \, d\Omega_m(p) \tag{6.13}$$

for any $\tilde{\phi}_1, \tilde{\phi}_2 \in L^2(\mathscr{V}_m^{\pm})$. Thus, we arrive at the following covariant expression for $\tilde{F}_\eta$:

$$\tilde{F}_\eta(B) = \frac{i\hbar}{Z_\eta} \int_B d^4q \, d\Omega_m(p) \, |\tilde{\eta}_{q,p}\rangle \, n^\mu \overleftrightarrow{\partial}_\mu \, \langle \tilde{\eta}_{q,p}| . \tag{6.14}$$

Using (4.20), we immediately arrive at (6.5), i.e. at the result that $\tilde{F}_\eta(B)$ and $\tilde{U}(a, \Lambda)$ constitute *Poincaré systems of covariance on the relativistic stochastic phase spaces* $\mathscr{M}_{m,\eta}^{\pm}$. Naturally, a dependence of $\tilde{F}_\eta$ on the normals n^μ to the hyperplanes σ_{q^0} constituting $\Sigma_m^{(q^0)\pm} = \sigma_{q^0} \times \mathscr{V}_m^{\pm}$ still remains.[15] As we shall discuss at greater length in Chapter 4, this is *de facto* a dependence on the direction of the chosen coherent flow of (stochastically extended) quantum test particles. However, since all such coherent flows are operationally equivalent, there is actually no frame-dependence, i.e. all inertial frames play totally equivalent roles. Thus the special relativity principle is satisfied.

The systems of covariance in (6.3) and (6.5) can be equally well expressed in each $L^2(\Sigma^{\pm}_{m,\eta})$ by using $U(a, \Lambda)$ in (3.9) and

$$\mathbb{P}_\eta(B) = \mathscr{W}_\eta \tilde{\mathbb{P}}_\eta(B) \mathscr{W}_\eta^{-1} = \int_B |\eta_\zeta\rangle \, \mathrm{d}\Sigma_m(\zeta) \, \langle \eta_\zeta|, \quad B \subset \Sigma^{\pm}_{m,\eta} \tag{6.15}$$

$$F_\eta(\mathring{B}) = \frac{i\hbar}{Z_\eta} \int_{\mathring{B}} \mathrm{d}^4 q \, \mathrm{d}\Omega_m(p) \, |\eta_\zeta\rangle \, n^\mu \overleftrightarrow{\partial}_\mu \, \langle \eta_\zeta|, \quad \mathring{B} \subset \mathscr{M}^{\pm}_{m,\eta}, \tag{6.16}$$

instead of (6.2) and (6.14), respectively, so that

$$U(a, R) \mathbb{P}_\eta(B) U^{-1}(a, R) = \mathbb{P}_\eta(a + RB), \; (a, R) \in \mathscr{E}(\sigma), \tag{6.17}$$

$$U(a, \Lambda) F_\eta(\mathring{B}) U^{-1}(a, \Lambda) = F_\eta(a + \Lambda\mathring{B}), \; (a, \Lambda) \in \mathscr{P}. \tag{6.18}$$

In (6.15) and (6.16) we have introduced by analogy with (1.6.7)

$$\eta_{q,p} = \mathscr{W}_\eta \tilde{\eta}_{q,p} = \mathscr{W}_\eta \tilde{U}(q, \Lambda_{p/m}) \tilde{\eta}. \tag{6.19}$$

In view of (4.23) and the unitarity of $\mathscr{W}_\eta$, we also have for $\zeta, \zeta' \in \Sigma^{\pm}_m$

$$\eta_\zeta(\zeta') = \langle \tilde{\eta}_{\zeta'} | \tilde{\eta}_\zeta \rangle_{\mathscr{V}^{\pm}_m} = \langle \eta_{\zeta'} | \eta_\zeta \rangle_{\Sigma^{\pm}_m}, \tag{6.20}$$

in complete analogy with (1.6.7) and (1.6.8). Pursuing these analogies further, but taking into account that in the relativistic regime – i.e. in the absence of absolute simultaneity – space and time degrees of freedom should be treated on an equal footing, we introduce as a counterpart of (1.6.9) the *relativistic free propagator* in $\mathscr{M}^{\pm}_{m,\eta}$:

$$\begin{aligned} K_\eta(q', p'; q, p) &= \langle \tilde{\eta}_{q',p'} | \tilde{\eta}_{q,p} \rangle_{\mathscr{V}^{\pm}_m} \\ &= \int_{\mathscr{V}^{\pm}_m} \exp\left[\frac{i}{\hbar}(q - q') \cdot k\right] \eta^*(p' \cdot k) \eta(p \cdot k) \, \mathrm{d}\Omega_m(k). \end{aligned} \tag{6.21}$$

We note that by (5.1) η_ζ constitutes a continuous resolution of the identity in $L^2(\Sigma^{\pm}_{m,\eta})$, i.e.

$$\int_{\Sigma^{\pm}_m} |\eta_\zeta\rangle \, \mathrm{d}\Sigma_m(\zeta) \, \langle \eta_\zeta| = \mathbb{1}^{(m,\eta)}_{\pm}, \tag{6.22}$$

where $\mathbb{1}_{\pm}^{(m,\eta)}$ are the identity operators in $L^2(\Sigma_{m,\eta}^{\pm})$, and that the values of all these resolutions of the identity for any fixed m and η can be extracted from (6.21):

$$\eta_\zeta(\zeta') = K_\eta(\zeta'; \zeta), \qquad \zeta', \zeta \in \Sigma_{m,\eta}^{\pm}. \tag{6.23}$$

The two properties that characterize $K_\eta(\zeta'; \zeta)$ as a propagator, namely

$$K_\eta^*(\zeta'; \zeta) = K_\eta(\zeta; \zeta'), \tag{6.24}$$

$$K_\eta(\zeta'; \zeta) = \int_{\Sigma_m^{\pm}} K_\eta(\zeta'; \zeta'')K_\eta(\zeta''; \zeta)\, \mathrm{d}\Sigma_m(\zeta''), \tag{6.25}$$

follow immediately from its definition (6.21), and from (5.1). Furthermore, in the same manner in which we have proved (6.11), we establish that the orthogonal projection operator $\mathbb{P}(\Sigma_{m,\eta}^{\pm})$ of $L^2(\Sigma_m^{\pm})$ onto $L^2(\Sigma_{m,\eta}^{\pm})$, for which therefore

$$\mathbb{P}(\Sigma_{m,\eta}^{\pm})L^2(\Sigma_m^{\pm}) = L^2(\Sigma_{m,\eta}^{\pm}), \tag{6.26}$$

acts as follows on any $\phi \in L^2(\Sigma_m^{\pm})$:

$$(\mathbb{P}(\Sigma_{m,\eta}^{\pm})\phi)(\zeta) = \int_{\Sigma_m^{\pm}} K_\eta(\zeta; \zeta')\phi(\zeta')\, \mathrm{d}\Sigma_m(\zeta'). \tag{6.27}$$

Hence, as a generalization of (6.22), we get

$$\mathbb{P}(\Sigma_{m,\eta}^{\pm}) = \mathbb{P}_\eta(\Sigma_m^{\pm}) = \int_{\Sigma_m^{\pm}} |\eta_\zeta\rangle\, \mathrm{d}\Sigma_m(\zeta)\, \langle\eta_\zeta|, \tag{6.28}$$

where $\mathbb{P}_\eta(B)$ is as defined in (6.15).

To provide some examples, let us consider the nonrelativistic optimal resolution generators (1.6.25), which in the momentum representation assume the form:

$$\tilde{\xi}^{(l)}(\mathbf{k}) = \left(\frac{2l^2}{\pi\hbar^2}\right)^{3/4} \exp\left(-\frac{l^2\mathbf{k}^2}{\hbar^2}\right). \tag{6.29}$$

Upon performing the identification (4.6) and (4.21), we get

$$\eta(p \cdot k) = (2mc)^{1/2} \left(\frac{2l^2}{\pi\hbar^2}\right)^{3/4} \exp\left[-\left(\frac{lp \cdot k}{mc\hbar}\right)^2 + \left(\frac{mcl}{\hbar}\right)^2\right], \tag{6.30}$$

so that the corresponding propagator (6.21) is

$$K_\eta(q',p';q,p) = \pm mc \left(\frac{2l^2}{\pi\hbar^2}\right)^{3/2} \int_{k\in\mathscr{V}_m^\pm} \frac{\mathrm{d}\mathbf{k}}{k^0} \times$$

$$\times \exp\left[\frac{i}{\hbar}(q-q')\cdot k - \left(\frac{lp\cdot k}{mc\hbar}\right)^2 - \left(\frac{lp'\cdot k}{mc\hbar}\right)^2 + 2\left(\frac{mcl}{\hbar}\right)^2\right]. \quad (6.31)$$

We do not label η and K by the proper radius l, since we reserve that notation for another class of l-dependent resolution generators and propagators, which we shall introduce in Section 2.9. In fact, the identification (4.21), albeit simple and therefore convenient, is not the only one allowed as an extrapolation from the nonrelativistic to the relativistic regime. Since any such extrapolation merely requires concordance between relativistic and nonrelativistic expressions at 3-velocities $|\mathbf{v}| \ll c$, any identification $\tilde{\xi} \mapsto \tilde{\eta}$ for which as $\mathbf{k} \longrightarrow \mathbf{0}$

$$\tilde{\eta}(k) = \text{const}\ \tilde{\xi}(\mathbf{k}) + O\left(\frac{|\mathbf{k}|}{mc}\right), \quad k = (k^0, \mathbf{k}) \in \mathscr{V}_m^\pm, \quad (6.32)$$

would fulfill this requirement. Following the group-contraction terminology that relates the Lie algebra of the Poincaré group to that of the Galilei group (Inönü and Wigner, 1953), we shall refer to any one-to-one mapping between sets of relativistic and nonrelativistic resolution generators as a *contraction of (relativistic) resolution generators*, provided that (6.32) is satisfied. As we shall see in Chapter 4, the choice of $\tilde{\eta}^{(l)}$ presented in Section 2.9 is dictated by the requirement that, in addition to possessing $\tilde{\xi}^{(l)}$ as contractions, the relativistic resolution generators $\tilde{\eta}^{(l)}$ should correspond to ground-state solutions of relativistic harmonic oscillators – as is the case with $\xi^{(l)}$ *vis-à-vis* nonrelativistic harmonic oscillators (cf. also Schroeck, 1978).

It is, nevertheless, instructive to consider the relativistic stochastic phase-space probability amplitudes

$$(\mathscr{W}_\eta\tilde{\phi})(q,p) = (2mc)^{1/2}\left(\frac{2l^2}{\pi\hbar^2}\right)^{3/4} \times$$

$$\times \int_{k\in\mathscr{V}_m^\pm} \exp\left[-\frac{i}{\hbar}q\cdot k - \left(\frac{lp\cdot k}{mc\hbar}\right)^2 + \left(\frac{mcl}{\hbar}\right)^2\right]\tilde{\phi}(k)\frac{\mathrm{d}\mathbf{k}}{2k^0} \quad (6.33)$$

in the sharp-point limit $l \longrightarrow +0$, as well as in the sharp-momentum limit $l \longrightarrow +\infty$. We easily get

$$\phi(q) = (2mc)^{-1/2} \lim_{l\to+0} (8\pi l^2)^{-3/4} (\mathscr{W}_\eta\tilde{\phi})(q,p), \quad (6.34)$$

where the function ϕ is defined by (1.2). By going to the stochastic rest frame of the test particle where $\mathbf{p}' = \mathbf{0}$, and then, after letting $l \to +\infty$, reverting to the laboratory frame, we obtain

$$\tilde{\phi}(p) = (2mc)^{-1/2} \exp\left(\frac{i}{\hbar}\, q \cdot p\right) \lim_{l \to +\infty} \left(\frac{l^2}{2\pi\hbar^2}\right)^{3/4} (\mathscr{W}_\eta \tilde{\phi})(q, p). \tag{6.35}$$

Thus, after performing suitable renormalizations, we can recover both the 'configuration' space as well as the momentum space wave functions in the sharp-point and sharp-momentum limits, respectively, of stochastic phase-space wave functions.

2.7. Relativistic Canonical Commutation Relations and Gauge Freedom

We have shown in Section 1.7 that the configuration space representation $\hat{U}(g)$, $g \in \mathscr{G}$, of the Galilei group $\mathscr{G}$ was not essential to the development and study of the stochastic phase-space representation $U(g)$ of $\mathscr{G}$ on $L^2(\Gamma_\xi)$, which could be carried out exclusively within the context of $L^2(\Gamma)$. Similarly, in this section we shall show that the momentum space representation $\tilde{U}(g)$, $g \in \mathscr{P}$, of the Poincaré group $\mathscr{P}$ is not at all essential to the theory of stochastic phase-space representations $U(g)$, $g \in \mathscr{P}$, in $L^2(\Sigma^{\pm}_{m,\eta})$, which can be carried out completely within the Hilbert spaces $L^2(\Sigma^{\pm}_m)$ (cf. Ali and Prugovečki, 1983, for harmonic analyses).

First of all, we can introduce the stochastic phase-space representative

$$\bar{\eta}(q, p) = (\mathscr{W}_\eta \tilde{\eta})(q, p) = \int_{\mathscr{V}^{\pm}_m} \exp\left(-\frac{i}{\hbar}\, q \cdot k\right) \eta^*(p \cdot k) \eta(mck^0)\, d\Omega_m(k) \tag{7.1}$$

of the resolution generator by defining it as equal to the right-hand side of (7.1) for some choice of a single-variable function η. Then, in accordance with (6.19) and (6.21), we can set by definition

$$K_{\bar{\eta}}(q', p'; q, p) = \langle \bar{\eta}_{q',p'} | \bar{\eta}_{q,p} \rangle_{\Sigma^{\pm}_m}, \tag{7.2a}$$

$$\bar{\eta}_{q,p} = U(q, \Lambda_{p/m}) \bar{\eta}, \qquad (q, p) \in \mathscr{M}^{\pm}_m, \tag{7.2b}$$

where $U(a, \Lambda)$ is given by (3.9). The spin-zero irreducible subspaces $L^2(\Sigma^{\pm}_{m,\bar{\eta}})$ of any $L^2(\Sigma^{\pm}_m)$ are then pinpointed as being equal to the ranges of the operators $\mathbb{P}_{\bar{\eta}}(\Sigma^{\pm}_m)$

$$(\mathbb{P}_{\bar{\eta}}(\Sigma^{\pm}_m)\phi)(\zeta) = \int_{\Sigma^{\pm}_m} K_{\bar{\eta}}(\zeta; \zeta') \phi(\zeta')\, d\Sigma_m(\zeta'), \tag{7.3}$$

which are projectors since, as a consequence of (7.1) and (7.2), we have

$$K_{\overline{\eta}}(\zeta;\zeta') = K^*_{\overline{\eta}}(\zeta';\zeta) = \int_{\Sigma_m^\pm} K_{\overline{\eta}}(\zeta;\zeta'')K_{\overline{\eta}}(\zeta'';\zeta')\,\mathrm{d}\Sigma_m(\zeta''). \tag{7.4}$$

The uniqueness for the given $U(a, \Lambda)$ of each resolution generator $\overline{\eta}$ in the subspaces $L^2(\Sigma^\pm_{m,\overline{\eta}})$ can be established essentially by the method used in Section 1.7 on nonrelativistic resolution generators. Indeed, we note that by (1.12b) and by the representation property

$$U(a_1, \Lambda_1)U(a_2, \Lambda_2) = U(a_1 + \Lambda_1 a_2, \Lambda_1\Lambda_2) \tag{7.5}$$

that follows from (3.9a), we get for $p \in \mathscr{V}^+_m$

$$U^{-1}(q, \Lambda_{p/m}) = U(-\Lambda_{\overline{p}/m}q, \Lambda_{\overline{p}/m}), \quad \overline{p} = (p^0, -\mathbf{p}). \tag{7.6}$$

On the other hand, by (7.3) and (7.4)

$$\overline{\eta}_\zeta(\zeta') = K_{\overline{\eta}}(\zeta';\zeta) \tag{7.7}$$

so that, in accordance with (7.2b),

$$\overline{\eta}(q, p) = K_{\overline{\eta}}(q, p; 0, (mc, \mathbf{0})). \tag{7.8}$$

Consequently, by (7.2a) and (7.6) we have

$$\overline{\eta}(\Lambda_{p/m}q; p) = \overline{\eta}^*(-q; \overline{p}) = \overline{\eta}^*(-\overline{q}; p), \tag{7.9}$$

and the counterpart of (1.7.8) is thus obtained. The uniqueness of $\overline{\eta}$ then follows from (1.7.9), and from (7.3) written in the form

$$(\mathbb{P}_{\overline{\eta}}(\Sigma^\pm_m)\phi)(q, p) = \langle\overline{\eta}_{q,p} | \phi\rangle_{\Sigma^\pm_m}, \tag{7.10}$$

by the sequence of steps used in (1.7.10).

Thus we can indeed argue, as we did for $U(b, \mathbf{a}, \mathbf{v}, R)$ in Section 1.7, that $U(a, \Lambda)$ and its structure can be deduced entirely within the phase-space context. However, since by Section 2.6 only $|\phi(\zeta)|^2$, and not $\phi(\zeta)$ themselves, are physically measurable quantities, we are also faced with the fact that the definition (3.9) of $U(a, \Lambda)$ is in part arbitrary from the physical point of view. Indeed, any unitarily equivalent representation

$$U_\omega(a, \Lambda) = \mathscr{U}_\omega U(a, \Lambda)\mathscr{U}_\omega^{-1}, \quad (a, \Lambda) \in \mathscr{P}, \tag{7.11}$$

corresponding to the gauge transformation

$$\mathscr{U}_\omega : \phi(q, p) \longmapsto \exp\left[\frac{i}{\hbar}\,\omega(q, p)\right]\phi(q, p) \tag{7.12}$$

would have done just as well: in conjunction with

$$\mathbb{P}_{\bar{\eta}}^{(\omega)}(B) = \mathcal{U}_{\omega}\mathbb{P}_{\bar{\eta}}^{(\omega)}(B)\mathcal{U}_{\omega}^{-1}, \tag{7.13}$$

$$F_{\bar{\eta}}^{(\omega)}(\mathring{B}) = \mathcal{U}_{\omega}F_{\bar{\eta}}(\mathring{B})\mathcal{U}_{\omega}^{-1}, \tag{7.14}$$

it would have given rise to counterparts of the systems of covariance in (6.17) and (6.18) on the subspaces

$$\mathcal{U}_{\omega}L^2(\Sigma_{m,\bar{\eta}}^{\pm}) = L^2(\Sigma_{m,\bar{\eta}^{\omega}}^{\pm}), \quad \bar{\eta}^{\omega} = \mathcal{U}_{\omega}\bar{\eta}, \tag{7.15}$$

of $\mathcal{U}_{\omega}L^2(\Sigma_m^{\pm})$ – which do not coincide with $L^2(\Sigma_m^{\pm})$ if ω is not constant along $\Sigma_m^{\pm}$, i.e. q^0-independent in stochastic rest frames for $\Sigma_m^{\pm}$.

It is interesting to see how the gauge transformation $\mathcal{U}_{\omega}$ in (7.12) affects the infinitesimal generators P^{μ} and

$$M^{\mu\nu} = Q^{\mu}P^{\nu} - Q^{\nu}P^{\mu} \tag{7.16}$$

of $U(a, \Lambda)$ in (3.9). Indeed, as can be established by purely algebraic methods (Brooke and Prugovečki, 1983b), the infinitesimal generators $M^{\mu\nu}$ of any unitary spin-zero representation of the Poincaré group can be written in the form (7.16), where Q^{μ} and P^{μ} satisfy the relativistic canonical commutation relations (RCCRs)

$$[Q^{\mu}, Q^{\nu}] = [P^{\mu}, P^{\nu}] = 0, \tag{7.17a}$$

$$[Q^{\mu}, P^{\nu}] = -i\hbar g^{\mu\nu}, \quad \mu, \nu = 0, \dots, 3. \tag{7.17b}$$

However, whereas the operators P^{μ}, as infinitesimal generators of spacetime translations (cf., e.g., (1.13) and (3.9b)) are unique and self-adjoint, that is not the case [16] with the Q^{μ} operators, which are neither, as evidenced by the fact that the addition of any functions $f^{\mu}(P)$ of P^{μ} to Q^{μ} leaves (7.16) as well as (7.17) unaltered. Of course, this ambiguity and general lack of self-adjointness reflects the fact that, as discussed in Sections 2.1 and 2.2, there are no (sharp) position operators for relativistic quantum particles.

For the representation $U(a, \Lambda)$ in (3.9) we immediately obtain in the standard manner (Schweber, 1961, Secs. 2–3)

$$M^{\mu\nu} = i\hbar\left(q^{\mu}\frac{\partial}{\partial q_{\nu}} - q^{\nu}\frac{\partial}{\partial q_{\mu}}\right) + i\hbar\left(p^{\mu}\frac{\partial}{\partial p_{\nu}} - p^{\nu}\frac{\partial}{\partial p_{\mu}}\right). \tag{7.18}$$

On the other hand, by (1.2), (1.5), (2.1)–(2.5) and (4.20),

$$M^{\mu\nu}\phi = (Q_{\eta}^{\mu}P^{\nu} - Q_{\eta}^{\nu}P^{\mu})\phi, \quad Q_{\eta}^{\mu} = \mathcal{W}_{\eta}\tilde{Q}^{\mu}\mathcal{W}_{\eta}^{-1}, \tag{7.19}$$

$$\tilde{Q}^{\mu} = \mathcal{W}_0^{-1}\hat{Q}^{\mu}\mathcal{W}_0 : \tilde{\phi}(k)\,\delta(k^2 - m^2c^2) \mapsto -i\hbar\frac{\partial}{\partial k_{\mu}}\tilde{\phi}(k)\,\delta(k^2 - m^2c^2), \tag{7.20}$$

when $\phi \in L^2(\Sigma^{\pm}_{m,\eta})$, so that by (5.5)

$$(Q^{\mu}_{\eta}\phi)(q, p)$$
$$= \int_{\mathscr{V}^{\pm}_{m}} \exp\left(-\frac{i}{\hbar}\, q \cdot k\right) \left[\left(q^{\mu} + i\hbar \frac{\partial}{\partial k_{\mu}}\right) \eta^*(p \cdot k)\right] \tilde{\phi}(k)\, \mathrm{d}\Omega_m(k). \tag{7.21}$$

In view of (5.16), $\eta^*(p \cdot k)$ can be regarded as a function of $(p - k)^2$, so that[17]

$$(Q^{\mu}_{\eta}\phi)(q, p) = \left(q^{\mu} - i\hbar \frac{\partial}{\partial p_{\mu}}\right) \phi(q, p), \quad \phi \in L^2(\Sigma^{\pm}_{m,\eta}). \tag{7.22}$$

Consequently, we can write

$$M^{\mu\nu} L^2(\Sigma^{\pm}_{m,\bar{\eta}}) = (Q^{\mu}P^{\nu} - Q^{\nu}P^{\mu}) L^2(\Sigma^{\pm}_{m,\bar{\eta}}) \tag{7.23}$$

if we adopt the following realization of the RCCRs:

$$Q^{\mu} = q^{\mu} - i\hbar \frac{\partial}{\partial p_{\mu}}, \qquad P^{\mu} = i\hbar \frac{\partial}{\partial q_{\mu}}. \tag{7.24}$$

In general, for $M^{\mu\nu}_{\omega}$ corresponding to $U_{\omega}(a, \Lambda)$ in (7.11) we should choose

$$Q^{\mu}_{\omega} = \mathscr{U}_{\omega} Q^{\mu} \mathscr{U}^{-1}_{\omega} = -i\hbar \frac{\partial}{\partial p_{\mu}} + q^{\mu} - \frac{\partial \omega}{\partial p_{\mu}}, \tag{7.25a}$$

$$P^{\mu}_{\omega} = \mathscr{U}_{\omega} P^{\mu} \mathscr{U}^{-1}_{\omega} = i\hbar \frac{\partial}{\partial q_{\mu}} + \frac{\partial \omega}{\partial q_{\mu}}. \tag{7.25b}$$

The 'pure gauge' term ω then manifests itself under 'parallel transport' (cf. Section 4.6) of $\bar{\eta}^{\omega}$ from stochastic rest at the origin to (q, p), since by (7.11)–(7.12) the general form of (7.10) is:

$$(\mathbb{P}_{\bar{\eta}^{\omega}}(\Sigma^{\pm}_{m})\phi)(q, p)$$
$$= \exp\left[\frac{i}{\hbar}\, \omega(q, p)\right] \langle \bar{\eta}^{\omega}_{q,p} \mid \phi \rangle_{\Sigma^{\pm}_{m}}, \quad \bar{\eta}^{\omega}_{q,p} = U_{\omega}(q, \Lambda_{p/m})\bar{\eta}^{\omega}. \tag{7.25c}$$

The realizations (7.25) of the RCCRs can be viewed as an extrapolation of the CCRs in (1.7.22), the latter being related to the former by Lie-algebra contractions (Inönü and Wigner, 1953). However, we must keep in mind that whereas (1.7.22a) hold globally on $L^2(\Gamma)$, as they are in fact uniquely determined by $U_{\omega}(0, \mathbf{0}, \mathbf{v}, I)$, this is not the case with (7.25a), which are merely the most convenient among the many choices for which

$$M^{\mu\nu}_{\omega}\phi = (Q^{\mu}_{\omega}P^{\nu}_{\omega} - Q^{\nu}_{\omega}P^{\mu}_{\omega})\phi, \quad \phi \in L^2(\Sigma^{+}_{m,\eta^{\omega}}), \tag{7.26}$$

On the other hand, (7.25b) are the unique (self-adjoint) infinitesimal generators of spacetime translations, for which therefore

$$U_\omega(a, I) = \exp\left(\frac{i}{\hbar} a_\mu P^\mu_\omega\right), \tag{7.27}$$

and as such they represent an unambiguous extrapolation of (1.7.22b) to the relativistic regime as well as to gauges that are time-dependent. Similarly to the situation in (1.7.28), the relativistic gauge $\bar{\omega} = q \cdot p/2$ will yield a reciprocally invariant realization of the RCCRs:

$$Q^\mu_{\bar{\omega}} = -i\hbar \frac{\partial}{\partial p_\mu} + \frac{q^\mu}{2}, \qquad P^\mu_{\bar{\omega}} = i\hbar \frac{\partial}{\partial q_\mu} + \frac{p^\mu}{2}. \tag{7.28}$$

In summary, the main conclusions that can be drawn from Sections 2.4–2.7 are totally analogous to the conclusions which we had drawn from Sections 1.4–1.7 at the end of Section 1.7. They are as follows:

For free quantum particles of zero spin there is a family of unitarily equivalent vector[18] (reducible) representations of the Poincaré group on each of the Hilbert spaces $L^2(\Sigma^\pm_m)$ over $\Sigma^\pm_m = \sigma \times \mathscr{V}^\pm_m$ (where σ is a spacelike hyperplane) which carry the inner product (3.8). Any two of these representations $U_{\omega_1}(g)$ and $U_{\omega_2}(g)$, $g \in \mathscr{P}$, are related by a gauge transformation

$$U_{\omega_1}(g) = \mathscr{U}_{\omega_1-\omega_2} U_{\omega_2}(g) \mathscr{U}^{-1}_{\omega_1-\omega_2}, \quad g \in \mathscr{P}, \tag{7.29}$$

with $\mathscr{U}_\omega$ for $\omega = \omega_1 - \omega_2$ defined by (7.12). The spin-zero subrepresentations of each $U_\omega(g)$, $g \in \mathscr{P}$, act on closed subspaces $L^2(\Sigma^\pm_{m,\bar{\eta}^\omega})$ of $L^2(\Sigma^\pm_m)$ determined by (unique in the chosen gauge) generators $\bar{\eta}^\omega \in L^2(\Sigma^\pm_{m,\bar{\eta}^\omega})$ of continuous resolutions

$$\bar{\eta}^\omega_{q,p} = U_\omega(q, \Lambda_{p/m})\bar{\eta}^\omega \tag{7.30}$$

for the projector $\mathbb{P}_{\bar{\eta}^\omega}(\Sigma^\pm_m)$ of $L^2(\Sigma^\pm_m)$ onto $L^2(\Sigma^\pm_{m,\bar{\eta}^\omega})$. In general

$$\mathbb{P}_{\bar{\eta}^\omega}(B) = \int_B |\bar{\eta}^\omega_\zeta\rangle \, d\Sigma_m(\zeta) \, \langle \bar{\eta}^\omega_\zeta| \tag{7.31}$$

is a POV measure on the Borel sets B in $\Sigma^\pm_m$. Each of these POV measures constitute a system of covariance

$$U_\omega(g)\mathbb{P}_{\bar{\eta}^\omega}(B)U^{-1}_\omega(g) = \mathbb{P}_{\bar{\eta}^\omega}(gB), \quad g \in \mathscr{E}(\sigma), \tag{7.32}$$

for the Euclidean group $\mathscr{E}(\sigma)$ in the stochastic rest frames of the given $\Sigma^\pm_{m,\bar{\eta}^\omega}$. Furthermore, the operators

$$F_{\bar{\eta}^\omega}(\mathring{B}) = i\hbar Z^{-1}_\eta \int_{\mathring{B}} |\bar{\eta}^\omega_{q,p}\rangle \, n^\mu \overleftrightarrow{\partial}_\mu \, \langle \bar{\eta}^\omega_{q,p}| \, d^4q \, d\Omega_m(p) \tag{7.33}$$

constitute a POV measure over the Borel sets $\mathring{B}$ in the relativistic phase spaces $\mathcal{M}_m^{\pm}$, which gives rise to a relativistic stochastic phase space system of covariance for $U_\omega(a, \Lambda)$:

$$U_\omega(g) F_{\bar{\eta}\omega}(\mathring{B}) U_\omega^{-1}(g) = F_{\bar{\eta}\omega}(g\mathring{B}), \; g \in \mathcal{P}. \tag{7.34}$$

Each two irreducible subrepresentations $U_\omega^{(i)}(g)$,

$$U_\omega^{(i)}(g) = U_\omega(g) \mathbb{P}_{\bar{\eta}_i^\omega}(\Sigma_m^{\pm}) \tag{7.35}$$

are unitarily equivalent, but that equivalence does not extend to the corresponding POV measures (7.32) or (7.33). This gives rise to distinct probability measures

$$p_\phi^{\bar{\eta}^\omega}(B) = \langle \phi | \mathbb{P}_{\bar{\eta}^\omega}(B) \phi \rangle_{\Sigma_m^{\pm}} \tag{7.36}$$

for distinct resolution generators $\bar{\eta}^\omega$. These probabilities therefore correspond to distinct (gauge-independent) relativistic stochastic phase spaces $\mathcal{M}_{m, \bar{\eta}^\omega}^{\pm}$ – that is, in operational terms, to probabilities measured with distinct quantum test particles.

2.8. Relativistically Covariant and Conserved Probability and Charge Currents

In Sections 2.3–2.7 of this chapter we have encountered a remarkable parallelism between nonrelativistic and relativistic quantum mechanics on stochastic phase space. This parallelism is all the more remarkable in view of the fundamental dissimilarities (discussed in Sections 2.1 and 2.2) displayed by these two disciplines when they are formulated in the 'configuration' representation. The existence of both probability as well as charge (covariant and conserved) currents in stochastic relativistic quantum mechanics extends even further the aforementioned parallelism, and at the same time throws some light on the mathematical reasons for the discrepancies that manifest themselves in the conventional approach.

The expressions (1.8.9) and (1.8.8) for the nonrelativistic probability 3-current $\mathbf{j}^\xi$ and the associated probability density ρ^ξ provide an unambiguous blueprint for the design of the expressions

$$j_\eta^\mu(q) = \pm 2 \int_{\mathcal{V}_m^{\pm}} \frac{p^\mu}{m} |\phi(q, p)|^2 \, d\Omega_m(p) \tag{8.1}$$

that should represent a relativistic probability 4-current. In view of the transformation law (3.9a) and the invariance of $d\Omega_m$ in (1.5), j_η^μ is obviously covariant under (a, Λ) in (1.8):

$$j_\eta^\mu(q) \mapsto j'^\mu_\eta(q') = \Lambda^\mu{}_\nu j_\eta^\nu(q), \qquad q' = a + \Lambda q, \tag{8.2}$$

Furthermore, by (1.19) and (5.11)

$$j_\eta^0(q) = \rho_\eta(q) = \int_{\mathbb{R}^3} |\phi(q, p)|^2 \, d\mathbf{p}, \quad p^0 = \pm(\mathbf{p}^2 + m^2c^2)^{1/2}, \tag{8.3}$$

so that $j_\eta^0(q) \geqslant 0$, and consequently j_η^μ is indeed a *probability* current, which in the nonrelativistic limit $c \to \infty$ obviously merges into ρ^ξ and $\mathbf{j}^\xi$ if the identification (4.21) is performed. We are left with the task of investigating whether this current is conserved or not.

To settle this very important point, let us use (5.5) to write:

$$\frac{\partial}{\partial q^\nu} j_\eta^\nu(q) = \frac{2}{\hbar} \operatorname{Im} \int_{\mathcal{V}_m^\pm} d\Omega_m(p)\, \phi^*(q, p) \int_{\mathcal{V}_m^\pm} d\Omega_m(k')\, (p \cdot k') \eta_{q,p}^*(k') \tilde{\phi}(k'). \tag{8.4}$$

Upon interchanging by Fubini's theorem orders of integration in p and k', we obtain, again by (5.5)

$$\frac{\partial}{\partial q^\nu} j_\eta^\nu(q) = \operatorname{Im} \int d\Omega_m(k)\, \tilde{\phi}_q^*(k) \int d\Omega_m(k')\, M_\eta(k, k') \tilde{\phi}_q(k'), \tag{8.5a}$$

$$\tilde{\phi}_q(k) = \exp\left(-\frac{i}{\hbar}\, q \cdot k\right) \tilde{\phi}(k) = (\tilde{U}(-q, I)\tilde{\phi})(k), \tag{8.5b}$$

$$M_\eta(k, k') = 2\hbar^{-1} \int (p \cdot k') \eta^*(p \cdot k') \eta(p \cdot k)\, d\Omega_m(p), \tag{8.5c}$$

where obviously $M_\eta(\Lambda k, \Lambda k') = M_\eta(k, k')$. Being invariant under Lorentz transformations, $M_\eta(k, k')$ as a function of $k, k' \in \mathcal{V}_m^\pm$ depends only on $k \cdot k'$, and therefore it is symmetric in k and k'. Since the necessary and sufficient condition for (8.5a) to equal zero for all $\tilde{\phi} \in L^2(\mathcal{V}_m^\pm)$ is that $M_\eta(k, k')$ in (8.5c) should be the kernel of a self-adjoint integral operator – i.e. that $M_\eta(k, k') = M_\eta^*(k', k)$ for (almost all) $k, k' \in \mathcal{V}_m^\pm$ – we arrive at the conclusion that

$$\frac{\partial}{\partial q^\nu} j_\eta^\nu(q) = 0, \quad q \in M(1,3) \tag{8.6}$$

if η in (4.6) is a real function of its argumental mck^0. Interestingly, we recall that in Section 1.8 we have arrived at the conclusion that the nonrelativistic current was conserved if $\xi(\mathbf{x})$, and therefore also $\tilde{\xi}(\mathbf{k})$, were real. With the identification (4.21) the conservation of nonrelativistic as well as the relativistic currents at stochastic configuration points is therefore secured.

In this context, it should be noted that the conservation (8.6) of j^ν_η is a necessary and sufficient condition in order that the free evolution of $\phi(q, p)$, governed by (3.9), remain unitary (and therefore probability conserving) even when σ' in $\Sigma'^{\pm}_m = \sigma' \times \mathscr{V}^{\pm}_m$ are general spacelike hypersurfaces in $M(1, 3)$, and not just hyperplanes. Indeed, if σ' is any spacelike deformation of a hyperplane, then by a standard application of Gauss' theorem we get that when $\Sigma'^{\pm}_m = \sigma' \times \mathscr{V}^{\pm}_m$

$$\langle \phi_1 | \phi_2 \rangle_{\Sigma'^{\pm}_m} = \int_{\Sigma'^{\pm}_m} \phi_1^*(q, p)\phi_2(q, p)\, \mathrm{d}\Sigma_m(q, p)$$

$$= 2 \int_\sigma \mathrm{d}\sigma_\nu(q) \int_{\mathscr{V}^{\pm}_m} \mathrm{d}\Omega_m(p)\, p^\nu \phi_1^*(q, p)\phi_2(q, p) \tag{8.7}$$

for $\phi_1, \phi_2 \in L^2(\Sigma^{\pm}_{m,\eta})$ if and only if

$$\int_R \mathrm{d}^4 q \frac{\partial}{\partial q_\nu} \int_{\mathscr{V}^{\pm}_m} \mathrm{d}\Omega_m(p)\, p^\nu \phi_1^*(q, p)\phi_2(q, p) = 0 \tag{8.8}$$

for the region $R \subset M(1, 3)$ enclosed by σ and σ'. Since (8.8) can always be expressed as a linear combination of integrals

$$\int_R \mathrm{d}^4 q \frac{\partial}{\partial q_\nu} j^\nu_\eta(q) = 0 \tag{8.9}$$

with ϕ in (8.1) assuming the values $\phi_1, \phi_2, \phi_1 \pm \phi_2$ and $\phi_1 \pm i\phi_2$, the imposition of (8.8) for all $\phi_1, \phi_2 \in L^2(\Sigma^{\pm}_{m,\eta})$ is equivalent to (8.8) being true for all $\phi \in L^2(\Sigma^{\pm}_{m,\eta})$. In turn, taking into account the fact that by (5.5) $j^\nu_\eta(q)$ and its partial derivatives are everywhere continuous, we see that (8.7) is true for all spacelike deformations σ' of all the spacelike hyperplanes σ if and only if (8.6) is true for all $q \in M(1, 3)$.

In later developments, and especially in Chapter 4 where we extend our considerations from special to general relativity, it will be essential that we should be able to carry out our considerations on Hilbert spaces $L^2(\Sigma'^{\pm}_{m,\eta})$ where σ' is in general a spacelike hypersurface, and not necessarily a hyperplane. Hence, *from now on we shall always consider only resolution generators for which*

$$\eta(p \cdot k) = \eta^*(p \cdot k), \tag{8.10}$$

where η is related to $\tilde{\eta} \in L^2(\mathscr{V}^{\pm}_m)$ by (4.6) and to $\overline{\eta} \in L^2(\Sigma^{\pm}_{m,\eta})$ by (7.1). The condition (8.10) greatly restricts the range of wave functions $\tilde{\eta}$ which are candidates for proper wave functions of spinless extended quantum particles. Yet in Chapter 4 it will transpire that this condition is satisfied when the results of this chapter are combined with those of Born's (1938, 1949) reciprocity theory.

The striking formal similarity between the expression (6.7a) for the inner product in $L^2(\Sigma^{\pm}_{m,\eta})$ and the expression (1.28) for the inner product in the 'configuration representation' spaces $L^2(\sigma^{\pm})$ suggests that the charge current (1.27) ought to have counterparts in the stochastic phase space $\mathcal{M}^{\pm}_{m,\eta}$, namely the *'charge' currents*

$$J^{\mu}_{\eta}(q) = \frac{i\hbar}{mZ_{\eta}} \int_{\mathcal{V}_m} \phi^*(q,p)\overleftrightarrow{\partial}_{\mu}\phi(q,p)\,\mathrm{d}\Omega_m(p). \tag{8.11}$$

Indeed, by (3.9) J^{μ}_{η} transform covariantly,

$$J^{\mu}_{\eta}(q) \longmapsto J'^{\mu}_{\eta}(q') = \Lambda^{\mu}{}_{\nu} J^{\nu}_{\eta}(q), \qquad q' = a + \Lambda q, \tag{8.12}$$

and since $\phi(q,p)$ satisfies the Klein–Gordon equation (3.11) for each $p \in \mathcal{V}^{\pm}_m$, the charge current (8.11) is also conserved:

$$\frac{\partial}{\partial q^{\mu}} J^{\mu}_{\eta}(q) = 0. \tag{8.13}$$

It is interesting to compare the conventional charge current $j^{\mu}(q)$ in (1.27) with $J^{\mu}_{\eta}(q)$ in (8.11) by reducing them to similar forms. Thus, by (1.2)

$$j^{\mu}(q) = m^{-1}(2\pi\hbar)^{-3} \int_{\mathcal{V}_m \times \mathcal{V}_m} \tilde{\phi}^*_q(k')\tilde{\phi}_q(k)(k'^{\mu} + k^{\mu})\,\mathrm{d}\Omega_m(k')\,\mathrm{d}\Omega_m(k) \tag{8.14}$$

for $\tilde{\phi}_q$ in (8.5b), whereas by (5.5), (6.12) and (8.11)

$$J^{\mu}_{\eta}(q) = m^{-1}(2\pi\hbar)^{-3} \int_{\mathcal{V}_m \times \mathcal{V}_m} \tilde{\phi}^*_q(k')\tilde{\phi}_q(k) N_{\eta}(k',k) \times$$

$$\times (k'^{\mu} + k^{\mu})\,\mathrm{d}\Omega_m(k')\,\mathrm{d}\Omega_m(k), \tag{8.15}$$

where $N_{\eta}(k',k)$ is the following invariant function of $k, k' \in \mathcal{V}_m$

$$N_{\eta}(k',k) = \Big[\int_{\mathcal{V}_m} \eta^2(mcp^0)\,\mathrm{d}\Omega_m(p)\Big]^{-1} \int_{\mathcal{V}_m} \eta(p\cdot k')\eta(p\cdot k)\,\mathrm{d}\Omega_m(p), \tag{8.16}$$

and therefore is in fact a function of $k \cdot k'$ only. If $\eta(mck^0) \geqslant 0$, as is the case with η in (6.30), then N_{η} is positive definite. In addition, setting $\tilde{\eta}_k(p) = \eta(p \cdot k)$, we see that (8.16) assumes the form

$$N_{\eta}(k',k) = \|\tilde{\eta}_{k'}\|^{-1}\|\tilde{\eta}_k\|^{-1}\langle\tilde{\eta}_{k'}|\tilde{\eta}_k\rangle_{\mathcal{V}_m} \leqslant 1, \tag{8.17}$$

and that $N_\eta(k, k) = 1$ for all $k \in \mathscr{V}_m$. However, (8.14) and (8.15) would coincide only if $N_\eta(k', k) = 1$ for all $k, k' \in \mathscr{V}_m$, and that is obviously not the case. For example, for η in (6.30) we get

$$N^{(l)}(k' \cdot k) = \left\{ \int_{\mathscr{V}_m} \exp\left[-2\left(\frac{lp^0}{\hbar}\right)^2\right] \mathrm{d}\Omega_m(p) \right\}^{-1} \times$$

$$\times \int_{\mathscr{V}_m} \exp\left\{-\frac{l^2}{(mc\hbar)^2} \left[(k' \cdot p)^2 + (k \cdot p)^2\right]\right\} \mathrm{d}\Omega_m(p). \quad (8.18)$$

In the sharp-point limit $l \to +0$ we therefore have

$$N^{(0)}(k' \cdot k) = \lim_{l \to +0} N^{(l)}(k' \cdot k) = \hat{N}(k' \cdot k)/\hat{N}(m^2c^2), \quad (8.19)$$

where $\hat{N}$ is obtained by performing the substitution $\mathbf{r} = l\mathbf{p}$ in (8.18), and then setting $r_0 = |\mathbf{r}|$:

$$\hat{N}(mck^0) = \int_{\mathbb{R}^3} \exp\left[-\hbar^{-2}\left(r_0^2 + \left(\frac{k \cdot r}{mc}\right)^2\right)\right] \frac{\mathrm{d}\mathbf{r}}{2r_0}. \quad (8.20)$$

Although $N^{(0)}(k' \cdot k) \leqslant 1$, and the equality holds when $k' = k$, it does not hold in general since (8.20) is obviously not a constant.[19] Thus, despite the strict equality of the inner products in (1.30) and (6.7a), which is due to the unitarity of $\mathscr{W}_0$ and $\mathscr{W}_\eta$, the charge currents that are implicit in these inner products do not coincide.

The probability current j^μ_η in (8.1) can be reduced to a form comparable to (8.14) and (8.15):

$$j^\mu(q) = \int_{\mathscr{V}_m^+ \times \mathscr{V}_m^+} \mathrm{d}\Omega_m(k)\, \mathrm{d}\Omega_m(k')\, \tilde{\phi}_q^*(k')\tilde{\phi}_q(k) \times$$

$$\times \int_{\mathscr{V}_m^+} \mathrm{d}\Omega_m(p)\, \frac{2p^\mu}{m}\, \eta(p \cdot k')\eta(p \cdot k). \quad (8.21)$$

Replacing k'^0 and k^0 in the expression (8.14) for $j^0(q)$ by integrals that result from (4.10), we get

$$j^0(q) - j^0_\eta(q) = \int_{\mathscr{V}_m \times \mathscr{V}_m} \tilde{\phi}_q^*(k') D_\eta(k', k) \tilde{\phi}_q(k)\, \mathrm{d}\Omega_m(k)\, \mathrm{d}\Omega_m(k') \quad (8.22)$$

where $D_\eta(k', k)$ is a positive-definite function:

$$D_\eta(k', k) = m^{-1} \int_{\mathbb{R}^3} [\eta(k' \cdot p) - \eta(k \cdot p)]^2 \, d\mathbf{p}. \tag{8.23}$$

Hence, if we insert (6.30) into (8.23), and then take the sharp-point limit $l \to +0$, we get

$$D^{(0)}(k', k) = \left(\frac{2}{\pi}\right)^{3/2} \frac{2c}{\hbar^3} \int_{r^0 = |\mathbf{r}|} \left\{ \exp\left[-\left(\frac{k' \cdot r}{mc\hbar}\right)^2 \right] - \right.$$

$$\left. - \exp\left[-\left(\frac{k \cdot r}{mc\hbar}\right)^2 \right] \right\}^2 d\mathbf{r}. \tag{8.24}$$

Thus, the probability current j_η^μ does not merge into the conventional charge current j^μ in the sharp-point limit. This was to be expected, since $j_\eta^0(q)$ is positive definite, whereas $j^0(q)$ is indefinite. However, when juxtaposed with the nonrelativistic result (1.8.11), the failure of both $j_\eta^\mu(q)$ and J_η^μ to approach the conventional charge current j^μ demonstrates that in the relativistic context the stochastic phase-space approach is capable of providing a new physical framework which might avoid the fundamental inconsistencies of the conventional one.

2.9. Propagators in Relativistic Stochastic Phase Spaces

On its most basic physical leve, the concept of propagator $\mathcal{K}_\eta(\zeta; \zeta')$ for a spinless extended quantum particle of proper wave function η in the relativistic stochastic phase space $\mathcal{M}^+_{m,\eta}$ should be the same as in the nonrelativistic context: the counterpart

$$\rho_\eta(\zeta'; \zeta'') = (2\pi\hbar)^3 \, |\mathcal{K}_\eta(\zeta''; \zeta')|^2 \tag{9.1}$$

of (1.9.17a) should represent the probability density that an extended particle known to have been at the stochastic point $\underset{\sim}{\zeta}' = (\underset{\sim}{q}', \underset{\sim}{p}')$ will reach the stochastic point $\underset{\sim}{\zeta}'' = (\underset{\sim}{q}'', \underset{\sim}{p}'') \in \mathcal{M}^+_{m,\eta}$. However, a basic difference between (9.1) and (1.9.17a) emerges as soon as we try to specify the probability measure

$$P^\eta_{\Sigma^+_m}(B) = \int_B \rho_\eta(\zeta'; \zeta) \, d\Sigma_m(\zeta) \tag{9.2}$$

to which ρ in (9.1) corresponds. Indeed, whereas in the nonrelativistic case we had absolute simultaneity, and therefore such a probability measure could indubitably be defined over the Borel sets of Γ, in the relativistic case (9.2) we have to specify the hypersurface Σ_m^+ to which B belongs, and over which the probability measure (9.2) is normalized. In fact, if the propagation takes place in an external field produced by some part of the laboratory apparatus (such as an electromagnet, a laser, a magnetic lens, etc.) then although $\mathscr{K}_\eta$ and ρ_η have to be covariant (i.e. transform as scalars under $(a, \Lambda) \in \mathscr{P}$) in order to satisfy the relativity principle, they will nevertheless display an intrinsic dependence on the laboratory frame, or in general on the frame to which the field generator is tied.

Such a dependence can be theoretically taken into account by envisaging a coherent flow of test particles (described in Section 2.3) in relation to which the laboratory frame is stochastically at rest. That coherent flow delineates a time-ordered family $\mathscr{S}^\uparrow$ of hypersurfaces Σ_m of operational simultaneity. These hypersurfaces constitute a foliation of the relativistic phase space $\mathscr{M}_m$. In other words, along each $\Sigma_m = \sigma \times \mathscr{V}_m$ all (classical or quantum) test particles in the flow are (deterministically or stochastically, respectively) at rest, so that each Σ_m can be labelled unambiguously by the proper time τ of these particles, with the time-ordering in $\mathscr{S}^\uparrow$ proceeding in the direction of increasing values of τ; furthermore, any two distinct hypersurfaces Σ_m' and Σ_m'' do not intersect, and the union of all $\Sigma_m \in \mathscr{S}^\uparrow$ equals $\mathscr{M}_m$. We shall call such a family $\mathscr{S}^\uparrow$ a *reference family*, and, naturally, each Σ_m (or Σ_m^+, if no antiparticles are present) will be called a *reference hypersurface*. As it stands, the definition is valid in special as well as general relativity, with the stipulation that in general relativity the particles in the coherent flo flow are in free fall. In special relativity, free fall becomes motion by inertia, and the hypersurfaces $\sigma \subset M(1, 3)$ of $\Sigma_m = \sigma \times \mathscr{V}_m \in \mathscr{S}^\uparrow$ are mutually parallel hyperplanes.

The introduction of reference families $\mathscr{S}^\uparrow$ with their implicit time-ordering enables us to extend to the relativistic regime the concept of a quantum stochastic process introduced in Section 1.10 in the nonrelativistic realm. Thus, given any finite sequence $\Sigma_m^{(0)} < \Sigma_m^{(1)} < \cdots < \Sigma_m^{(N)}$ of reference hypersurfaces in some $\mathscr{S}^\uparrow$, which are ordered in accordance with the time-ordering in $\mathscr{S}^\uparrow$, we define the counterparts

$$K_\eta(\zeta''; B_{N-1}, \ldots, B_1; \zeta') = \int_{B_{N-1}} d\Sigma_m(\zeta_{N-1}) \mathscr{K}_\eta(\zeta''; \zeta_{N-1}) \times$$

$$\times \int_{B_{N-2}} d\Sigma_m(\zeta_{N-2}) \ldots \int_{B_1} d\Sigma_m(\zeta_1)\, \mathscr{K}_\eta(\zeta_2; \zeta_1)\, \mathscr{K}_\eta(\zeta_1; \zeta'') \tag{9.3}$$

of (1.10.7) as the *relativistic probability amplitudes for a quantum chain* from $\underset{\sim}{\zeta}' \in \Sigma_{m,\eta}^{(0)}$ to $\underset{\sim}{\zeta}'' \in \Sigma_{m,\eta}^{(N)}$, in which the filtration procedures through B_j have been performed along $\Sigma_m^{(j)}$ for $j = 1, \ldots, N-1$. Of course, in order that this definition be consistent, we have to impose the condition

$$\mathscr{K}_\eta(\zeta''; \zeta') = \int_{\Sigma_m^\pm} \mathscr{K}_\eta(\zeta''; \zeta) \mathscr{K}_\eta(\zeta; \zeta')\, d\Sigma_m(\zeta) \tag{9.4a}$$

along any $\Sigma_m \in \mathscr{S}^\uparrow$. Together with

$$\mathscr{K}^*_\eta(\zeta';\zeta'') = \mathscr{K}_\eta(\zeta'';\zeta') \tag{9.4b}$$

this condition will constitute hereafter the key ingredient of the mathematical definition of all propagators in relativistic stochastic phase spaces.

The conditions (9.4) are expected to be obeyed in the free case, as well as in the presence of an external field. In addition, relativistic covariance should be satisfied, i.e. in the present special relativity context, in a new inertial frame $\mathscr{L}'$ we should have:

$$\mathscr{K}'_\eta(\zeta';\hat{\zeta}') = \mathscr{K}_\eta(\zeta;\hat{\zeta}), \quad \zeta' = (q', p') = (a + \Lambda q, \Lambda p). \tag{9.5}$$

The fact that the free propagators K_η in (6.21) satisfy (9.4) has been established in (6.24)–(6.25) for the chosen gauge. Of course that implies right away the validity of (9.4) in arbitrary gauges. The covariance condition (9.5) is also an immediate consequence of (6.21) when $\tilde{U}(a, \Lambda)$ in (1.9) is applied, and obviously also remains valid in arbitrary gauges. For propagators $\mathscr{K}_\eta$ in external fields these same properties can be derived from those of the free propagators K_η through the use of path integrals that are the relativistic counterparts of (1.10.11) and (1.10.12). Specific examples of such derivations will be encountered in the next section.

It should be emphasized that in operationally interpreting such covariance conditions as (9.5), we have to think of the coherent flow (that is, mathematically, of the reference family $\mathscr{S}^\uparrow$) as fixed while the change of inertial frame is executed. In other words, the kinematical operations of translation, rotation and boosting take place against the background supplied by the coherent flow.[20] Since our universe of discourse operationally consists of the system and its environment, which in turn consists of frames of reference and coherent flows of test particles (i.e. of microdetectors), an application of any of these operations simultaneously to all these elements is operationally meaningless, and if mathematically implemented, it would bring back into the picture absolute space and time. The key point is, of course, that these operations are deemed to be those of some of the abovementioned components of the universe of discourse relative to the others. For example, a rotation R of the frame of reference in relation to the coherent flow should have exactly the same effect as a rotation R^{-1} of the coherent flow (whose test particles are used to established where the system actually is) in relation to the frame.

Leaving aside for the time being these epistemological questions (which will be further discussed in Chapter 4), we note that the dependence of the probabilities (9.2) on $\mathscr{S}^\uparrow$ (i.e. on the choice of coherent flow) cannot be avoided even if we resort to relative probabilities $P_\eta(\mathring{B}/\mathring{D})$ (cf. (1.3.8)) over Borel set $\mathring{B} \subset \mathring{D}$ in $\mathscr{M}^+_m$. Indeed, by (6.16) or (7.33) in the free case we have

$$P_\eta(\mathring{B}/\mathring{D}) = \mathscr{P}_\eta(\mathring{B})/\mathscr{P}_\eta(\mathring{D}), \tag{9.6a}$$

where $\mathscr{P}_\eta$, as given by the covariant expression

$$\mathscr{P}_\eta(\mathring{B}) = i\hbar \int_{\mathring{B}} K_\eta(\zeta'; \zeta) n^\mu \overleftrightarrow{\partial}_\mu K_\eta(\zeta; \zeta')\, d^4 q\, d\Omega_m(p), \tag{9.6b}$$

is the relative probability of the particle propagating from $\underset{\sim}{\zeta}' \in \mathscr{M}^+_{m,\eta}$ to the region $\mathring{B}$ in $\mathring{D}$ – if it had propagated to $\mathring{D}$ at all. The intrinsic dependence[21] of (9.6) on $\mathscr{S}^\uparrow$ is then manifested by the presence in (9.6b) of the normal n^μ to the hyperplanes σ for $\Sigma_m = \sigma \times \mathscr{V}_m \in \mathscr{S}^\uparrow$.

This dependence of probabilities on the coherent flow of the test particles – i.e. in practical terms, on the state of motion of the preparatory and detecting components of the apparatus – does not necessarily entail the dependence of the propagators themselves on $\mathscr{S}^\uparrow$. To demonstrate this point, and to supply at the same time a specific class of explicit examples of free propagators, let us compute $K_\eta(\zeta; \zeta')$ for

$$\eta^{(l)}(mck^0) = N_{m,l} \exp\left(-\frac{2l^2}{\hbar^2}\, mck^0\right), \tag{9.7}$$

where $N_{m,l}$ is a normalization constant to be computed from (4.10), i.e. from the condition that

$$\int_{p \in \mathscr{V}^+_m} |\eta^{(l)}(mcp^0)|^2\, d\mathbf{p} = 2mc(2\pi\hbar)^{-3}. \tag{9.8}$$

In Chapter 4 the resulting propagators

$$K^{(l)}(q, p; q', p') = \int_{\mathscr{V}^+_m} \tilde{\eta}^{(l)*}_{q,p}(k) \tilde{\eta}^{(l)}_{q',p'}(k)\, d\Omega_m(k) \tag{9.9}$$

will be playing a key role in the formulation of reciprocity theory in quantum spacetime, since $\eta^{(l)}$ in (9.7) will turn out to be directly related to the ground-state solution of relativistic harmonic oscillators describing minimal fluctuations in the metric of quantum spacetime.

From (9.7) and (9.8) we obtain that

$$N^{-2}_{m,l} = \frac{(2\pi\hbar)^3}{2mc} \int_{\mathbb{R}^3} \exp\left[-\frac{4l^2}{\hbar^2}\, mc(\mathbf{p}^2 + m^2c^2)^{1/2}\right] d\mathbf{p}. \tag{9.10}$$

The above integral can be evaluated by employing the following Fourier sine transform formula (Erdélyi, 1954, p. 75) for Re α, Re $\beta > 0$:

$$F(y, \alpha, \beta) = \int_0^\infty \exp[-\beta(\alpha^2 + x^2)^{1/2}] \sin xy \frac{x\,dx}{(\alpha^2 + x^2)^{1/2}}$$

$$= \alpha y(y^2 + \beta^2)^{-1/2} K_1[\alpha(y^2 + \beta^2)^{1/2}]. \tag{9.11}$$

Here we use the customary notation K_i, $i = 1, 2, \ldots$, for the modified Bessel functions. We get

$$\frac{\alpha}{\beta} K_1(\alpha\beta) = \lim_{y\to+0} y^{-1} F(y, \alpha, \beta)$$

$$= \int_0^\infty \exp[-\beta(\alpha^2 + x^2)^{1/2}] \frac{x^2\,dx}{(\alpha^2 + x^2)^{1/2}}, \tag{9.12}$$

and therefore from (9.10)

$$N_{m,l} = \left\{ -(2\pi\hbar)^3 \frac{2\pi}{mc} \frac{\partial}{\partial\beta} \frac{\alpha}{\beta} K_1(\alpha\beta) \right\}^{-1/2}$$

$$= (2\pi\hbar)^{-2} \left\{ \frac{\hbar}{4l^2} K_2\left(\left[\frac{2lmc}{\hbar} \right]^2 \right) \right\}^{-1/2}. \tag{9.13}$$

To compute $K^{(l)}$ in (9.9), which by (4.1), (4.8) and (9.7) equals

$$K^{(l)}(q, p; q', p')$$

$$= N_{m,l}^2 \int_{\mathscr{V}_m^+} \exp\left[-\frac{i}{\hbar}(q - q') \cdot k - \frac{2l^2}{\hbar^2}(p + p') \cdot k \right] d\Omega_m(k), \tag{9.14}$$

let us introduce the new variables $k' = \Lambda_v^{-1} k$, where

$$v_\mu = [(p + p') \cdot (p + p')]^{-1/2} (p_\mu + p'_\mu). \tag{9.15}$$

Since $d\Omega_m(k)$ is left invariant by Λ_v^{-1}, we have

$$K^{(l)}(q, p; q', p') = N_{m,l}^2 \int_{\mathscr{V}_m^+} \exp\left(-\frac{i}{\hbar} q'' \cdot k' - \frac{2l^2}{\hbar^2} p'' \cdot k' \right) d\Omega_m(k'), \tag{9.16a}$$

$$q'' = \Lambda_v^{-1}(q - q'), \qquad p'' = \Lambda_v^{-1}(p + p') \tag{9.16b}$$

$$\mathbf{p}'' = 0, \qquad p_0'' = [(p + p') \cdot (p + p')]^{1/2} = (2m^2c^2 + 2p \cdot p')^{1/2} \tag{9.16c}$$

If we introduce, by analogy with (1.9.23b)

$$l^2(q_0'') = l^2 + i\frac{\hbar}{2}\frac{q_0''}{p_0''} = l^2 + i\frac{\hbar}{4m}t'' + O\left(\frac{|\mathbf{p}''|}{mc}\right), \quad q_0'' = ct'', \tag{9.17}$$

then we are able to cast (9.16a) into the form

$$K^{(l)}(q, p; q', p') = \frac{N_{m,l}^2}{2}\int_{\mathbb{R}^3} \exp\left[\frac{i}{\hbar}\mathbf{q}''\cdot\mathbf{k} - \frac{2l^2(q_0'')}{\hbar^2}p_0''(\mathbf{k}^2 + m^2c^2)^{1/2}\right]\frac{d\mathbf{k}}{(\mathbf{k}^2 + m^2c^2)^{1/2}}. \tag{9.18}$$

In terms of spherical coordinates, the integration in (9.18) can be performed immediately over the angular coordinates, whereas the remaining integration in $|\mathbf{k}|$ yields an integral of the type (9.11). Hence we obtain

$$K^{(l)}(q, p; q', p') = \frac{2\pi m^2c^2N_{m,l}^2}{f_l(q, p; q', p')}K_1(f_l(q, p; q', p')) \tag{9.19}$$

where in terms of q'' and p'' in (9.16b)

$$f_l = \mathbf{q}''^2 + \left[\frac{2l^2(q_0'')}{\hbar}\right]^2 p_0''^2 = \frac{4l^4}{\hbar^2}p_0''^2 + \frac{4il^2}{\hbar}p_0''q_0'' - q''\cdot q''. \tag{9.20}$$

Reverting to the original variables we arrive at the following relativistically as well as reciprocally[22] invariant expression for f_l:

$$f_l(q, p; q', p') = \frac{lmc}{\hbar}\left\{\left[\frac{i}{l}(q - q') + \frac{2l}{\hbar}(p + p')\right]^2\right\}^{1/2}. \tag{9.21}$$

Hence, as a function of the above invariant quantity (9.21), the free propagator $K^{(l)}$ in (9.19) is itself a relativistically invariant quantity. Thus indeed, as asserted earlier, there is no explicit or implicit dependence of these free particle propagators either on the inertial frame or on the coherent flow.

In conclusion, we note that $\tilde{\xi}^{(l)}$, given by (1.6.30) for $\mathbf{q} = \mathbf{p} = \mathbf{0}$, is a nonrelativistic approximation of $\tilde{\eta}^{(l)}(k)$ supplied by (9.7) in accordance with (6.32):

$$\eta^{(l)}(mck^0) = N_{m,l}(2mc)^{1/2}\exp\left[-2\left(\frac{lmc}{\hbar}\right)^2\right]\left[\tilde{\xi}^{(l)}(\mathbf{k}) + O\left(\frac{|\mathbf{k}|}{mc}\right)\right]. \tag{9.22}$$

Furthermore, both $\tilde{\xi}^{(l)}$ and $\tilde{\eta}^{(l)}$ are related to ground states of nonrelativistic and relativistic harmonic oscillators, respectively. Consequently, the propagator $K^{(l)}$ in

(9.19) can be justifiably regarded as the relativistic counterpart of the propagator $K^{(l)}$ in (1.9.23).

* 2.10. Relativistic Extended Quantum Particles in External Electromagnetic Fields

In Sections 2.3–2.9 we have studied the main kinematical aspects of the motion of extended relativistic quantum particles which are free, and therefore move by inertia. In this section we intend to outline the fundamental aspects of consistent relativistically covariant theories of motion of such particles in the presence of external electromagnetic fields. For further details and references the reader is directed to the papers by Prugovečki (1978f, 1981b) and by Ali and Prugovečki (1980, 1981).

The conventional theory of a spin-zero quantum particle in an external electromagnetic field described by the 4-potential $A^\mu(x)$ is based on the Klein–Gordon equation with minimal coupling (Bjorken and Drell, 1964; Berestetskiĭ *et al.*, 1979):

$$\left[\left(i\hbar\partial_\nu - \frac{e}{c}A_\nu(x)\right)\left(i\hbar\partial^\nu - \frac{e}{c}A^\nu(x)\right) - m^2c^2\right]\phi(x) = 0. \tag{10.1}$$

This model, however, suffers from two fundamental inconsistencies. First of all, it shares all the difficulties of the free model based on (1.1), since the latter is obviously a special case of (10.1). Hence, despite formal appearances, the values of the variables x^μ do not refer to true spacetime points at which the quantum (point) particle might be found when $A^\mu(x)$ acts upon it. In addition, there is a second difficulty, namely that many solutions of (1.1) for initial conditions $\phi \in L^2(\sigma^+)$ corresponding to positive-energy solutions in accordance with (1.30) (i.e. to a particle rather than antiparticle initial state) do not leave the space $L^2(\sigma^+)$ of these positive-energy solutions invariant. In a one-particle model, such as (1.1), this is unacceptable[23] since such behavior violates the conservation of the total probability (equal to 1) of the particle being somewhere (to the extent this concept makes sense in the absence of a consistent configuration representation) and of having at each instant some 4-momentum (which for (1.1) is a consistent concept, since the momentum representation is consistent in the free special relativistic case).

The construction of relativistically covariant models for extended particles which avoid both these pitfalls essentially requires the introduction of the relativistic stochastic phase space $\mathcal{M}^+_{m,\eta}$ corresponding to the proper wave function η of the chosen particle. Indeed, we can then work with *bona fide* probability amplitudes $\phi(q, p)$, which for a chosen reference family $\mathcal{S}^\uparrow$ (i.e. chosen coherent flow of test particles) assume values in $L^2(\Sigma^+_{m,\eta})$ for $\Sigma^+_m = \sigma \times \mathcal{V}^+_m \in \mathcal{S}^\uparrow$. This avoids the first inconsistency, but we are then faced with the problem of deciding on the mathematical expression that would appropriately describe the action of the external field upon an extended particle at the stochastic location $\underset{\sim}{q}^\nu$ and of stochastic 4-momentum $\underset{\sim}{p}^\nu$.

Extrapolating from (1.8.30)–(1.8.33), we introduce the 4-potential operators

$$A^\mu_\eta(\sigma) = \mathbb{P}_\eta(\Sigma^+_m)A^\mu(Q_{\text{st}})\mathbb{P}_\eta(\Sigma^+_m) \tag{10.2}$$

where Q^μ_{st} are defined as in (5.23), but are acting on $\Phi \in L^2(\Sigma^+_m)$ which are not necessarily restrictions to Σ^+_m of some ϕ in (3.9). Hence

$$(A^\mu(Q_{\text{st}})\Phi)\,(q, p) = A^\mu(q)\Phi(q, p) \tag{10.3}$$

defines $A^\mu(Q_{\text{st}})$ as a family of self-adjoint operators in $L^2(\Sigma^+_m)$ for all $\Sigma^+_m \in \mathscr{S}^\uparrow$. In (5.25) we have seen that in the laboratory frame, where Σ^+_m corresponds to the q^0 = const hypersurface in $\mathscr{M}^+_m$, the operators Q^j_η, $j = 1, 2, 3$, when restricted to $L^2(\Sigma^+_{m,\eta})$ actually coincide with the unitary transforms (under $\mathscr{W}_\eta$) of the Newton–Wigner operators X^j_{NW} in $L^2(\mathscr{V}^+_m)$. Hence A^μ_η in (10.2), which by (7.3) act upon $\phi \in L^2(\Sigma^+_{m,\eta})$ as follows,

$$(A^\mu_\eta(\sigma)\phi)\,(q, p) = \int_{\sigma \times \mathscr{V}^+_m} K_\eta(q, p; q', p')A^\mu(q')\phi(q', p')\,\mathrm{d}\Sigma_m(q', p'), \tag{10.4}$$

are plausible candidates for the 4-potential acting on an extended particle at $\underset{\sim}{\zeta} = (\underset{\sim}{q}, \underset{\sim}{p}) \in \mathscr{M}^+_{m,\eta}$. Since, however, there is no consistent configuration representation for relativistic quantum particles, the justification of this assumption can be achieved only by first carrying out a contraction $\eta \mapsto \xi$ of the relativistic resolution generators into nonrelativistic ones in accordance with (6.32), and then writing down in the configuration representation the counterparts

$$(W^{-1}_\xi A^\mu_\xi W_\xi \psi)\,(\mathbf{x}, \mathbf{t}) = A^\mu_\xi(\mathbf{x}, t)\psi(\mathbf{x}, t), \tag{10.5a}$$

$$A^\mu_\xi(\mathbf{x}, t) = (2\pi\hbar)^3 \int_{\mathbb{R}^3} |\xi(\mathbf{x} - \mathbf{q})|^2 A^\mu((ct, \mathbf{q}))\,\mathrm{d}\mathbf{q}, \tag{10.5b}$$

of (1.8.32) and (1.8.33) for the external 4-potential A^μ acting upon a nonrelativistic extended particle. We see that (10.5b) provides the appropriate 'smearing', which in the sharp-point limit (1.8.1) merges into the conventional expression $A^\mu(x)$.

In a relativistic model for a quantum particle moving in an external field the dynamics can be formulated in a manifestly covariant manner in two ways: by means of a relativistically invariant equation, such as in (1.1), (3.11) and (3.30), or by means of a Schrödinger equation

$$ic\hbar \frac{\partial \Phi}{\partial q_0} = H\Phi, \qquad q_0 = ct, \tag{10.6}$$

in which $\hat{P}^0 = c^{-1}H$ is the zeroth component of an operator-valued 4-vector $\hat{P}^\mu$. In the free case the two methods are equivalent since the four components P^μ commute. However, when a nonvanishing external field is present that is no longer the case, and if H in (10.6) is to play the dual role of generator of time translations and of energy operator, then only the second option is available.

In the classical theory of a charged particle minimally coupled to an electromagnetic field there is a great variety of Hamiltonians H that give rise to the correct canonical equations of motion, but there is only one expression that also represents the total energy of the particle, namely (Barut, 1964):

$$H^{\mathrm{cl}} = [(c\mathbf{p} - e\mathbf{A}(x))^2 + m^2c^4]^{1/2} + eA^0(x). \tag{10.7}$$

Of course, if we try to quantize (10.7) via the usual substitution $\mathbf{p} \mapsto -i\hbar\nabla_{\mathbf{x}}$ we are left with a mathematically ill-defined and physically unacceptable outcome on account of the form of the expression between square brackets, which is not a positive-definite operator since, due to (2.9), multiplication by $A^j(x)$ does not yield a symmetric operator in $L^2(\sigma^+)$. Consequently, the square root of this expression cannot possibly be self-adjoint, as it is required if H in (10.6) were to represent the total energy operator. For that reason, in the conventional approach one is forced to proceed via an invariant equation, such as the Klein–Gordon equation (10.1).[24]

On the other hand, we encounter no similar difficulties in $L^2(\Sigma^+_{m,\eta})$ with

$$H_\eta = [(ic\hbar\nabla_{\mathbf{q}} + e\mathbf{A}_\eta)^2 + m^2c^4]^{1/2} + eA^0_\eta \tag{10.8}$$

since A^μ_η are self-adjoint operators in $L^2(\Sigma^+_{m,\eta})$, whereas $P = -i\hbar\nabla_{\mathbf{q}}$ are self-adjoint on $L^2(\Sigma^+_m)$ and leave $L^2(\Sigma^+_{m,\eta})$ invariant since $U(a, I)$ in (3.9b) does so for all $a \in M(1, 3)$. Thus, if we label the initial-data hypersurface $\hat{\Sigma}^+_m$ by $q^0 = 0$ in the laboratory frame, then the evolution governed by

$$U_\eta(q_0'', q_0') = T\exp\left[-\frac{i}{\hbar c}\int_{q_0'}^{q_0''} H_\eta(q^0)\,\mathrm{d}q^0\right] \tag{10.9}$$

is unitary in $L^2(\hat{\Sigma}^+_{m,\eta})$, and the total probability is conserved. It is not yet evident, however, that the resulting model is relativistically covariant and gauge invariant.

To establish relativistic covariance, we have to prove first of all that

$$P^0_\eta(\sigma) = c^{-1}H_\eta\mathbb{P}_\eta(\Sigma^+_m) = c^{-1}\{[(c\mathbf{P} - e\mathbf{A}_\eta(\sigma))^2 + m^2c^4]^{1/2} + eA^0_\eta\}\mathbb{P}_\eta(\Sigma^+_m) \tag{10.10}$$

combined with $\mathbf{P} = -i\hbar\nabla_{\mathbf{q}}$ gives rise to a 4-vector of operators in $L^2(\Sigma^+_{m,\eta})$. Since in the classical theory $p^0 = c^{-1}H^{\mathrm{cl}}$ in (10.7) and $\mathbf{p}$ constitute a vector field if $A^\mu(x)$ is a vector field, we shall indeed have that

$$P^\mu_\eta \mapsto P'^\mu_\eta = \Lambda^\mu{}_\nu P^\nu_\eta, \qquad P^j_\eta(\sigma) = P^j\mathbb{P}_\eta(\Sigma^+_m), \tag{10.11}$$

under all $(a, \Lambda) \in \mathscr{P}$ if under such transformations

$$A_\eta^\mu \mapsto A_\eta'^\mu = \Lambda^\mu{}_\nu A_\eta^\nu. \tag{10.12}$$

In view of the 4-vector field property of $A^\mu(x)$,

$$A^\mu(x) \mapsto A'^\mu(x) = \Lambda_\nu^\mu A^\nu(\Lambda^{-1}(x-a)), \tag{10.13}$$

the fact that (10.12) is indeed true is established by writing A_η^μ as an integral operator,

$$(A_\eta^\mu(\sigma)\phi)(\zeta) = \int_{\sigma \times \mathscr{V}_m^+} A_\eta^\mu(\zeta;\zeta')\phi(\zeta')\, \mathrm{d}\Sigma_m(\zeta'), \tag{10.14}$$

whose kernel is obtained from (7.3) and (10.2):

$$A_\eta^\mu(\zeta';\zeta'') = \int_{\Sigma_m^+} K_\eta(\zeta'; q, p) A^\mu(q) K_\eta(q, p; \zeta'')\, \mathrm{d}\Sigma_m(q, p). \tag{10.15}$$

Since the free propagator K_η as well as $\mathrm{d}\Sigma_m(\zeta)$ are covariant, we get from (10.13) that

$$A_\eta'^\mu(q', p'; q'', p'') = \Lambda^\mu{}_\nu A_\eta^\nu(\Lambda^{-1}(q'-a), \Lambda^{-1}p'; \Lambda^{-1}(q''-a), \Lambda^{-1}p''), \tag{10.16}$$

and therefore (10.12) follows. Hence (10.9) regarded as an evolution operator from $\Sigma_m' = \sigma' \times \mathscr{V}_m^+ \in \mathscr{S}^\uparrow$ to $\Sigma_m'' = \sigma'' \times \mathscr{V}_m^+ \in \mathscr{S}^\uparrow$ can be written in a manifestly covariant form:

$$U_\eta(\sigma''; \sigma') = T \exp\left(-\frac{i}{\hbar} \int_{\sigma'}^{\sigma''} P_\eta^\mu\, \mathrm{d}q_\mu\right). \tag{10.17}$$

It should be emphasized, however, that in the above time-ordered (Bochner) integral the ordering in time does not change with the change of inertial frame, but it is always determined by the background coherent flow (i.e. in practical terms, by the field-producing apparatus), and it is specified by the straight worldlines followed by the classical test particles in that flow in spacetime regions where the external field vanishes or is switched off.

The covariance of U_η in (10.17) and of η_ζ in (6.19) immediately implies the covariance (in accordance with (9.5)) of the propagator

$$\mathscr{K}_\eta(\zeta''; \zeta') = \langle \eta_{\zeta''} \mid U_\eta(\sigma''; \sigma')\eta_{\zeta'}\rangle_{\Sigma_m^+} \tag{10.18}$$

from $\underset{\sim}{\zeta}' \in \Sigma'^{+}_{m,\eta}$ to $\underset{\sim}{\zeta}'' \in \Sigma''^{+}_{m,\eta}$ for an extended particle of proper wave function η and electric charge e that is moving in the external electromagnetic field

$$F^{\mu\nu}_{\eta} = \mathbb{P}_{\eta}(\Sigma^{+}_{m}) \, [\partial^{\mu} A^{\nu}(Q_{\mathrm{st}}) - \partial^{\nu} A^{\mu}(Q_{\mathrm{st}})] \, \mathbb{P}_{\eta}(\Sigma^{+}_{m}). \tag{10.19}$$

The fact that $\mathcal{K}_{\eta}$ is indeed a propagator – i.e. that it satisfies the relation (9.4) – follows directly from (6.28) and its definition (10.18).

The above propagator can be expressed also as a path integral in $\mathcal{M}^{+}_{m,\eta}$ of the general form (10.11). Indeed, the same method by which (10.12) was derived yields when we work in the laboratory frame,

$$\mathcal{K}_{\eta}(\zeta''; \zeta') = \lim_{N \to \infty} \int \mathrm{d}\Sigma_{m}(\zeta'_{N}) \int \prod_{j=N-1}^{1} \mathrm{d}\Sigma_{m}(\zeta_{j}) \, \mathrm{d}\Sigma_{m}(\zeta'_{j}) \times$$
$$\times \prod_{j=N}^{1} \left\langle \eta_{0,q_{j};p_{j}} \,\middle|\, \exp\left[-\frac{i}{\hbar} (P^{0}_{\eta}(\sigma_{j}) - P^{0}) \, (q^{0}_{j} - q^{0}_{j-1}) \right] \times \right.$$
$$\left. \times \; \eta_{0,q'_{j};p'_{j}} \right\rangle_{\Sigma^{+}_{m}} K_{\eta}(\zeta'_{j}; \zeta_{j-1}). \tag{10.20}$$

To recast this expression in a manifestly covariant form we first note that $q^{0}_{0} = q'^{0} < q^{0}_{1} < \cdots < q^{0}_{N} = q''^{0}$ determine a time-ordered family

$$\Sigma^{(0)+}_{m} < \Sigma^{(1)+}_{m} < \cdots < \Sigma^{(N)+}_{m} \tag{10.21}$$

of reference hypersurfaces $\Sigma^{(j)+}_{m} = \sigma_{j} \times \mathcal{V}^{+}_{m}$. The inner product in (10.20) can be written on $L^{2}(\Sigma^{(j)+}_{m})$ as

$$\left\langle \eta_{q_{j}, p_{j}} \,\middle|\, \exp\left[-\frac{i}{\hbar} (P^{0}_{\eta}(\sigma_{j}) - P^{0}) \, (q^{0}_{j} - q^{0}_{j-1}) \right] \eta_{q'_{j}, p'_{j}} \right\rangle_{\Sigma^{(j)}_{m}} \tag{10.22}$$

since by (6.21) and (6.23) we have in general

$$\eta_{a+q,p}(a + q', p') = \eta_{q,p}(q', p'), \quad a \in M(1, 3). \tag{10.23}$$

Working still in the laboratory frame (where (3.7) is valid) and using (1.19), we can write

$$\langle \eta_{\zeta_{j}} \,|\, (P^{0}_{\eta}(\sigma_{j}) - P^{0}) \eta_{\zeta_{j-1}} \rangle_{\Sigma^{(j)}_{m}}$$
$$= 2 \int_{\Sigma^{(j)+}_{m}} \eta^{*}_{\zeta_{j}}(q, p) \, (p_{\mu} \, [P^{\mu}_{\eta}(\sigma_{j}) - P^{\mu}] \, \eta_{\zeta_{j-1}}) \, (q, p) \, \mathrm{d}\mathbf{q} \, \mathrm{d}\Omega_{m}(p) \tag{10.24}$$

since $P^j_\eta(\sigma)\phi = P^j\phi$, $j = 1, 2, 3$, for $\phi \in L^2(\sigma \times \mathcal{V}^+_m)$. Let us introduce by analogy with (1.10.13)

$$\overline{p \cdot (P_\eta - P)}\,(j; j-1) K_\eta(\zeta_j, \zeta_{j-1})$$

$$= \int_{(j-1)}^{(j)} K_\eta(\zeta_j; q, p)\,(p_\mu P^\mu_\eta - p_\mu P^\mu) K_\eta(q, p; \zeta_{j-1})\, \mathrm{d}^4 q\, \mathrm{d}\Omega_m(p), \quad (10.25)$$

where the integration with respect to the invariant measure $\mathrm{d}^4 q\, \mathrm{d}\Omega_m(p)$ extends over the entire region of $\mathcal{M}^+_m$ that lies between $\Sigma^{(j-1)+}_m$ and $\Sigma^{(j)+}_m$. Then we see that (10.20) assumes the following manifestly covariant form:[25]

$$\mathcal{K}_\eta(\zeta''; \zeta') = \lim_{N\to\infty} \int \exp\left[-\frac{2i}{\hbar}\, \overline{p \cdot (P_\eta - P)}\,(j, j-1)\right] \times$$

$$\times \prod_{j=1}^{N-1} K_\eta(\zeta_j; \zeta_{j-1})\, \mathrm{d}\Sigma_m(\zeta_j). \quad (10.26)$$

Albeit it is exact and manifestly covariant, the above path integral does not contain an explicit action term that is recognizably related to the classical action (Infeld, 1957) of a relativistic point particle, which can be expressed in terms of its classical Lagrangian (Barut, 1964; Jackson, 1975):

$$L^{\mathrm{cl}} = L^{\mathrm{cl}}_0 + L^{\mathrm{cl}}_I = -(k^\nu k_\nu)\frac{\mathrm{d}\tau}{\mathrm{d}s} + \frac{e}{mc^2}\, k_\nu A^\nu(x(\tau))\, \frac{\mathrm{d}\tau}{\mathrm{d}s}\,. \quad (10.27)$$

Let us show, however, that if we expand $p \cdot (P_\eta - P)$ in a perturbation series by writing

$$P^0_\eta - P^0 = \frac{e}{c} A^0 - \frac{1}{4}\frac{e}{c}\, \{(P^0)^{-1}, \mathbf{P}\cdot\mathbf{A} + \mathbf{P}\cdot\mathbf{A}\} + \ldots \quad (10.28)$$

then a counterpart of L^{cl}_I emerges.[26]

In frames where Σ_m corresponds to $q^0 = \text{const}$, (6.7) assumes the form

$$\langle \phi_1 | \phi_2 \rangle_{\Sigma^\pm_{m,\eta}} = \frac{2}{Z_\eta} \int_{\Sigma^\pm_\eta} \phi_1^*(q, p)\,(P^0 \phi_2)(q, p)\, \mathrm{d}\mathbf{q}\, \mathrm{d}\Omega_m(p). \quad (10.29)$$

Consequently by (10.28) we can rewrite (10.24) to the first order in e/c as

$$\frac{e}{2Z_\eta} \int_{\Sigma^{(j)+}_m} \eta^*_{\zeta_j}(q, p)\,(P_\mu A^\mu_\eta + A^\mu_\eta P_\mu)\eta_{\zeta_{j-1}}(q, p)\, \mathrm{d}\mathbf{q}\, \mathrm{d}\Omega_m(p). \quad (10.30)$$

Thus, if we introduce the renormalized charge

$$e' = \frac{mc}{2Z_\eta}\, e = \frac{e}{2(2\pi\hbar)^3}\left[\int_{\mathbb{R}^3} |\tilde{\xi}(\mathbf{k})|^2 \frac{\mathrm{d}\mathbf{k}}{(\mathbf{k}^2 + m^2c^2)^{1/2}}\right]^{-1} \tag{10.31}$$

we obtain, to the first order in e/c,

$$\overline{p \cdot (P_\eta - P)}\,(j, j-1) = \frac{e'}{mc^2}\,\frac{1}{2}\,\overline{(P \cdot A_\eta + A_\eta \cdot P)} + \ldots \tag{10.32}$$

where, on account of (10.30),

$$\overline{(P \cdot A_\eta)} K_\eta(\zeta_j, \zeta_{j-1}) = \int_{(j-1)}^{(j)} K_\eta(\zeta_j; \zeta)(P \cdot A_\eta) K_\eta(\zeta; \zeta_{j-1})\, \mathrm{d}^4 q\, \mathrm{d}\Omega_m(p) \tag{10.33}$$

with P^μ and A_η^μ acting on the $\zeta = (q, p)$ variables. Thus a quantum analogue of L_I^{cl}, with a renormalized charge e', makes its appearance in (10.26) in the limit of weak fields or small couplings.

For the sake of simplicity, we have chosen to work in this section with the gauge corresponding to (7.24) – i.e. to $\omega \equiv 0$ in (7.25) – but our considerations are equally valid in all gauges. In fact, the model is gauge invariant since the kinetic 4-momentum operators

$$\Pi_{\overline{\eta}}^\mu = P_{\overline{\eta}}^\mu - \frac{e}{c} A_{\overline{\eta}}^\mu, \quad \overline{\eta} = \mathscr{W}_\eta \eta, \tag{10.34}$$

with $P_{\overline{\eta}}^\mu = P_\eta^\mu$ and $A_{\overline{\eta}}^\mu = A_\eta^\mu$, undergo a unitary transformation corresponding to an unobservable change of phase factors under the combined gauge transformation

$$\phi \longmapsto \phi' = \mathscr{U}_\omega \phi, \tag{10.35a}$$

$$A^\mu(q) \longmapsto A'^\mu(q) = A^\mu(q) - \frac{c}{e}\,\partial^\mu \omega, \tag{10.35b}$$

with $\mathscr{U}_\omega$ defined by (7.12). Indeed, in that case

$$\overline{\eta} \longmapsto \overline{\eta}' = \mathscr{U}_\omega \overline{\eta} = \overline{\eta}_\omega \tag{10.36}$$

and therefore we shall have

$$A_{\overline{\eta}}^\mu \longmapsto A_{\overline{\eta}'}^{\prime\mu} = A_{\overline{\eta}'}^\mu - \frac{c}{e}\,\mathbb{P}_{\overline{\eta}'}(\Sigma_m^+)\partial^\mu \omega \mathbb{P}_{\overline{\eta}'}(\Sigma_m^+), \tag{10.37a}$$

$$P_{\overline{\eta}}^\mu \longmapsto P_{\overline{\eta}'}^{\prime\mu} = P^\mu \mathbb{P}_{\overline{\eta}'}(\Sigma_m^+). \tag{10.37b}$$

Hence, by (7.25b) we get

$$\Pi^{\mu}_{\bar{\eta}} \mapsto \Pi'^{\mu}_{\bar{\eta}'} = P'^{\mu}_{\bar{\eta}'} - \frac{e}{c} A'^{\mu}_{\bar{\eta}'}$$

$$= P_{\omega} \mathbb{P}_{\bar{\eta}\omega}(\Sigma^{+}_{m}) - \frac{e}{c} A^{\mu}_{\eta\omega} = \mathscr{U}_{\omega} \Pi^{\mu}_{\bar{\eta}} \mathscr{U}^{-1}_{\omega}. \tag{10.38}$$

Thus upon executing the gauge transformations (10.35), the new particle propagation governed by (10.6) written in a manifestly gauge-invariant manner, namely as

$$\Pi^{0}_{\bar{\eta}\omega} \Phi^{\omega} = \left[\sum_{j=1}^{3} (\Pi^{j}_{\bar{\eta}\omega})^2 \right]^{1/2} \Phi^{\omega}, \tag{10.39}$$

will be physically indistinguishable from the old one:

$$\Phi^{\omega}(\zeta) = (\mathscr{U}_{\omega}\Phi)(\zeta) = \exp\left[\frac{i}{\hbar}\,\omega(\zeta)\right] \Phi(\zeta). \tag{10.40}$$

In particular, the new propagators in the external field (10.19) will be physically equivalent to the old ones:

$$\mathscr{K}_{\bar{\eta}\omega}(\zeta''; \zeta') = \exp\left[\frac{i}{\hbar}\,\omega(\zeta'')\right] \mathscr{K}_{\bar{\eta}}(\zeta''; \zeta') \exp\left[-\frac{i}{\hbar}\,\omega(\zeta')\right]. \tag{10.41}$$

We see that in general terms the gauge invariance on stochastic phase space for models based on (10.39) exhibits the same qualitative features as in models based on (10.1). It might appear, however, that in case of (10.1) we are working on the space $L^2(\sigma^+)$ for a single irreducible representation of $\mathscr{P}$, so that the counterpart of (10.37b) is the identity transformation $\hat{P}_{\mu} \mapsto \hat{P}_{\mu} = i\hbar\partial_{\mu}$, and that therefore the counterpart of (10.38) assumes an apparently simpler form:

$$\hat{\Pi}^{\mu} = i\hbar\partial^{\mu} - \frac{e}{c} A^{\mu}(x) \mapsto \hat{\mathscr{U}}_{\omega} \hat{\Pi}^{\mu} \hat{\mathscr{U}}^{-1}_{\omega} = i\hbar\partial^{\mu} + \partial^{\mu}\omega - \frac{e}{c} A^{\mu}(x). \tag{10.42}$$

This first impression is, however, misleading since the operator $\hat{\mathscr{U}}_{\omega}$, where

$$(\hat{\mathscr{U}}_{\omega}\Phi)(x) = \exp\left[\frac{i}{\hbar}\,\omega(x)\right] \Phi(x) \tag{10.43}$$

is not in general unitary in the space $L^2(\sigma^+)$ with inner product (1.28). To make it unitary we have to replace in (1.28) ordinary by covariant derivatives – i.e. $i\hbar\partial^{\mu}$ by $\hat{\Pi}^{\mu}$ – so that we are not actually working any longer in $L^2(\sigma^+)$, but rather in a family of gauge-dependent Hilbert spaces $L^2(\sigma^{+}_{\omega})$ related by the unitary transforms $\hat{\mathcal{U}}_\omega$, just as the Hilbert spaces $L^2(\Sigma^+_{m,\bar{\eta}\omega})$ (with the gauge-independent inner product (3.8)) are related by $\mathcal{U}_\omega$.

It is interesting that the above external-field model on $\mathcal{M}^+_{m,\eta}$, although based on the equation of motion (10.39), can nevertheless be extrapolated to the spin-½ case. In fact, it turns out that a Dirac theory of bispinor wave functions $\psi(\zeta)$, $\underset{\sim}{\zeta} \in \mathcal{M}_{m,\eta}$, can be formulated, but that the introduction of a minimal coupling to an external electromagnetic field does not leave the space of positive-energy solutions invariant (Prugovečki, 1980). This is not that surprising since, as we shall see in Section 5.4, the Foldy–Wouthuysen type of transformations required to take us from spinor probability amplitudes to the bispinor Dirac-type wave functions $\psi(\zeta)$ does not allow the retention of a direct probability-amplitude interpretation for $\psi(\zeta)$. Hence, in order to develop a consistent dynamics for extended spin-½ particles in external fields, we must work in the context of two-component fermion theories reminiscent of those considered in the late Fifties by Feynman and Gell-Mann (1958), and by Brown (1958).

The Hamiltonian representing the energy of a classical point particle with intrinsic magnetic momentum μ and dipole electric moment $\mathbf{d}$ in an external electromagnetic field $F^{\mu\nu}(x)$ is obtained (Barut, 1964, pp. 73–83) by adding to the rest mass m in (10.7) the term $\frac{1}{2}\sigma^{\mu\nu}_{\text{cl}}F_{\mu\nu}$, where $\sigma^{\mu\nu}_{\text{cl}} = -\sigma^{\nu\mu}_{\text{cl}}$ and

$$\sigma^{0j}_{\text{cl}} = d^j, \quad \sigma^{ij}_{\text{cl}} = \sum_{k=1}^{3} \epsilon^{ijk}\mu^k, \qquad i, j = 1, 2, 3. \tag{10.44}$$

Hence, in the stochastic quantization of this model[27] the Hamiltonian that replaces (10.8) and acts on two-component spinor wave functions should be

$$H_\eta = [(c\mathbf{P} - e\mathbf{A}_\eta)^2 + (m + \tfrac{1}{2}\sigma_{\mu\nu}F^{\mu\nu}_\eta)^2 c^4]^{1/2} + eA^0_\eta. \tag{10.45}$$

It turns out (Ali and Prugovečki, 1981) that the only nontrivial choice for $\sigma^{\mu\nu}$ which makes $\sigma_{\mu\nu}F^{\mu\nu}_\eta$ invariant is

$$\sigma^{\mu\nu} = \frac{i}{2}\,(|\mu|\, g^{\mu\lambda}g^{\nu\kappa} + |\mathbf{d}|\, \epsilon^{\mu\nu\lambda\kappa})\,(\sigma_\kappa\sigma_\lambda - \sigma_\lambda\sigma_\kappa), \tag{10.46}$$

where σ^μ act on the spin degrees of freedom and can be realized by means of Pauli matrices, so that

$$\tfrac{1}{2}\sigma_{\mu\nu}F^{\mu\nu}_\eta = |\mathbf{d}|\, \sigma\cdot\mathbf{E}_\eta + |\mu|\, \sigma\cdot\mathbf{B}_\eta. \tag{10.47}$$

This model for stochastically extended spin-½ relativistic particles in external fields retains all the covariance and invariance features of the earlier discussed spin-zero model. The derivation of these features is in fact very much the same in both cases. Furthermore, in the presence of an external Coulomb field, the above spin-½ model displays an energy spectrum in general agreement with up-to-date experimental results (Ali and Prugovečki, 1981, Section 4), exhibiting doublet-splitting energies that contribute to the Lamb shift terms[28] similar to those discovered in other recent models (Reuse, 1978; Schwebel, 1978) of electrons in Coulomb fields.

Notes

1 For a discussion of most of the attempts initiated until the beginning of the seventies see the review article by Kálnay (1971), which also contains an extensive bibliography.

2 In curved spacetime, momentum is no longer conserved and there is no counterpart to the Poincaré group. Hence, the physical as well as the mathematical meaning of the momentum representation becomes questionable. This point will be discussed in Section 4.4, where a solution will also be found in the context of quantum spacetime.

3 In view of the later discussed indefiniteness of $j^\mu(x)$, it superficially appears as if $ej^\mu(x)$ $d\sigma_\mu(x)$ might represent a probability for charge detection since charge can be positive or negative. This conclusion is, of course, fallacious since a charged-point particle carries *either* positive *or* negative total charge, so that $ej^0(x)$ would have to be either positive definite or negative definite if it represented a *bona fide* charge density. The misnomer has, however, persisted primarily because the interpretation of $e\phi^* \overleftrightarrow{\partial}^\mu \phi$ as a charge current can be retained (superficially speaking) when $\phi(x)$ is a 'local' quantum field. More about that in Chapter 5!

4 In reviewing the history of the Dirac equation on the occasion of Dirac's seventieth birthday, A. Pais (1972) reminds us that this equation had resulted from a "search for a positive-definite one-particle probability density with the right covariance and conservation properties". However, as Wightman (1972) points out on the same occasion, the assumption that $\bar{\psi}\gamma^0\psi$ is a probability density "cannot be correct because multiplication by **x** does not carry positive-energy solutions into positive-energy wave functions".

5 Reviews of these results and of the difficulties they spell out for the conventional interpretation of solutions for such equations can be found in Wightman (1971) and Velo and Wightman (1978). Rañada and Rodero (1980) have shown that in the presence of derivative couplings to external fields even the Dirac equation can display acausal behavior.

6 For example, for spin-½ we can work in the Foldy–Wouthuysen representation in which (1.35) replaces (1.32), and our subsequent remarks apply to each component of ψ_{FW}.

7 One way is to argue (Kálnay and Toledo, 1967) that the complex values $\lambda = \lambda_1 + i\lambda_2$ in the spectrum of such operators represent an ordered pair (λ_1, λ_2) of measurable quantities. Another way is to give an interpretation to *complex* probabilities (Dirac, 1945; Prugovečki, 1966, 1967) that can be associated with such operators. However, neither of these possibilities has led to a successful generalization of conventional quantum mechanics.

8 Thus, we can combine (Prugovečki, 1976c) the concept of extension intrinsic to the nonrelativistic stochastic phase-space approach of Chapter 1 with the subsequently discussed Newton–Wigner operators, and arrive at a purportedly relativistic formalism for extended quantum particles which, however, does not display true relativistic covariance since space and time coordinates are not treated on an equal footing.

9 In fact, in the context of localization in stochastic phase space (rather than of sharp localization), this will turn out to be a true statement (cf. the concluding paragraph in Section 2.5) if the microdetectors are extended quantum test particles.

10 Upon analyzing basic mathematical aspects of the localization problems in nonrelativistic quantum mechanics in search of a solution (cf. also Castrigiano and Mutze, 1982) to the relativistic problem, Wightman (1962, p. 851) concludes that in the later case " . . . a sensible notion of localizability in space-time does not exist". However, this observation is not accurate without the qualifying term 'sharp': as the later results of this chapter will

demonstrate, once we drop the restrictions characterizing the naive notion of sharp localizability discussed in Section 1.2, very sensible solutions to the localizability problem in relativistic quantum mechanics turn out to be feasible.

11 The textbooks which openly, and from the outset, stress the fact that *there is no consistent* local relativistic quantum mechanics are, unfortunately, the exception rather than the rule. One of these remarkable exceptions can be found in the Landau series of physics textbooks, where it has already been emphasized in the introduction that: "There is as yet no logically consistent and complete relativistic quantum theory" (Berestetskiĭ *et al.*, 1979, p. 4).

12 It is the arbitrary form of 'removal' of these divergences that has spurred the acknowledged founder of 'local' quantum field theory and quantum electrodynamics to severely and repeatedly criticize the main developments in this area since the inception of the 'renormalization' program (Dirac, 1951, 1965, 1973, 1977, 1978). For example, Dirac has made the following unequivocal statement: " . . . I find the present quantum elepctrodynamics quite unsatisfactory. One ought not to be complacent about its faults. The agreement with observation is presumably a coincidence, just like the original calculations of the hydrogen spectrum with Bohr orbits . . . Quantum electrodynamics is rather like the Klein–Gordon equation. It was built up from physical ideas that were not correctly incorporated into the theory and it has no sound mathematical foundation" (Dirac, 1978b, p. 5).

13 This alternative method was subsequently used by Ali and Giovannini (1982) in deriving related phase-space K-representations (Giovannini, 1981) of the Poincaré and Einstein groups.

14 For example, in Chapter 5, Z_η^{-1} will be seen to be related to the infinite renormalization constants in LQFT, obtained by taking the sharp-point limit (1.8.1) in which Z_η^{-1} becomes infinite, as seen from (1.6.30), (4.21) and (6.12) (cf. Z_A in (5.1.29)).

15 In fact, the conventional inner product (1.28) is also n^μ-dependent. Hence, even if (1.23) were a *bona fide* probability (which it is not!) on $\sigma \subset M(1, 3)$, the resulting probabilities over Borel sets B in the Minkowski space $M(1, 3)$ would be of the form $\int_B n^\mu j_\mu(x)\, d^4x$, and consequently n^μ-dependent, i.e. dependent on the direction of the chosen coherent flow of classical test particles.

16 In the nonrelativistic case Q^j, $j = 1, 2, 3$, were uniquely pinpointed as generators of velocity boosts. As such, they were also self-adjoint (since the considered representations of $\mathscr{G}$ were unitary) and therefore could be interpreted as position operators.

17 Note that η in (5.5) certainly has the set $(-\infty, -mc] \cup [mc, +\infty)$ in its domain of definition. Hence $\psi \in L^2(\Sigma^\pm_{m,\eta})$ is defined for p on $\mathscr{V}^\pm_m$ as well as in some neighborhood of the mass hyperboloid $\mathscr{V}_m$, so that in (7.22), (7.24) and (7.25) $\partial/\partial p_\mu$ can be treated as derivatives with respect to independent variables.

18 It should be recalled that the Poincaré group as opposed to the Galilei group, possesses no nontrivial unitary ray representations, i.e. no ray representations that are not equivalent to vector representations (Bargmann, 1947, 1954; Inönü and Wigner, 1952). Hence (7.11) is as general as required.

19 This is the case asymptotically as $m \to \infty$, i.e. when test particles begin to display semiclassical behavior. Hence the statement that (8.11) can approach (1.27) in the sharp-point limit (Prugovečki, 1978d) has to be qualified accordingly.

20 From the cosmological perspective, that will be eventually adopted in Chapter 4, the most universal coherent flow is fixed by the average motion of matter in the universe, whose consideration is required in the operational interpretation of Mach's principle in general relativity (cf., e.g., Adler *et al.*, 1975, pp. 441 and 448).

21 Recall that this dependence would be present also in the conventional approach based on (1.23) (since $d\sigma^\mu$ in (1.21) always points in the direction of n^μ) if (1.23) were a true probability measure. On the other hand, since a unique mass hyperboloid is associated with each particle, no analogous explicit dependence manifests itself in the momentum representation, which provides the basis of contemporary computations of probabilities. Nevertheless, an implicit operational dependence on $\mathscr{S}^\uparrow$ exists even then, as will be discused in Chapter 4.

22 We shall give Born's (1949) formal definition of reciprocal invariance in Section 4.5, where we review and discuss reciprocity theory.

23 An interpretation of this phenomenon as due to pair creation or annihilation is unacceptable in a one-particle model, when $\phi(x)$ is the probability amplitude for that particle alone, and it is not related to systems of particles and antiparticles, as is the case with a quantum field (see Chapter 5).

24 Whereas $i\hbar\nabla_x$ are self-adjoint in $L^2(\sigma^+)$, that is not the case with multiplication of $\phi(x)$ by $\mathbf{A}(x)$, as easily seen by using (1.28). On the physical level, this feature reflects the fact that multiplication by x^μ has no direct relation to localization in (classical) Minkowski spacetime – as discussed in Section 2.1.

25 If the particle is very massive, so that its propagation is nearly classical, then (3.30) is valid and the term $\overline{p \cdot P}$ can be dropped.

26 For a relativistic classical particle, the free Lagrangian L_0^{cl} is a constant of motion, and therefore its appearance in the action integral is only formal (Infeld, 1957).

27 In stochastic quantization, the particle extension manifests itself only at the quantum level, since it is due to the purely quantum concept of proper wave function. Hence, whereas models of an extended electron (Dirac, 1962; Gnadig *et al.*, 1978) based on conventional quantization have to start from models that are extended already at the classical level, and therefore exhibit severe instabilities, that is not the case in the present stochastic type of models. The latest experimental upper bound on lepton r.m.s. radii is 10^{-16} cm (Barber *et al.*, 1979).

28 The fact that the mean-square deviation of stochastic values is related to the Lamb shifts when external electromagnetic fields act in stochastic spaces has been previously noted and commented upon by Blokhintsev (1975, p. 247). This feature is not surprising in view of Welton's (1948) semiclassical interpretation of the Lamb shift (Itzykson and Zuber, 1980, p. 81).

Chapter 3

Statistical Mechanics on Stochastic Phase Spaces

"I maintain that the mathematical concept of a point in a continuum has no direct physical significance . . . . This attitude is generally accepted in quantum mechanics. But it has actually a more fundamental significance and is only indirectly connected to the special features characteristic of quantum mechanics. It ought to be applied to classical mechanics as well. . . . It is misleading to compare quantum mechanics with deterministically formulated classical mechanics; instead, one should first reformulate the classical theory, even for a single particle, in an indeterministic, statistical manner. Then some of the distinctions between the two theories disappear, others emerge with great clarity. Amongst the first is the feature of quantum mechanics, that each measurement interrupts the automatic flow of events and introduces new initial conditions (so-called 'reduction of probability'); this is true just as well for a statistically formulated classical theory."

BORN (1955c, pp. 3–4, 25–26)

Classical statistical mechanics, as well as quantum statistical mechanics, are both acknowledgedly statistical theories even if one subscribes to the conventional point of view about the intrinsically disparate natures of classical and quantum mechanics, respectively. Yet, the conventional mathematical frameworks for these two theories are remarkably dissimilar even at the most fundamental level of description of states of ensembles: in the classical case states are described by distribution functions $f(q, p)$ on phase space, whereas in the quantum case they are given by density operators ρ.

However, according to Born (1955, 1956), there is no sharp dividing line between classical and quantum physics on the most fundamental epistemological level. Born's argument (restated in the language of stochastic spaces used in this monograph) is that all measurements produce only nonsharp stochastic values, and, therefore, deterministically formulated classical theories reflect mathematical fiction rather than physical reality.

In this chapter we shall demonstrate that upon reinstating in the mathematical formalisms for classical and quantum theories a closer link with physical reality via the medium of stochastic phase spaces, we can achieve a remarkable unification of classical and quantum statistical mechanics. Thus, on one hand, as shown in Sections 3.3 and 3.6, we can formulate quantum statistical mechanics in the language of phase-space distribution functions that is traditionally employed in classical statistical mechanics. On the other hand, as shown in Section 3.5, we can equally well formulate classical statistical mechanics in the framework of Liouville spaces used in quantum statistical mechanics. Hence indeed, in complete accord with Born's fundamental thesis, there is no essential conceptual distinction between classical and quantum theories: the differences are the ones in dynamics, and in the lower bound for the spreads of stochastic phase values imposed by the uncertainty principle in the quantum realm – as opposed to the nonexistence of such bounds from the classical point of view.

The analogy between the classical and quantum realm becomes even more striking once we dispose of another kind of fiction often perpetrated in textbooks on quantum mechanics, namely that any self-adjoint (loosely called 'Hermitian') operator in a Hilbert space $\mathcal{H}$ of pure states of some quantum mechanical system represents an observable for that system. Not only does the existence of superselection rules (Wick *et al.*, 1952; Hegerfeldt *et al.*, 1968) disprove the general validity of this assumption, but as implicit in remarks by Wigner (1963, p. 14), the truth of the matter is precisely the opposite: most self-adjoint operators (such as those we obtain by appropriately symmetrizing polynomials in the position and momentum operators) are *not known* to be measurable. The arbitrary assumption that they might be measurable introduces a restrictive structural component into quantum theories which renders trivial an otherwise deep physical question: When are quantum theories informationally complete? In other words, when can we pinpoint a physical state unambiguously by measuring *known* physical quantities (as opposed to quantities that are merely *claimed* to be measurable, without any evidence being presented as to their measurability – e.g. the aforementioned polynomials of position and momentum operators, such as $X_jP_j^2 + P_j^2X_j$, $j = 1, 2, 3$)?

As shown in Section 3.2, conventional quantum mechanics without the *ad hoc* assumptions as to the measurability of all self-adjoint operators is not an informationally complete theory: there are distinct quantum mechanical states which cannot be empirically distinguished by any measurement procedure if we restrict ourselves to sharp measurements of position and momentum. However, as soon as simultaneous measurements of stochastic position and momentum that comply with the uncertainty principle are taken into consideration, informational completeness is fully achieved by Theorem 4 in Section 3.3. Thus, stochastic phase spaces are more than just convenient devices for the solution of the localizability problem in relativistic quantum mechanics: stochastic phase space emerges as a key ingredient for deriving the informational completeness of quantum theories, without which redundancy in the quantum description of physical reality would be present even in the most elementary quantum models.

Informational completeness and related foundational topics are discussed in Sections 3.1–3.3. The remainder of this chapter deals with questions of primary interest in statistical mechanics, and it can be skipped at a first reading. The proofs of many of the results which we shall present and discuss in this chapter are rather technical and lengthy. Hence, very often we shall limit ourselves to describing the key results and discussing their significance without presenting the proofs – directing instead the interested reader to the original source.

3.1. Phase-Space Representations of Quantum Statistical Mechanics

In this chapter we shall consider systems with N degrees of freedom, such as $N_0 = N/3$ particles without spin [1] moving in three dimensions. We shall denote by

$$q = (q_1, \ldots, q_N) \in \mathbb{R}^N, \qquad p = (p_1, \ldots, p_N) \in \mathbb{R}^N \tag{1.1}$$

the configuration and momentum variables, so that throughout this chapter (q, p) assumes values in $\Gamma^N = \mathbb{R}^{2N}$. Hence, the state of an ensemble of such systems will be described in classical statistical mechanics by a distribution function $\rho^{\text{cl}}(q, p)$ on the phase space Γ^N. We shall denote by $\mathscr{P}(\Gamma^N)$ the family of all such distribution functions, i.e. $\mathscr{P}(\Gamma^N)$ is the set of all probability densities on Γ^N for which

$$P^{\text{cl}}(\Delta) = \int_\Delta \rho^{\text{cl}}(q, p)\, \text{d}q\, \text{d}p \tag{1.2}$$

represents the probability that a system within the ensemble will display values (q, p) within the Borel set $\Delta \subset \Gamma^N$.

On the other hand, a state of an ensemble of such systems is described in quantum statistical mechanics by a density operator ρ from the Liouville space $\mathscr{L}(\mathbb{R}^N)$, which coincides with the Hilbert–Schmidt class [2] $\mathscr{B}_2(\mathscr{H})$ over the configuration representation Hilbert space $\mathscr{H} = L^2(\mathbb{R}^N)$. This Liouville space is itself a Hilbert space that consists of all operators A on $\mathscr{H}$ that can be written in the form

$$A = \sum_{i=1}^{\infty} |e_i'\rangle\, \lambda_i\, \langle e_i''|, \qquad \lambda_1 \geqslant \lambda_2 \geqslant \cdots \geqslant 0, \tag{1.3a}$$

where $\{e_i'\}$ and $\{e_i''\}$ are any two orthonormal bases in $\mathscr{H}$ and

$$\sum_{i=1}^{\infty} \lambda_i^2 = \text{Tr}(A^* A) < \infty. \tag{1.3b}$$

The inner product of any two $A_1, A_2 \in \mathscr{B}_2(\mathscr{H})$ is defined to be:

$$\langle A_1 | A_2 \rangle_2 = \text{Tr}(A_1^* A_2). \tag{1.3c}$$

The set $\mathscr{R}(\mathscr{H})$ of all density operators is a subset of the trace-class $\mathscr{B}_1(\mathscr{H})$, and it consists of all positive-definite operators of trace 1, i.e. of all Hilbert–Schmidt operators ρ obtained by the following specialization of (1.3):

$$\rho = \sum_{i=1}^{\infty} |e_i\rangle\, \lambda_i\, \langle e_i|, \qquad \sum_{i=1}^{\infty} \lambda_i = \text{Tr}\, \rho = 1. \tag{1.4}$$

The question with which we shall be concerned in this section is whether these quantum mechanical descriptions of the state of an ensemble can be recast in a form resembling the classical description in terms of distribution functions $\rho^{\mathrm{cl}}(q, p)$ in some of the stochastic phase spaces Γ_ξ in (1.5.5), or some more general stochastic phase spaces that might be compatible with the uncertainty principle.

With this motivation in mind, let us introduce the general concept of *phase-space representation of the quantum statistical mechanics* of ensembles described in the Liouville space $\mathscr{L}(\mathbb{R}^N)$ as a one-to-one affine[3] map

$$\pi : \rho \mapsto P_\rho, \quad \rho \in \mathscr{R}(\mathscr{H}), \tag{1.5}$$

of the set $\mathscr{R}(\mathscr{H})$ of density operators onto a set of probability measures P_ρ on Γ, for which

$$P_{U_g^* \rho U_g}(\Delta) = P_\rho(g\Delta), \quad g \in \mathscr{G}'. \tag{1.6}$$

In the relation (1.6) U_g is the tensor product of $N_0 = N/3$ representations $U(g)$, $g \in \mathscr{G}$, in (1.4.5), corresponding to the representation of the Galilei group on the Hilbert space $L^2(\mathbb{R}^N)$ of states of N_0-particle systems, and $g\Delta$ is the appropriate generalization of (1.3.19b) to N degrees of freedom. Hence (1.6) ensures that the covariance under the isochronous Galilei group $\mathscr{G}'$ is preserved upon making the transition (1.5) from density operators to probability measures on the Borel sets in Γ^N.

Due to Galilei covariance the measures in (1.5) are absolutely continuous with respect to the Lebesgue measure $\mathrm{d}q\, \mathrm{d}p$ on Γ, so that they can be written by analogy with (1.2) in the form (cf. (A.26) in Appendix A)

$$P_\rho(\Delta) = \int_\Delta \rho(q, p)\, \mathrm{d}q\, \mathrm{d}p, \quad \Delta \in \mathscr{B}^{2N}. \tag{1.7}$$

The ensuing mappings $\rho \mapsto \rho(q, p)$ are special cases of more general types of phase space representations considered[4] in the past. Of all the various phase-space transforms, the best known is the Wigner (1932) transform (cf. O'Connell (1983) for a review and further references)

$$w_\rho(q, p) = (2\pi\hbar)^{-N} \int_{\mathbb{R}^N} \exp\left(-\frac{i}{\hbar} p \cdot x\right) \left\langle q + \frac{x}{2} \middle| \rho \middle| q - \frac{x}{2} \right\rangle \mathrm{d}x \tag{1.8}$$

of the density operator ρ, obtained from the density matrix (in the configuration representation) of ρ in (1.4):

$$\langle x' | \rho | x'' \rangle = \sum_{i=1}^{\infty} \lambda_i e_i^*(x') e_i(x''). \tag{1.9}$$

However, the Wigner transform $w_\rho(q, p)$, as well as most of the transforms considered in the past, are not positive definite, and therefore cannot be interpreted as probability densities. It is the requirement that P_ρ should be a probability measure, and consequently that in (1.7) $\rho(q, p) \geqslant 0$, that fundamentally distinguishes the present definition of phase-space representation $\rho(q, p)$ of a density operator from those considered in the past.

It is quite obvious that the required kind of phase-space representations for one-particle systems might be obtained from the (isochronous) Galilei systems of covariance $\hat{\mathbb{P}}_\xi$ in (1.6.2) and (1.6.4) by setting

$$P_\rho^{(\xi)}(\Delta) = \mathrm{Tr}[\rho \hat{\mathbb{P}}_\xi(\Delta)], \quad \rho \in \mathscr{R}(L^2(\mathbb{R}^3)), \tag{1.10}$$

provided that the one-to-one property of the mapping $\rho \mapsto P_\rho^\xi$ can be established. On the other hand, it is equally clear that (1.10) will not be the most general type of phase-space representation of ρ compatible with (1.6). Indeed, let us define in general a *phase-space system of covariance* as a normalized POV measure $E(\Delta)$ (of operators on $\mathscr{H} = L^2(\mathbb{R}^N)$), over the Borel sets Δ in $\Gamma^N = \mathbb{R}^{2N}$ which is such that

$$U_g E(\Delta) U_g^{-1} = E(g\Delta), \quad g \in \mathscr{G}'. \tag{1.11}$$

Then we can obviously extrapolate (1.10) into

$$P_\rho(\Delta) = \mathrm{Tr}[\rho E(\Delta)], \quad \rho \in \mathscr{R}(\mathscr{H}). \tag{1.12}$$

However, the one-to-one property required of phase-space representations (1.5) of quantum statistical mechanics has to be postulated separately. Since the affine mapping (1.5) can be extended in an evident manner to a linear mapping $A \mapsto \mu_A$ of the entire trace-class $\mathscr{B}_1(\mathscr{H})$ into the set of complex measures $\mu_\rho(\Delta)$ on the Borel sets $\Delta \subset \Gamma^N$, this last request is equivalent to the property imposed on this mapping by the following definition: A phase-space system of covariance is *informationally complete* if and only if the equation

$$\mathrm{Tr}[AE(\Delta)] \equiv 0, \quad A \in \mathscr{B}_1(\mathscr{H}), \tag{1.13}$$

can be satisfied for all Borel sets Δ in Γ^N only by $A = 0$.

The following theorem (whose proof can be found in Appendix A of Ali and Prugovečki, 1977a) establishes the intimate relationship between phase-space representations of quantum statistical mechanics and phase-space systems of covariance.

THEOREM 1. *Every phase-space representation* (1.5)–(1.7) *of quantum statistical mechanics determines a unique informationally complete phase-space system of covariance for which* (1.12) *is true, and conversely to each informationally complete phase-space system of covariance we can assign by* (1.12) *a unique phase-space representation* (1.5)–(1.7) *of quantum statistical mechanics.*

Guided by the considerations in Section 1.5, we might seek to interpret $P_\rho(\Delta)$ as a probability measure on stochastic phase spaces $\underset{\sim}{\Gamma}^N$ described by (1.1.7)–(1.1.12), provided that for some choice of confidence functions χ_q' and χ_p'' we get

$$P_\rho(\Delta_1 \times \mathbb{R}^N) = \int_{\Delta_1} \mathrm{d}q \int_{\mathbb{R}^N} \mathrm{d}x\, \chi_q'(x) \langle x | \rho | x \rangle, \tag{1.14a}$$

$$P_\rho(\mathbb{R}^N \times \Delta_2) = \int_{\Delta_2} \mathrm{d}p \int_{\mathbb{R}^N} \mathrm{d}k\, \chi_p''(k) \langle k | \rho | k \rangle, \tag{1.14b}$$

where $\langle x | \rho | x \rangle$ is obtained by setting $x = x' = x''$ in (1.9), and $\langle k | \rho | k \rangle$ is obtained by setting $k = k' = k''$ in

$$\langle k' | \rho | k'' \rangle = \sum_{i=1}^{\infty} \lambda_i \tilde{e}_i^*(k') \tilde{e}_i(k''). \tag{1.15}$$

It is easily seen that the POV measures

$$E(\Delta) = \int_\Delta \chi_q'^{1/2}(\hat{Q}) \chi_p''(\hat{P}) \chi_q'^{1/2}(\hat{Q})\, \mathrm{d}q\, \mathrm{d}p \tag{1.16}$$

constructed from positive-definite functions of the position and momentum operators

$$(\hat{Q}_\alpha \psi)(x) = x_\alpha \psi(x), \; (\hat{P}_\alpha \psi)(x) = -i\hbar \frac{\partial}{\partial x_\alpha} \psi(x) \tag{1.17}$$

in $L^2(\mathbb{R}^N)$, lead to phase-space representations which satisfy one but not both marginality conditions (1.14). Thus, if we define a *stochastic phase-space representation* of quantum statistical mechanics as a phase-space representation which obeys both marginality conditions (1.14) for some choice

$$\chi_q'(x) = \chi_0'(x - q), \; \chi_p''(k) = \chi_0''(k - p), \tag{1.18}$$

of confidence function on Γ^N, then we have to conclude that not all phase-space representations are stochastic. In the next section we shall search for the essential ingredient that distinguishes the stochastic case from the general case.

3.2. Informational Equivalence and Completeness of POV Measures

The spectral measures $E^{\hat{Q}}(\Delta_1)$ and $E^{\hat{P}}(\Delta_2)$ of the position and momentum operators in (1.17), namely

$$(E^{\hat{Q}}(\Delta_1)\psi)(x) = \chi_{\Delta_1}(x)\psi(x), \; \Delta_1 \in \mathscr{B}^N, \tag{2.1a}$$

$$(E^{\hat{P}}(\Delta_2)\psi)^\sim(k) = \chi_{\Delta_2}(k)\tilde{\psi}(k), \; \Delta_2 \in \mathscr{B}^N, \tag{2.1b}$$

are POV measures on $\mathscr{H} = L^2(\mathbb{R}^N)$. These PV measures give rise to systems of covariance under the unitary ray representation U_g of the isochronous Galilei group $\mathscr{G}'$,

$$U_g E^{\hat{Q}}(\Delta_1) U_g^{-1} = E^{\hat{Q}}(g\Delta_1), \qquad \Delta_1 \subset \mathbb{R}^N_{\text{conf}} \tag{2.2a}$$

$$U_g E^{\hat{P}}(\Delta_2) U_g^{-1} = E^{\hat{P}}(g\Delta_2), \qquad \Delta_2 \subset \mathbb{R}^N_{\text{mom}} \tag{2.2b}$$

on the configuration and momentum spaces of the $N_0 = N/3$ particles, respectively. On the other hand, by (1.11), the POV measures

$$E'(\Delta_1) = E(\Delta_1 \times \mathbb{R}^N), \qquad \Delta_1 \subset \underset{\sim}{\mathbb{R}}^N_{\text{conf}} \tag{2.3a}$$

$$E''(\Delta_2) = E(\mathbb{R}^N \times \Delta_2), \qquad \Delta_2 \subset \underset{\sim}{\mathbb{R}}^N_{\text{mom}} \tag{2.3b}$$

also give rise to systems of covariance on $\underset{\sim}{\mathbb{R}}^N_{\text{conf}}$ and $\underset{\sim}{\mathbb{R}}^N_{\text{mom}}$, respectively. If these systems of covariance were to possess a physical interpretation that is related to those for $E^{\hat{Q}}$ and $E^{\hat{P}}$ in (2.1), we would expect that E' and E'' would be as informationally effective as their conventional counterparts $E^{\hat{Q}}$ and $E^{\hat{P}}$. In this context, the most natural gauge of informational effectiveness is provided by the ability of pinpointing the state ρ of the ensemble as accurately as possible on the basis of measuring all related probabilities, which in case of E' and E'' are:

$$P'_\rho(\Delta_1) = \mathrm{Tr}[\rho E'(\Delta_1)], \quad P''_\rho(\Delta_2) = \mathrm{Tr}[\rho E''(\Delta_2)]. \tag{2.4}$$

With this criterion in mind, we shall say that two POV measures $E_1(\Delta)$ and $E_2(\Delta)$ over the same family of Borel sets Δ are *informationally equivalent* if and only if whenever for $\rho', \rho'' \in \mathscr{R}(\mathscr{H})$,

$$\mathrm{Tr}[\rho' E_1(\Delta)] = \mathrm{Tr}[\rho'' E_1(\Delta)] \tag{2.5}$$

for all Δ, then we also have

$$\mathrm{Tr}[\rho' E_2(\Delta)] = \mathrm{Tr}[\rho'' E_2(\Delta)] \tag{2.6}$$

for the same pair of density operators, and vice versa: if (2.6) is true for some ρ', $\rho'' \in \mathscr{R}(\mathscr{H})$, then (2.5) is also true. In other words, two informationally equivalent POV measures possess precisely the same ability to differentiate between the states of the ensemble. We note that according to the definition of informational completeness of systems of covariance given in the preceding section, we are entitled to say that $E_1(\Delta)$ is *informationally complete* if and only if (2.5) is satisfied for all Δ only when $\rho' = \rho''$. In that case, any informationally equivalent POV measure $E_2(\Delta)$ shall also be informationally complete.

The following theorem (proven in Appendix B of Ali and Prugovečki, 1977a) provides via the medium of informational equivalence the key ingredient that distinguishes stochastic representations from all the other phase-space representations of quantum statistical mechanics of N degrees of freedom defined by (1.5).

THEOREM 2. *The marginal components* (2.3) *of the phase-space system of covariance in* (1.11) *are informationally equivalent to the sharp position and momentum systems of covariance* (2.2), *respectively, if and only if there is a stochastic phase space*

$$\underset{\sim}{\Gamma}^N = \{(q, \mu_q') \times (p, \mu_p'') \mid (q, p) \in \Gamma^N\} = \underset{\sim}{\mathbb{R}}^N_{\mathrm{conf}} \times \underset{\sim}{\mathbb{R}}^N_{\mathrm{mom}} \tag{2.7}$$

such that for all wave functions ψ *from the Schwartz space* $\mathscr{S}(\mathbb{R}^N)$ (*dense in* $L^2(\mathbb{R}^N)$) *we have*

$$\langle \psi \mid E(\Delta_1 \times \mathbb{R}^N) \psi \rangle = \int_{\Delta_1} \mathrm{d}q \int_{\mathbb{R}^N} \mathrm{d}\mu_q'(x) \mid \psi(x) \mid^2, \tag{2.8a}$$

$$\langle \psi \mid E(\mathbb{R}^N \times \Delta_2) \psi \rangle = \int_{\Delta_2} \mathrm{d}p \int_{\mathbb{R}^N} \mathrm{d}\mu_p''(k) \mid \tilde{\psi}(k) \mid^2. \tag{2.8b}$$

We note that (2.8a) and (2.8b) lead directly to (1.14a) and (1.14b), respectively, since due to Galilei covariance the confidence measure μ_q' and μ_p'' are absolutely continuous with respect to the Lebesgue measure on $\mathbb{R}^N$, so that (cf. (A.26) in Appendix A)

$$\mu_q'(\Delta') = \int_{\Delta'} \chi_q'(q')\, \mathrm{d}q', \quad \mu_p''(\Delta'') = \int_{\Delta''} \chi_p''(p')\, \mathrm{d}p'. \tag{2.9}$$

This is the situation encountered in the next section, where we shall arrive at a generalized form of the stochastic phase-space formalism of Section 1.5. In the meantime, it is important to point out that the informational equivalence of the marginal components of some phase-space POV measures to both the sharp position and the sharp momentum spectral measures $E^{\hat{Q}}$ and $E^{\hat{P}}$ in (2.1) does not secure informational completeness, since even when considered together, the probability measures

$$\mathrm{Tr}\,[\rho E^{\hat{Q}}(\Delta_1)] = \int_{\Delta_1} \langle x \mid \rho \mid x \rangle\, \mathrm{d}x \tag{2.10a}$$

$$\mathrm{Tr}\,[\rho E^{\hat{P}}(\Delta_2)] = \int_{\Delta_2} \langle k \mid \rho \mid k \rangle\, \mathrm{d}k \tag{2.10b}$$

are not capable of identifying a unique density operator ρ.

This somewhat surprising conclusion can be reached by constructing examples (Prugovečki, 1977; Vogt, 1978; Corbett and Hurst, 1978) of distinct density operators ρ_i, $i = 1, 2$, for which the probability measures in (2.10) are equal. Simple instances of $\rho_i = |\psi_i\rangle\langle\psi_i|$ for one-particle systems can be arrived at by choosing

$$\psi_1(\mathbf{x}) = |\psi_1(\mathbf{x})| \exp[i\theta(\mathbf{x})], \qquad |\psi_1(-\mathbf{x})| = |\psi_1(\mathbf{x})| \tag{2.11}$$

in such a way that $\theta(\mathbf{x}) + \theta(-\mathbf{x})$, $\theta(x) = \arg \psi_1(\mathbf{x})$, does not equal a constant modulo 2π almost everywhere in $\mathbb{R}^3$. In that case

$$\psi_2(\mathbf{x}) = |\psi_1(\mathbf{x})| \exp[-i\theta(-\mathbf{x})] = \psi_1^*(-\mathbf{x}) \tag{2.12}$$

represents a state that is distinct from the one represented by $\psi_1(\mathbf{x})$. On the other hand, by (2.12)

$$\tilde{\psi}_2(\mathbf{k}) = (2\pi\hbar)^{-3/2} \int_{\mathbb{R}^3} \exp\left(\frac{i}{\hbar}\,\mathbf{k}\cdot\mathbf{x}\right) \psi_1^*(\mathbf{x})\,d\mathbf{x} = \tilde{\psi}_1^*(\mathbf{k}), \tag{2.13}$$

so that $|\psi_1(\mathbf{x})|^2 = |\psi_2(\mathbf{x})|^2$ and $|\tilde{\psi}_1(\mathbf{k})|^2 = |\tilde{\psi}_2(\mathbf{k})|^2$, and therefore

$$\langle\psi_1 | E^{\hat{Q}}(\Delta_1)\psi_1\rangle = \langle\psi_2 | E^{\hat{Q}}(\Delta_1)\psi_2\rangle, \Delta_1 \in \mathscr{B}^3, \tag{2.14a}$$

$$\langle\psi_1 | E^{\hat{P}}(\Delta_2)\psi_1\rangle = \langle\psi_2 | E^{\hat{P}}(\Delta_2)\psi_2\rangle, \Delta_2 \in \mathscr{B}^3. \tag{2.14b}$$

The fact that $E^{\hat{Q}}$ and $E^{\hat{P}}$ lack informational completeness even when considered together has profound implications for the quantum theory of measurement: this fact implies that in general the state of an ensemble of quantum systems cannot be pinpointed unambiguously by performing exclusively sharp position and sharp momentum measurements on any parts of the ensemble (even if we assume that such measurement were feasible), and therefore that the conventional theory of quantum measurement is incomplete. Of course, one could always argue that there might be other, as yet unknown, procedures for measuring all other self-adjoint operators, but such an *ad hoc* assumption is totally unwarranted.[5] In fact, not only do superselection rules (Wick *et al.*, 1952) provide a counterexample to such *ad hoc* assumptions, but fundamental inconsistencies (Wigner, 1952; Araki and Yanasi, 1961) are reached if one assumes that sharp measurements of such very conventional observables as spin components are feasible.[6]

As we shall see in the next section, informationally complete stochastic phase-space systems of covariance do exist so that one can always pinpoint any state ρ of a quantum ensemble by measuring the probabilities $P_\rho(\Delta)$ in (1.12). Thus, stochastic phase space turns out to be as essential to the unambiguous measurement of the state of a quantum ensemble as ordinary phase space is to the measurement of a state of a classical ensemble.

By Theorem 2, the marginal components (2.3) of stochastic phase-space POV measure $E(\Delta)$ are informationally equivalent to $E^{\hat{Q}}$ and $E^{\hat{P}}$, despite the fact that the former relate to the measurement of spread-out stochastic values of position and momentum, whereas the later relate to the measurement of sharp values of the same quantities. On first sight, this might appear paradoxical. To illustrate, however, the naturalness of this feature, let us consider the probability density

$$\bar{\rho}^{\text{cl}}(q, p) = \int_{\Gamma^N} \chi_{q,p}(q', p')\rho^{\text{cl}}(q', p')\, dq'\, dp' \tag{2.15}$$

corresponding to a classical distribution function on the stochastic phase space $\underset{\sim}{\Gamma}^N$ in (1.1.7). Despite the fact that $\bar{\rho}^{\text{cl}}(q, p)$ corresponds to measurements of the stochastic spread-out values $\underset{\sim}{q} = (q, \chi_q')$ and $\underset{\sim}{p} = (p, \chi_p'')$, where $\rho^{\text{cl}}(q, p)$ is the distribution function in (1.2) corresponding to the measurement of the sharp values (q, δ^3) and (p, δ^3), the two are informationally equivalent in the sense that one can be deduced from the other. Indeed, the fact that $\rho^{\text{cl}}(q, p)$ can be derived from $\bar{\rho}^{\text{cl}}(q, p)$ for known confidence functions $\chi_{q,p}$ given by (1.1.8) and (1.1.12) can be immediately established by taking Fourier transforms of both sides of (2.15), which yield:

$$\tilde{\bar{\rho}}^{\text{cl}}(x, y) = \tilde{\chi}_{0,0}(x, y)\tilde{\rho}^{\text{cl}}(x, y), \quad x, y \in \mathbb{R}^N. \tag{2.16}$$

Naturally, in practice one cannot measure either $\rho^{\text{cl}}(q, p)$ or $\bar{\rho}^{\text{cl}}(q, p)$ exactly and at more than a limited number of sample points, and in those circumstances $\rho^{\text{cl}}(q, p)$, if actually known at those points, would provide more information about the state of the ensemble than the knowledge of $\bar{\rho}^{\text{cl}}(q, p)$ at the same values of (q, p) might be able to supply.

3.3. Generalized Stochastic Phase-Space Systems of Covariance

In Theorem 2 of the preceding section we have seen that phase-space systems of covariance, whose marginal components (2.3) are informationally equivalent to the sharp position and sharp momentum spectral measures, do indeed lead to phase-space probability measures (1.12) in stochastic phase spaces (2.7). However, these stochastic phase spaces are of a much more general type than those in Section 1.5, not only because they involve N degrees of freedom rather than just three, but also because the confidence measures in (2.7) are not necessarily related to the spectral resolution generator ξ via such relations as (1.5.2) and (2.9). In fact, if we consider the expression (1.6.2) for $\hat{\mathbb{P}}_\xi(B)$, and take note of (1.4.6b), we observe that its extrapolation to $L^2(\mathbb{R}^N)$ assumes the form

$$\mathbb{P}_\gamma(\Delta) = \int_\Delta \gamma_{q,p}\, dq\, dp, \qquad \gamma_{q,p} = U_{q,p}\gamma U_{q,p}^{-1}, \tag{3.1}$$

$$\gamma = |\xi\rangle\langle\xi|, \quad \xi = \xi_{(1)} \otimes \ldots \otimes \xi_{(N_0)}, \tag{3.2}$$

where $\xi_{(1)}, \ldots, \xi_{(N_0)} \in L^2(\mathbb{R}^3)$ satisfy (1.4.9) and (1.4.16) which equals U_g for g consisting of space translations $q = (\mathbf{q}_1, \ldots, \mathbf{q}_{N_0})$ and velocity boosts

$$(\mathbf{p}_1/m_1, \ldots, \mathbf{p}_{N_0}/m_{N_0}), \qquad p = (\mathbf{p}_1, \ldots, \mathbf{p}_{N_0}), \tag{3.3}$$

but no rotations or time translations. Thus, in accordance with (1.2.7)–(1.2.9) and (1.17):

$$U_{q,p} = \exp\left(\frac{i}{\hbar} q \cdot p\right) \exp\left(\frac{i}{\hbar} p \cdot Q\right) \exp\left(-\frac{i}{\hbar} q \cdot P\right)$$

$$= \exp\left(-\frac{i}{\hbar} q \cdot P\right) \exp\left(\frac{i}{\hbar} p \cdot Q\right) = \exp\left[\frac{i}{\hbar}\left(p \cdot Q - q \cdot P + \frac{1}{2} q \cdot p\right)\right], \tag{3.4}$$

$$q \cdot p = \sum_{\alpha=1}^{N} q_\alpha p_\alpha,\ q \cdot P = \sum_{\alpha=1}^{N} q_\alpha P_\alpha,\ p \cdot Q = \sum_{\alpha=1}^{N} p_\alpha Q_\alpha. \tag{3.5}$$

It is easy to check that in fact $\mathbb{P}_\gamma(\Delta)$ in (3.1) satisfies the covariance condition (1.11) if the *generalized resolution generator* γ is rotationally invariant, i.e. commutes with U_g for all $g \in SO(3)$. Hence, by definition, a *generalized stochastic phase-space system of covariance* is a phase-space system of covariance whose POV measure $E(\Delta)$ is of the form (3.1), where γ is a rotationally invariant positive trace-class operator of trace $(2\pi\hbar)^{-N}$. The restriction to trace-class operators is physically desirable, since by (1.4) we can then write γ in the form

$$\gamma = \sum_{i=1}^{\infty} |\xi_i\rangle\, \lambda_i\, \langle \xi_i|, \qquad \|\xi_i\| = (2\pi\hbar)^{-N/2}, \tag{3.6}$$

where $\xi_i \in L^2(\mathbb{R}^N)$, $i = 1, 2, \ldots$, are mutually orthogonal, so that

$$P_\rho(\Delta) = \sum_{i=1}^{\infty} \lambda_i \int_\Delta \langle U_{q,p}\xi_i | \rho U_{q,p}\xi_i\rangle\, dq\, dp. \tag{3.7}$$

In other words, in this case the probability of a reading with $(q, p) \in \Delta$ is a superposition of probabilities (1.5.11) and (1.6.1) for measurements with microdetectors of proper wave function ξ. This kind of superposition is exactly what we expect to obtain if the microdetectors are extended objects with an internal structure (e.g. atoms or molecules), and λ_i are the probabilities for the various internal states in which the microdetector could be found prior to the detection process.

The POV measures (3.1) belong to the general class of POV measures $E(\Delta)$ with a *spectral density*

$$F(q, p) = U^*_{q,p} F(0, 0) U_{q,p}, \tag{3.8}$$

i.e. of POV measures which are Bochner integrals of the following form:

$$E(\Delta) = \int_{\Delta} F(q, p)\, dq\, dp. \tag{3.9}$$

The next theorem (whose proof can be found in Appendix B of Ali and Prugovečki, 1977a) reveals that generalized phase-space systems of covariance can be unambiguously characterized by the existence of a spectral density (3.8), and by the informational equivalence of their marginal components to $E^{\hat{Q}}$ and $E^{\hat{P}}$ in (2.1).

THEOREM 3. *A phase-space system of covariance* $E(\Delta)$ *with a spectral density* (3.8) *is a generalized stochastic phase-space system of covariance if and only if its marginal components* (2.3) *are informationally equivalent to the respective sharp position and sharp momentum spectral measures* (2.2). *When that is the case, then*

$$P_\rho(\Delta_1 \times \mathbb{R}^N) = \mathrm{Tr}[\rho E'(\Delta_1)] = \int_{\Delta_1} dq \int_{\mathbb{R}^N} \chi_q'(x) \langle x | \rho | x \rangle, \tag{3.10a}$$

$$P_\rho(\mathbb{R}^N \times \Delta_2) = \mathrm{Tr}[\rho E''(\Delta_2)] = \int_{\Delta_2} dp \int_{\mathbb{R}^N} \chi_p''(k) \langle k | \rho | k \rangle, \tag{3.10b}$$

where both families of confidence functions χ_q' *and* χ_p'' *are translationally invariant,*

$$\chi_q'(x) = \chi_0'(x - q),\ \chi_p''(k) = \chi_0''(k - p), \tag{3.11}$$

and are related to the trace-class operators $\gamma_{q,p} = F(q, p)$ *as follows:*

$$\chi_q'(x) = (2\pi\hbar)^N \langle x | \gamma_{q,p} | x \rangle,\ \chi_p''(k) = (2\pi\hbar)^N \langle k | \gamma_{q,p} | k \rangle. \tag{3.12}$$

The above theorem can be restated from the point of view of associating with each density operator ρ the probability density

$$\rho(q, p) = \mathrm{Tr}[\rho F(q, p)]. \tag{3.13}$$

Pending such an association, we can ask the question as to which are the families of operator-valued functions (3.8), that for some choice of confidence functions $\chi_q'(x)$ and $\chi_p''(k)$ satisfying (3.11) yield densities (3.13) for which

$$\int_{\mathbb{R}^N} \rho(q, p)\, dp = \int_{\mathbb{R}^N} \chi_q'(x) \langle x | \rho | x \rangle\, dx, \tag{3.14a}$$

$$\int_{\mathbb{R}^N} \rho(q, p)\, dq = \int_{\mathbb{R}^N} \chi_p''(k) \langle k | \rho | k \rangle\, dk. \tag{3.14b}$$

Wigner's (1932, 1979) theorem answers the question for the case that

$$\chi'_q(x) = \delta^N(x-q),\ \chi''_p(k) = \delta^N(k-p), \tag{3.15}$$

by affirming that there are no operator-valued functions (3.8) with such properties. However, Theorem 3 implicitly confirms that there are no such $F(q, p)$ even if only one of the two conditions in (3.15) is imposed. In effect, Theorem 3 states that (3.14) can be satisfied if and only if $F(0, 0)$ is an operator of trace $(2\pi\hbar)^{-N}$, and furthermore

$$\chi'_0(x) = (2\pi\hbar)^N \langle x|\, F(0,0)\, |x\rangle,\ \chi''_0(k) = (2\pi\hbar)^N \langle k|\, F(0,0)\, |k\rangle. \tag{3.16}$$

This last observation sheds some light on an otherwise puzzling dichotomy (Prugovečki, 1967) in the interpretation of the uncertainty relations: on one hand, in Heisenberg's (1930) *gedanken* experiments, Δx^j and Δk^j are interpreted as uncertainties in the simultaneous measurement of the position and momentum of a *single* particle; whereas, on the other hand, the textbook derivation (cf., e.g., Prugovečki, 1981a, p. 330) of these relations from the CCRs is predicated on the interpretation of Δx^j and Δk^j as the mean square standard deviations of outcomes of, respectively, sharp position and sharp momentum measurements on respective *ensembles* of particles that are in one and the same quantum state. In other words, in the first instance we are dealing with limitations on the spreads of confidence functions resulting from the accuracy calibration of any instrument for the simultaneous measurement of position and momentum (i.e. with the interpretation (1.5.9) of Δx^j and Δk^j), whereas in the second case we are dealing with properties displayed by ensembles of identically prepared quantum systems when each of its members is subjected to perfectly accurate measurements of position *or* momentum. That the same uncertainty relations have to hold in both cases becomes clear from (3.16). Indeed, due to the fact that $F(0, 0)$ is of trace class, the minimal spreads of $\chi'_q(x)$ and $\chi''_p(k)$ occur when $F(0, 0)$ is of the form $|\xi\rangle\langle\xi|$. But then (1.5.2) is in effect, and the aforementioned *mathematical* derivation of the uncertainty relations from the CCRs applies. However, in this application Δx^j and Δk^j have a *physical* meaning that is different from the one in case of sharp measurement on ensembles, namely that of standard deviations for confidence functions that describe uncertainties in every single measurement of position and momentum – such as those examined by Busch (1982).

Although a generalized stochastic phase-space system of covariance supplies probability measures (3.7) which can be physically interpreted as probability measures on the stochastic phase space

$$\underset{\sim}{\Gamma}^N = \{(q, \chi'_q) \times (p, \chi''_p) \mid (q, p) \in \Gamma^N\}, \tag{3.17}$$

we still require the informational completeness of its POV measure $E(\Delta)$ in order to arrive at a phase-space representation (1.5) of quantum statistical mechanics on the

Liouville space $\mathscr{L}(\mathscr{H})$ for N degrees of freedom. The following theorem (proven in Appendix A of Ali and Prugovečki, 1977b) provides a necessary and sufficient condition for this to be the case.

THEOREM 4. *A POV measure on the Borel sets Δ in Γ^N which is of the form*

$$\mathbb{P}_\gamma(\Delta) = \int_\Delta U^*_{q,p}\gamma U_{q,p}\, dq\, dp, \qquad \mathrm{Tr}\,\gamma = (2\pi\hbar)^{-N} \tag{3.18}$$

is informationally complete if and only if the Weyl transform[7]

$$\gamma_W(q,p) = \mathrm{Tr}[\gamma U_{q,p}] \tag{3.19}$$

of the generalized resolution generator γ is different from zero for almost all $(q,p)\in\Gamma$.

We note that if $\gamma = |\xi\rangle\langle\xi|$, $\xi\in L^2(\mathbb{R}^N)$, then

$$\gamma_W(q,p) = \langle\xi \mid U_{q,p}\xi\rangle, \tag{3.20}$$

so that $\gamma^*_W(q,p)$ coincides with the $L^2(\Gamma^N)$ counterpart $\bar{\xi}(q,p)$ of the stochastic phase-space representative (1.7.1) of a resolution generator (i.e. proper wave function in the gauge $\omega = 0$ of Section 1.7). The entire theory of Sections 1.6–1.12 can be then generalized in a most straightforward manner to N degrees of freedom. For example, in place of (1.6.9) we shall be dealing with

$$K_\xi(q,p;q',p') = \langle U_{q,p}\xi \mid U_{q',p'}\xi\rangle, \tag{3.21}$$

where by (3.4) we have in general

$$(U_{q,p}\psi)(q,p) = \exp\left[\frac{i}{\hbar}\, p\cdot(x-q)\right]\psi(x-q). \tag{3.22}$$

Thus, if ξ is as in (3.2), then

$$K_\xi(q,p;q',p') = \prod_{\beta=1}^{N_0} K_{\xi(\beta)}(\mathbf{q}_\beta,\mathbf{p}_\beta;\mathbf{q}'_\beta,\mathbf{p}'_\beta), \tag{3.23}$$

where $K_{\xi(\beta)}$ are the one-particle reproducing kernels of Section 1.6. For example, in the optimal case each $\xi_{(\beta)}(\mathbf{x}_\beta)$ is of the form (1.6.25) for some value of $l_\beta > 0$, and therefore each factor in (3.23) is of the form (1.6.31). Thus, we see immediately that in this case $\gamma_W(q,p) \neq 0$ at all $(q,p)\in\Gamma^N$, and we have informational completeness for $\mathbb{P}_\gamma(\Delta)$ in (3.18). Hence, each of the stochastic phase-space probability densities

$$\rho^{(l)}(q,p) = \langle U_{q,p}\xi^{(l)} | \rho U_{q,p}\xi^{(l)}\rangle, \tag{3.24}$$

corresponding to some specific choices of $\xi^{(l_\beta)} \in L^2(\mathbb{R}^3)$ in

$$\xi^{(l)}(x) = \xi^{(l_1)}(x_1) \ldots \xi^{(l_{N_0})}(\mathbf{x}_{N_0}) \in L^2(\mathbb{R}^N), \tag{3.25}$$

unambiguously characterizes a density operator ρ as the one and only $\rho \in \mathscr{R}(\mathscr{H})$, $\mathscr{H} = L^2(\mathbb{R}^N)$, for which, when $\gamma = |\xi^{(l)}\rangle\langle\xi^{(l)}|$, we have

$$\operatorname{Tr}[\rho \mathbb{P}_\gamma(\Delta)] = \int_\Delta \rho^{(l)}(q, p)\, dq\, dp \tag{3.26}$$

for all Borel sets $\Delta \subset \Gamma$. In this context, it is interesting to note that if we set, in terms of (1.6.27),

$$\chi_{q,p}^{(l)}(q', p') = \prod_{\beta=1}^{N} \chi_{\mathbf{q}_\beta}^{(l_\beta)}(\mathbf{q}'_\beta)\overline{\chi}_{\mathbf{p}_\beta}^{(l_\beta)}(\mathbf{p}'_\beta), \tag{3.27}$$

then we can verify by explicit computation that

$$\rho^{(l)}(q, p) = \int_{\Gamma^N} \chi_{q,p}^{(l)}(q', p') w_\rho(q', p')\, dq'\, dp', \tag{3.28}$$

where w_ρ is the Wigner transform (1.8). The formal analogy between (2.15) and (3.28) is unmistakable, but the difference at a deeper level is just as great: in (2.15) the classical distribution function $\rho^{\text{cl}}(q, p)$ is a true probability density, whereas in (3.28) the Wigner transform is not positive definite, and therefore it is certainly not a probability density.

The POV measures (3.18) constructed from $\xi^{(l)}$ in (3.25) are by no means the only informationally complete ones. For example, using the techniques of Section 3.6, one can compute K_ξ in (3.23) for a ξ of the form (3.2) for the case that $\xi_{(1)}, \ldots, \xi_{(N_0)}$ are excited states of the nonrelativistic harmonic oscillator, by starting with K_ξ corresponding to (3.25), and therefore constructed out of the ground states of the same harmonic oscillator. Such a computation would reveal that $\gamma_W(q, p) \neq 0$ for almost all $(q, p) \in \Gamma^N$, so that in these cases, as well as in many others, we would have informational completeness. By Theorem 2, in all such cases we would also have informational equivalence between the marginal components (2.3) of these POV measures, and the sharp position and momentum measures in (2.1). Thus, since, as shown in Section 3.2, the latter cannot pinpoint all quantum states unambiguously, neither can the former. Hence we arrive at the general conclusion that by *separate* measurements of (sharp or spread-out) values of stochastic position and stochastic momentum on the elements of an ensemble we cannot pinpoint in general the state ρ of that ensemble, but that ρ can be always determined unambiguously by the *simultaneous* measurement of stochastic position and momentum, such as the values in the spaces

$$\Gamma_\xi^N(l) = \Gamma^{(l_1)} \times \cdots \times \Gamma^{(l_{N_0})} \tag{3.29}$$

corresponding to the optimal confidence functions in (1.6.27). Of course, this determination of ρ presupposes that probabilities, such as those in (3.26), can be inferred with total exactitude from measurements on finite samples – but this kind of question is part of a general problem that faces the probability theory as a whole (Fine, 1973).

3.4. Classical and Quantum Mechanics in $L^2(\Gamma^N)$

In the preceding section we arrived at the conclusion that there are stochastic phase spaces $\underset{\sim}{\Gamma}^N = \Gamma_\gamma^N$, such as those in (3.29), which afford a unique description of the state $\rho \in \mathscr{R}(\mathscr{H})$ of any quantum ensemble by means of a distribution function

$$\rho^\gamma(q, p) = \mathrm{Tr}(\rho\gamma_{q,p}), \quad \gamma_{q,p} = U_{q,p}^* \gamma U_{q,p}, \tag{4.1}$$

on Γ_γ^N. This description of states of quantum ensembles is therefore of the same mathematical and physical nature as that of classical ensembles by means of the distribution functions $\bar{\rho}^{\mathrm{cl}}(q, p)$ in (2.15) on $\underset{\sim}{\Gamma}^N$ – with the distribution functions $\rho^{\mathrm{cl}}(q, p)$ on Γ^N being special cases corresponding to confidence functions $\chi_{q,p}$ that are δ-functions at $(q, p) \in \Gamma^N$. However, in order to arrive at a truly unified mathematical framework for classical and quantum statistical mechanics, we also have to bridge the gap between the disparate manners on which the observables and the dynamics are specified in these two theories. Thus, in the classical case the observables are presumed to be functions $A^{\mathrm{cl}}(q, p)$ of the phase-space variables in (1.1), and their expectation (i.e. mean) values in a state ρ^{cl} are given by:

$$\langle A^{\mathrm{cl}} \rangle = \int_{\Gamma^N} A^{\mathrm{cl}}(q, p)\rho^{\mathrm{cl}}(q, p)\, \mathrm{d}q\, \mathrm{d}p. \tag{4.2}$$

On the other hand, in the quantum case the observables are represented by self-adjoint operators A in $\mathscr{H} = L^2(\mathbb{R}^N)$, and their expectation values are:

$$\langle A \rangle = \mathrm{Tr}(\rho A), \quad \rho \in \mathscr{R}(\mathscr{H}). \tag{4.3}$$

For example, for one and the same type of model with potential interaction, the classical Hamiltonian is

$$H^{\mathrm{cl}}(q, p) = \sum_{\alpha=1}^{N} \frac{p_\alpha^2}{2m_\alpha} + V(q), \tag{4.4}$$

where m_α is the mass of the βth particle for $\alpha = 3\beta - 2, 3\beta - 1, 3\beta$, whereas the quantum Hamiltonian is the Schrödinger operator

$$\hat{H} = \sum_{\alpha=1}^{N} \frac{\hat{P}_\alpha^2}{2m_\alpha} + V(\hat{Q}), \tag{4.5}$$

with $\hat{Q}_\alpha$ and $\hat{P}_\alpha$ defined by some representations of the CCRs, such as that in (1.17). Despite the formal similarities of (4.4) and (4.5), which are primarily due to the choice of notation, their mathematical natures as, respectively, real function and self-adjoint operator, are rather different. That fact is reflected by the roles they play in the dynamics.

Thus, in the classical case (4.4) gives rise via the solution $q_\alpha(t)$ and $p_\alpha(t)$ of the Hamilton equations

$$\dot{q}_\alpha = \frac{\partial H^{\text{cl}}}{\partial p_\alpha} = \frac{p_\alpha}{m_\alpha}, \qquad \dot{p}_\alpha = -\frac{\partial H^{\text{cl}}}{\partial q_\alpha} = -\frac{\partial V}{\partial q_\alpha}, \tag{4.6}$$

to a time-dependent distribution function

$$\rho_t^{\text{cl}}(q, p) = \rho_0^{\text{cl}}(q(-t), p(-t)), \qquad q = q(0), p = p(0), \tag{4.7}$$

that obeys the Liouville equation

$$\partial_t \rho_t(q, p) = L^{\text{cl}} \rho_t(q, p), \quad L^{\text{cl}} = L_0^{\text{cl}} + L_I^{\text{cl}}. \tag{4.8}$$

In this equation Liouville operators L_0^{cl} and L_I^{cl} for the free and interaction part make their appearance:

$$L_0^{\text{cl}} = -\sum_\alpha \frac{p_\alpha}{m_\alpha} \frac{\partial}{\partial q_\alpha}, \qquad L_I^{\text{cl}} = \sum_\alpha \frac{\partial V(q)}{\partial q_\alpha} \frac{\partial}{\partial p_\alpha}. \tag{4.9}$$

On the other hand, in the quantum case (4.5) gives rise via the Schrödinger equation to a time-dependent density operator

$$\rho_t = \exp\left(-\frac{i}{\hbar}\hat{H}t\right) \rho_0 \exp\left(\frac{i}{\hbar}\hat{H}t\right) \tag{4.10}$$

that obeys the von Neumann equation

$$i\hbar\partial_t \rho_t = [\hat{H}, \rho_t], \quad \hat{H} = \hat{H}_0 + \hat{H}_I, \tag{4.11}$$

in which no new operators distinct from the Hamiltonian appear, but rather

$$\hat{H}_0 = \sum_\alpha \frac{\hat{P}_\alpha^2}{2m_\alpha}, \quad \hat{H}_I = V(\hat{Q}). \tag{4.12}$$

Thus, at this level almost all the formal similarities between classical and quantum statistical mechanics seem to have been lost. In the remainder of this chapter we shall show how they can be recovered by working in a master Liouville space for N degrees of freedom (Ali and Prugovečki, 1977b, c; Prugovečki, 1978a, b, c) which coincides with the Hilbert–Schmidt class over $L^2(\Gamma^N)$. In fact, the structural similarities between classical and quantum statistical mechanics that emerge from this approach are rather remarkable: classical expectation values can be written in the form (4.3), and quantum values in the form (4.2), for suitable choices of operators A^{cl} in (4.3) and functions $A(q, p)$ in (4.2).

The first step in achieving this structural unification consists of embedding both classical and quantum mechanics for N degrees of freedom into the same Hilbert space $L^2(\Gamma^N)$ of wave function $\psi(q, p)$ with inner product

$$\langle \psi_1 | \psi_2 \rangle_\Gamma = \int_{\Gamma^N} \psi_1^*(q, p) \psi_2(q, p) \, \mathrm{d}q \, \mathrm{d}p. \tag{4.13}$$

For the quantum case, this is achieved by extending to N degrees of freedom the unitary transformation W_ξ in (1.4.6). Thus, working with an arbitrary gauge ω, we define the unitary mapping

$$\begin{aligned} W_\xi^{(\omega)} &: \psi(x) \mapsto \psi(q, p) \\ &= \int_{\mathbb{R}^N} \exp\left\{\frac{i}{\hbar}\,[p \cdot (q - x) - \omega(q, p)]\right\} \xi(x - q) \psi(x) \, \mathrm{d}x \end{aligned} \tag{4.14}$$

of $L^2(\mathbb{R}^N)$ onto $L^2(\Gamma^N_{\xi\omega})$. In the Hilbert space $L^2(\Gamma^N_{\xi\omega})$ the position and momentum operators will be represented in a fashion similar to that in (1.7.22), namely by restrictions to $L^2(\Gamma^N_{\xi\omega})$ of the following operators (densely) defined on all of $L^2(\Gamma^N)$:

$$Q_\alpha^{(\omega)} = i\hbar \frac{\partial}{\partial p_\alpha} + q_\alpha + \frac{\partial \omega}{\partial p_\alpha}, \tag{4.15a}$$

$$P_\alpha^{(\omega)} = -i\hbar \frac{\partial}{\partial q_\alpha} - \frac{\partial \omega}{\partial q_\alpha}. \tag{4.15b}$$

Consequently, the total Hamiltonian in (4.5) will be represented by the corresponding restriction of

$$H^{(\omega)} = \sum_\alpha \frac{P_\alpha^{(\omega)2}}{2m_\alpha} + V(Q^{(\omega)}). \tag{4.16}$$

If we assume that $V(q)$ in (4.4) is an entire function, so that at all $q \in \mathbb{R}^N$

$$V(x) = \sum_{|n|=0}^{\infty} \frac{1}{n!} \frac{\partial^n V(q)}{\partial q^n} (x - q)^n, \tag{4.17a}$$

$$(x - q)^n = (x_1 - q_1)^{n_1} \ldots (x_N - q_N)^{n_N}, \; n = (n_1, \ldots, n_N), \tag{4.17b}$$

$$\frac{\partial^0 V}{\partial q^0} = V(q), \frac{\partial^n V}{\partial q^n} = \frac{\partial^{|n|} V}{\partial q_1^{n_1} \ldots \partial q_N^{n_N}}, \; |n| > 0, \tag{4.17c}$$

$$|n| = n_1 + \ldots + n_N, \; n! = n_1!, \ldots, n_N!, \tag{4.17d}$$

then in the gauge $\omega = 0$ we get

$$H^{(0)} = -\sum_{\alpha} \frac{\hbar^2}{2m_\alpha} \frac{\partial^2}{\partial q_\alpha^2} + \sum_{|n|=0}^{\infty} \frac{(i\hbar)^{|n|}}{n!} \frac{\partial^n V(q)}{\partial q^n} \frac{\partial^n}{\partial p^n}. \tag{4.18}$$

Thus, to the zeroth order in $|n|$, (4.18) has the formal appearance of a Schrödinger operator.

Viewed in global terms, a pure quantum state is represented on $L^2(\Gamma)$ by an equivalence class of wave functions: in each $L^2(\Gamma_{\xi\omega})$, this equivalence class has (in the absence of superselection rules) one representative modulo a multiplicative constant of absolute value 1, and the operators $W_\xi^{(\omega_2)} W_\xi^{(\omega_1)-1}$ map a representative $\psi_1 \in L^2(\Gamma_{\xi\omega_1})$ into another $\psi_2 \in L^2(\Gamma_{\xi\omega_2})$.

An equivalence-class type of description in $L^2(\Gamma^N)$ can be also achieved for a classical distribution function $\rho^{\text{cl}}(q, p)$ if we assign to it all $\psi^{\text{cl}} \in L^2(\Gamma^N)$ such that

$$|\psi^{\text{cl}}(q, p)|^2 = \rho^{\text{cl}}(q, p), \quad q, p \in \Gamma. \tag{4.19}$$

It is then very easy to check that if we set

$$\psi_t^{\text{cl}} = \exp\left(-\frac{i}{\hbar} H^{\text{cl}} t\right) \psi_0^{\text{cl}}, \quad H^{\text{cl}} = H_0^{\text{cl}} + H_I^{\text{cl}}, \tag{4.20}$$

where H^{cl} is given by the self-adjoint operators

$$H_0^{\text{cl}} = -i\hbar \sum_{\alpha} \frac{p_\alpha}{m_\alpha} \frac{\partial}{\partial q_\alpha}, \qquad H_I^{\text{cl}} = i\hbar \sum_{\alpha} \frac{\partial V}{\partial p_\alpha} \frac{\partial}{\partial p_\alpha}, \tag{4.21}$$

then the time evolution of $|\psi_t^{\text{cl}}(q, p)|^2$ coincides with that of $\rho_t^{\text{cl}}(q, p)$ in (4.7). Indeed, by (4.20)

$$\partial_t \psi_t^{\text{cl}}(q, p) = \sum_{\alpha} \left(\frac{\partial V}{\partial q_\alpha} \frac{\partial}{\partial p_\alpha} - \frac{p_\alpha}{m_\alpha} \frac{\partial}{\partial q_\alpha} \right) \psi_t^{\text{cl}}(q, p), \tag{4.22}$$

and (4.22) implies that $|\psi_t^{\text{cl}}(q, p)|^2$ satisfies (4.8) with L^{cl} given by (4.9). Naturally, the appearance of $\hbar$ in (4.20) and (4.21) is a matter of notational convention, and it is aimed at facilitating comparisons of classical and quantum models, in which the classical counterparts of certain expressions can be expected to be a kind of first approximations in powers of $\hbar$ of their quantum counterpart.

A good example of this feature is obtained if we adopt in (4.15) and (4.16) the gauge $\omega = -q \cdot p$, and assume that $V(x)$ is given by (4.17). We then immediately get

$$H^{(\omega)} = H^{\text{cl}} + \left[\sum_{\alpha} \frac{p_\alpha^2}{2m_\alpha} + V(q) \right] - \sum_{\alpha} \left(\frac{\hbar^2}{2m_\alpha} \frac{\partial^2}{\partial q_\alpha} + q_\alpha \frac{\partial V}{\partial q_\alpha} \right) + \\ + \sum_{|n| \geqslant 2} \frac{1}{n!} \frac{\partial^n V}{\partial q^n} \left(i\hbar \frac{\partial}{\partial p} - q \right)^n. \tag{4.23}$$

We note that in (4.23) H^{cl}, as given by (4.21), contains for $|n| \leqslant 1$ the lowest power of $\hbar$, except for the term in the square bracket and the second term in the first round bracket. It is easy to see, however, that these last two terms do not contribute to the time evolution of $|\psi_t(q, p)|^2$. Rather, they contribute only a physically insignificant phase factor to $\psi_t(q, p)$, since they contain no partial derivatives in q or p.

In view of (4.19), we see that the classical probability (1.2) can be written as an expectation value

$$P^{\text{cl}}(\Delta) = \langle \psi | E^{\text{cl}}(\Delta) \psi \rangle, \tag{4.24}$$

$$(E^{\text{cl}}(\Delta)\psi)(q, p) = \chi_\Delta(q, p)\psi(q, p), \tag{4.25}$$

of the joint spectral measure of the classical position and momentum operators Q^{cl}_α and P^{cl}_α:

$$(Q^{\text{cl}}_\alpha \psi)(q, p) = q_\alpha \psi(q, p), \; (P^{\text{cl}}_\alpha \psi)(q, p) = p_\alpha \psi(q, p). \tag{4.26a}$$

Thus, once the classical theory has been embedded in $L^2(\Gamma^3)$, the stochastic position operators in (1.8.29) assume a new meaning: they represent sharp position in the classical approximation of the quantum theory. This result is, of course, completely in accord with the results in Section 1.11 on the semiclassical approximation of quantum propagation. We note also that the classical observables in (4.2) can be now represented by the operators $A^{\text{cl}}(Q^{\text{cl}}, P^{\text{cl}})$, since by (4.13), (4.19) and (4.26a) we can rewritten (4.2) as

$$\langle A^{\text{cl}} \rangle = \langle \psi^{\text{cl}} \,|\, A^{\text{cl}}(Q^{\text{cl}}, P^{\text{cl}})\, \psi^{\text{cl}} \rangle_\Gamma. \tag{4.26b}$$

It might appear that the present unified framework of classical and quantum mechanics in $L^2(\Gamma^N)$ still confronts us with a conceptual dichotomy, since on the quantum level it deals with wave functions that can describe[8] individual systems, and not just ensembles, whereas it might appear that due to (4.19) the wave functions $\psi^{\text{cl}}_t(q, p)$ can describe only classical ensembles represented by a distribution function $\rho^{\text{cl}}_t(q, p)$. However, it should be recalled that the quantum wave functions in (4.14) refer to spread-out stochastic values (q, χ'_q) and (p, χ''_p). Hence, let us consider instead of distributions at sharp values $(q, p) \in \Gamma^N$, the distribution function (2.15) at spread-out stochastic values $(q, p, \chi_{q,p}) \in \underset{\sim}{\Gamma}^N$, where the choices of confidence functions (1.1.12) are not restricted – at least in our minds – by the imposition of uncertainty relations. A classical system following the trajectory $(\bar{q}(t), \bar{p}(t))$ in phase space can then be equally well described by

$$\rho^{\text{cl}}_t(q', p') = \delta^{2N}(q' - \bar{q}(t), \quad p' - \bar{p}(t)), \tag{4.27}$$

or, alternatively,[9] by the distribution function

$$\bar{\rho}^{\text{cl}}_t(q, p) = \chi_{0,0}(q - \bar{q}(t), \quad p - \bar{p}(t)) \tag{4.28}$$

obtained by inserting (4.27) into (2.15). Since $\bar{q}_\alpha(t)$ and $\bar{p}_\alpha(t)$ satisfy Hamilton's equations (4.6), $\bar{\rho}_t^{\mathrm{cl}}(q, p)$ satisfies the Liouville equation (4.8). Consequently, any $\bar{\psi}_t^{\mathrm{cl}}(q, p)$ which evolves according to (4.20) and satisfies the relation

$$\bar{\rho}_t^{\mathrm{cl}}(q, p) = |\bar{\psi}_t^{\mathrm{cl}}(q, p)|^2, \tag{4.29}$$

at $t = 0$, will satisfy it at all times t, and therefore can be viewed as the wave function of a single classical system in the stochastic phase space $\underset{\sim}{\Gamma}^N$ in (1.1.7). This conclusion is very much in keeping with that reached by Born (1955c), namely that many apparent epistemic barriers separating classical and quantum descriptions of the physical universe melt away, once the basic fact that no sharp (deterministic) values of observable quantities with continuous spectra ever result from *any* measurement process is injected into the classical framework.

3.5. Classical and Quantum Statistical Mechanics in Master Liouville Space $\mathscr{L}(\Gamma^N)$

In the preceding section we have shown that classical as well as quantum systems with N degrees of freedom can be represented in $L^2(\Gamma^N)$. Their states are then described by equivalence classes of wave functions $\psi(q, p)$ in $L^2(\Gamma^N)$, for which $|\psi(q, p)|^2$ represents the probability density for observing stochastic values $(\underset{\sim}{q}, \underset{\sim}{p})$ in some stochastic phase space $\underset{\sim}{\Gamma}^N$. Consequently, it naturally follows that density operators $\rho \in \mathscr{R}(\mathscr{H})$, $\mathscr{H} = L^2(\underset{\sim}{\Gamma}^N)$, can be used to described the states of ensembles of such (classical or quantum) systems. The general representation (3.4) of these density operators indicates that the diagonal values

$$\langle q, p|\, \rho\, |q, p\rangle = \sum_{i=1}^{\infty} \lambda_i\, |\bar{e}_i(q, p)|^2, \qquad \bar{e}_i \in L^2(\Gamma^N), \tag{5.1}$$

of density 'matrices' (Prugovečki, 1981a, pp. 401–404)

$$\langle q, p|\, \rho\, |q', p'\rangle = \sum_{i=1}^{\infty} \lambda_i \bar{e}_i(q, p) \bar{e}_i^*(q', p') \tag{5.2}$$

constructed from wave functions $\bar{e}_i(q, p)$ belonging to the aforementioned equivalence classes are the actual objects of physical significance. Hence we can embed the classical statistical mechanics for N degrees of freedom into the Liouville space $\mathscr{L}(\Gamma^N)$ (i.e. the Hilbert–Schmidt class $\mathscr{B}_2(\mathscr{H})$ over $\mathscr{H} = L^2(\Gamma^N)$) by allowing a state of a classical ensemble described by the distribution function $\rho^{\mathrm{cl}}(q, p)$ to be represented by the equivalence class of all $\rho^{\mathrm{cl}} \in \mathscr{R}(\mathscr{H})$ for which

$$\langle q, p|\, \rho^{\mathrm{cl}}\, |q, p\rangle = \rho^{\mathrm{cl}}(q, p). \tag{5.3}$$

We note that each such equivalence class will then contain all $\rho^{\mathrm{cl}} = |\psi^{\mathrm{cl}}\rangle\langle\psi^{\mathrm{cl}}|$ with $\psi^{\mathrm{cl}}(q, p)$ that satisfies (4.19), as well as many additional density operators which are not of this form. Moreover, the classical observables in (4.26b) can be represented now by superoperators $\underset{\sim}{A}^{\text{cl}} = \frac{1}{2}\{A^{\text{cl}}, \cdot\}$ in $\mathscr{L}(\Gamma^N)$, since we have

$$\langle A^{\text{cl}} \rangle = \text{Tr}(\underset{\sim}{A}^{\text{cl}} \rho), \tag{5.4a}$$

$$\underset{\sim}{A}^{\text{cl}} \rho = \frac{1}{2} [A^{\text{cl}}(Q^{\text{cl}}, P^{\text{cl}})\rho + \rho A^{\text{cl}}(Q^{\text{cl}}, P^{\text{cl}})]. \tag{5.4b}$$

To arrive at the equivalence class of density operators in the master Liouville space $\mathscr{L}(\Gamma^N)$ that represent a state of a quantum ensemble, we have to construct the representatives $\underline{W}_\gamma^{(\omega)}\rho$ of the density operator ρ in the configuration Liouville space $\mathscr{L}(\mathbb{R}^N)$ for all generalized resolution generators γ that give rise to generalized stochastic phase-space systems of covariance. Since any $A \in \mathscr{L}(\Gamma^N)$ can be represented in a manner analogous to (1.3a), namely as

$$\bar{A} = \sum_{i=1}^{\infty} |\bar{e}_i'\rangle \lambda_i \langle \bar{e}_i''|, \quad \bar{e}_i', \bar{e}_i'' \in L^2(\Gamma^N), \tag{5.5}$$

we immediately conclude that if $\gamma = |\xi\rangle\langle\xi|$, then we shall have

$$\underline{W}_\gamma^{(\omega)} : A \mapsto \bar{A} = W_\xi^{(\omega)} A W_\xi^{(\omega)-1}, \quad A \in \mathscr{L}(\mathbb{R}^N), \tag{5.6}$$

where $W_\xi^{(\omega)}$ is given by (4.14). However, in order to deal with the general type of γ considered in Section 3.3, it is best to construct $\underline{W}_\gamma^{(\omega)}$ directly at the level of Liouville space by Hilbert–Schmidt methods.

To do that, we have to recall that the Liouville spaces $\mathscr{L}(\mathbb{R}^N)$ and $\mathscr{L}(\Gamma^N)$ are Hilbert spaces in their own right, with inner products given by (1.3c). Furthermore, they can be identified (Prugovečki, 1981a, p. 397) with the spaces $L^2(\mathbb{R}^N \times \mathbb{R}^N)$ and $L^2(\Gamma^N \times \Gamma^N)$, respectively, by assigning to the Hilbert–Schmidt operators (1.3a) and (5.5) the respective L^2–functions

$$\langle x'|A|x''\rangle = \sum_{i=1}^{\infty} \lambda_i e_i'(x') e_i''^*(x''), \tag{5.7a}$$

$$\langle q', p'|\bar{A}|q'', p''\rangle = \sum_{i=1}^{\infty} \lambda_i \bar{e}_i'(q', p') \bar{e}_i''^*(q'', p''). \tag{5.7b}$$

Let us introduce now in $\mathscr{L}(\mathbb{R}^N)$ the superoperators $\underline{U}_\omega(q', p'; q'', p'')$ by setting

$$\underline{U}_\omega(q', p'; q'', p'')A = U_\omega(q', p')AU_\omega^*(q'', p''), \tag{5.8a}$$

$$U_\omega(q, p) = \exp\left(\frac{i}{\hbar} p \cdot Q\right) \exp\left(-\frac{i}{\hbar} q \cdot P\right). \tag{5.8b}$$

It is easy to check, using (5.6) and the $L^2(\mathbb{R}^N)$ equivalent of (1.2.9), that

$$\langle x'|\, \underline{U}_\omega(q', p'; q'', p'')A\, |x''\rangle$$
$$= \exp\left[\frac{i}{\hbar}(p' \cdot x' - p'' \cdot x'')\right] \langle x' - q'|\, A\, |x'' - q''\rangle. \tag{5.9}$$

Hence, an explicit computation in $L^2(\mathbb{R}^N \times \mathbb{R}^N)$ which is totally analogous to that for (1.4.11), yields the resolution

$$\int_{\Gamma^N \times \Gamma^N} |\underline{U}_\omega(q', p'; q'', p'')Y\rangle\, dq'\, dp'\, dq''\, dp''\, \langle \underline{U}_\omega(q', p'; q'', p'')Y| = \mathbb{1} \tag{5.10}$$

for any $Y \in \mathscr{L}(\mathbb{R}^3)$ with

$$\langle Y | Y\rangle_2 = \mathrm{Tr}(Y^* Y) = (2\pi\hbar)^{-2N}, \tag{5.11}$$

where $\mathbb{1}$ denotes the identity superoperator in $\mathscr{L}(\mathbb{R}^N)$. Hence the mapping

$$\underline{W}_\gamma^{(\omega)}\colon\ \mathscr{L}(\mathbb{R}^N) \to \mathscr{L}(\Gamma^N) \tag{5.12}$$

which assigns to $A \in \mathscr{L}(\mathbb{R}^N)$ the element of $\bar{A}$ of $\mathscr{L}(\Gamma^N)$ for which

$$\langle q', p'\, |\bar{A}|\, q'', p''\rangle = \frac{(2\pi\hbar)^{-N}}{\mathrm{Tr}\,\gamma^2}\ \mathrm{Tr}[U_\omega(q', p')\gamma U_\omega^*(q'', p'')A] \tag{5.13}$$

is isometric. Thus, its range

$$\mathscr{L}(\Gamma_{\gamma\omega}^N) \overset{\text{def}}{=} W_\gamma^{(\omega)}\, \mathscr{L}(\mathbb{R}^N) \subset \mathscr{L}(\Gamma^N) \tag{5.14}$$

is a closed subspace of $\mathscr{L}(\Gamma^N)$, and $\underline{W}_\gamma^{(\omega)}$ is a unitary mapping of $\mathscr{L}(\mathbb{R}^N)$ onto $\mathscr{L}(\Gamma_{\gamma\omega}^N)$. It is easy to check that if $\gamma = |\xi\rangle\langle\xi|$, then $\underline{W}_\gamma^{(\omega)}$ in (5.12) coincides with (5.5) in the gauge $\omega = -q \cdot p$. By changing the gauge in (5.8), we can arrive at the representatives (5.13) of $A \in \mathscr{L}(\mathbb{R}^N)$ in other gauges, but the diagonal elements of (5.13), which are the objects of exclusive physical interest, are obviously the same in all gauges.

It is now possible to recapitulate the entire theory of Section 1.7 from the point of view of the Liouville space $\mathscr{L}(\Gamma^N)$, with many of its technical features remaining almost unaltered. For example, the kernel of the projector $\mathbb{P}_\gamma(\Gamma^N)$ onto $\mathscr{L}(\Gamma_\gamma^N)$ equals by (5.10) and (5.11)

$$K_\gamma^{(\omega)}(q, \ldots, p''') = \frac{(2\pi\hbar)^{-2N}}{\mathrm{Tr}\,\gamma^2}\ \langle \underline{U}_\omega(q, p; q', p')\gamma\, |\, \underline{U}_\omega(q'', p''; q''', p''')\gamma\rangle_2. \tag{5.15}$$

The above expression is the analogue of (1.6.9), but an alternative expression with inner product in $\mathscr{L}(\Gamma^N)$, which is the analogue of (1.7.2a), can be easily obtained by working with the counterparts of $\underline{U}_\omega$ in $\mathscr{L}(\Gamma^N)$.

When working on the Liouville space $\mathscr{L}(\Gamma^N)$, rather than the Hilbert space $L^2(\Gamma^N)$, basic differences emerge in physical interpretation, and therefore also in those technical aspects intimately related to this interpretation. Thus, whereas the generators of various kinematical operations (translations, boosts, etc.) represent quantum mechanical observables in $L^2(\mathbb{R}^N)$, that is no longer the case in $\mathscr{L}(\Gamma^N)$ (or, for that matter, in $\mathscr{L}(\mathbb{R}^N)$). A good example of this fact is provided by the quantum mechanical Hamiltonian (4.16), in its dual role of generator of time translations and of energy operator. In the master Liouville space $\mathscr{L}(\Gamma^N)$ the time evolution of a density operator is given by

$$\rho_t = \exp\left(-\frac{i}{\hbar} Ht\right) \qquad \rho_0 \exp\left(\frac{i}{\hbar} Ht\right), \tag{5.16}$$

where $H = H_0 + H_I$ is given by (4.21) if ρ_t is a density operator for a classical ensemble, and by (4.16) if it represents the state if a quantum ensemble. In either case ρ_t obeys the equation

$$i\partial_t \rho_t = \underline{H}\rho_t, \quad \underline{H} = \hbar^{-1}[H, \cdot] \tag{5.17}$$

where $\underline{H}$ is a Liouville (self-adjoint) superoperator, namely

$$\underline{H}A = \hbar^{-1}(HA - AH)^{**}, \tag{5.18}$$

which is in general defined for all A from a dense set in $\mathscr{L}(\Gamma^N)$ (Prugovečki and Tip, 1974, 1975). On the other hand, it is the superoperator

$$\underset{\sim}{H} = \frac{1}{2}\{H, \cdot\}, \tag{5.19}$$

and not $\underline{H}$, that provides the correct expectation value for the total energy of the system:

$$\mathrm{Tr}(\rho H) = \frac{1}{2}\mathrm{Tr}(H\rho + \rho H) = \mathrm{Tr}(\underset{\sim}{H}\rho). \tag{5.20}$$

Thus, at the level of Liouville space, the roles of generator of time translation and of energy superoperator are played by distinct entities not only in the classical, but also in the quantum context. In fact, in the last stage of the unification process of classical and quantum statistical mechanics reached in the next section – where we shall represent quantum observables by distributions on stochastic phase space to achieve complete analogy with the classical situation in (4.2) – the classical Hamiltonian and the classical Liouville operator will emerge as the lowest order approximations of their quantum counterparts.

* 3.6. Quantum Observables as Density Distributions on Stochastic Phase Spaces

In the last section we have shown that classical as well as quantum states of ensembles can be represented by equivalence classes of density operators ρ in master Liouville spaces $\mathscr{L}(\Gamma^N)$, and that classical as well as quantum observables can be represented by superoperators in $\mathscr{L}(\Gamma^N)$. This establishes that a unified formalism for classical and quantum mechanics exists, in which the classical theory has been recast into a Liouville space framework traditionally associated only with quantum statistical mechanics. In this section we shall show, however, that this procedure can be reversed by recasting quantum statistical mechanics in terms of the space $\mathscr{P}(\Gamma^N)$ of distribution functions on Γ^N, traditionally deemed to be the domain of classical statistical mechanics.

Half of this task has already been accomplished in Theorems 3 and 4 in Section 3.3, which establish that for certain stochastic phase spaces Γ_γ^N each state of a quantum ensemble can be uniquely represented by the distribution function $\rho^\gamma(q, p)$ in (4.1). Hence, for each such Γ_γ^N there is a convex subset

$$\mathscr{P}(\Gamma_\gamma^N) = \{\mathrm{Tr}(\rho\gamma_{q,p}) \mid \rho \in \mathscr{R}(\mathscr{H})\} \tag{6.1}$$

of $\mathscr{P}(\Gamma^N)$, whose elements unambiguously describe states of quantum ensembles. It is obvious that each $\mathscr{P}(\Gamma_\gamma^N)$ is a proper subset of $\mathscr{P}(\Gamma^N)$ since, first of all, all elements $\rho^\gamma(q, p)$ of (6.1) are continuous functions (whereas $\mathscr{P}(\Gamma^N)$ also contains many discontinuous functions), and second, $\mathscr{P}(\Gamma^N)$ contains functions $\rho(q, p)$ of arbitrarily small supports, which cannot possibly represent quantum states when the uncertainty principle is violated by the size of the phase-space cells containing those supports.

To complete the formulation of quantum statistical mechanics on $\mathscr{P}(\Gamma_\gamma^N)$, we have to show that each quantum observable, represented thus far by a self-adjoint operator A in $L^2(\mathbb{R}^N)$, can also be represented by a density distribution $A^\gamma(q, p)$ on Γ_γ^N in such a manner that

$$\mathrm{Tr}(A\rho) = \int_{\Gamma^N} A^\gamma(q, p)\rho^\gamma(q, p)\, dq\, dp. \tag{6.2}$$

Furthermore, we have to establish the existence of Liouville operators L^γ on $\mathscr{P}(\Gamma_\gamma^N)$ such that the dynamics governed by the von Neumann equation (4.11) can be recast in the form of a Liouville equation

$$\partial_t \rho_t^\gamma(q, p) = L^\gamma \rho_t^\gamma(q, p), \qquad L^\gamma = L_0^\gamma + L_I^\gamma, \tag{6.3}$$

that is formally analogous to (4.8).

Both these facts have been established on a most general level by Ali and Prugovečki (1977c), but the proofs are somewhat technical and lengthy. We shall therefore concentrate in this section on the case of the optimal stochastic phase spaces Γ_γ^N in (3.29), and for the sake of notational simplicity we shall choose $l_1 = \ldots = l_{N_0} = l$, and denote the resulting stochastic phase space by $\Gamma_{(l)}^N$.

The computations of the density distribution $A^{(l)}(q, p)$ corresponding to the present choice of $\gamma = |\xi^{(l)}\rangle\langle\xi^{(l)}|$ in (6.2) can be carried out explicitly (Prugovečki, 1978b) when $\hat{A}$ is a polynomial or a power series Φ in $\hat{Q}_\alpha$ and $\hat{P}_\alpha$ – as is the case with observables of practical interest; for example, for the Hamiltonians in (4.12) and for $V(x)$ given by (4.17). Choosing then $\omega = -q \cdot p$ in (4.14), we can re-express the action of $W_{(l)}$ on $\psi \in L^2(\mathbb{R}^N)$ by

$$(W_{(l)}\psi)(q, p) = \int_{\mathbb{R}^N} \xi_{q,p}^{(l)*}(x)\psi(x)\,\mathrm{d}x, \tag{6.4a}$$

$$\xi_{q,p}^{(l)}(x) = (8\pi^3\hbar^2 l^2)^{-N/4} \exp\left[-\frac{(x-q)^2}{4l^2} + \frac{i}{\hbar}p \cdot x\right]. \tag{6.4b}$$

Let us denote by Q_α and P_α the operators (4.15) for this choice of ω:

$$Q_\alpha = i\hbar\frac{\partial}{\partial p_\alpha}, \qquad P_\alpha = p_\alpha - i\hbar\frac{\partial}{\partial q_\alpha}. \tag{6.5}$$

Then the observable represented by $\hat{A} = \Phi(\hat{Q}, \hat{P})$ in $L^2(\mathbb{R}^N)$ is represented by

$$A^{(l)} = W_{(l)}\hat{A}W_{(l)}^{-1} = \Phi(Q, P)\mathbb{P}_{(l)} \tag{6.6}$$

in $L^2(\Gamma_{(l)}^N)$, if we agree to limit in (6.6) the domain of the projector $\mathbb{P}_{(l)}$ onto $L^2(\Gamma_{(l)}^N)$ to $L^2(\Gamma_{(l)}^N)$ itself. With this convention in effect we can also write

$$Q_\alpha^{(l)} = W_{(l)}\hat{Q}_\alpha W_{(l)}^{-1} = Q_\alpha\mathbb{P}_{(l)}, \tag{6.7a}$$

$$P_\alpha^{(l)} = W_{(l)}\hat{P}_\alpha W_{(l)}^{-1} = P_\alpha\mathbb{P}_{(l)}. \tag{6.7b}$$

Using (1.17) and (6.4), we can then easily verify that when applied to $\psi \in L^2(\Gamma_{(l)}^N)$ the above operators can also be exhibited in an alternative form as

$$Q_\alpha^{(l)} = q_\alpha + 2l^2\frac{\partial}{\partial q_\alpha}, \qquad P_\alpha^{(l)} = p_\alpha + \frac{i\hbar}{2l^2}q_\alpha + \frac{\hbar^2}{2l^2}\frac{\partial}{\partial p_\alpha}. \tag{6.8}$$

Comparing (6.5) and (6.8) we see that

$$P_\alpha^{(l)} = \left[p_\alpha - \frac{i\hbar}{2l^2}(Q_\alpha - q_\alpha)\right]\mathbb{P}_{(l)}. \tag{6.9}$$

When we use (6.9) in (6.6), we arrive at the conclusion that $A^{(l)}$ can always be written in the form

$$A^{(l)} = \sum_n [R_n(q, p) + iI_n(q, p)]\, Q^n \mathbb{P}_{(l)}, \tag{6.10a}$$

$$Q^n = Q_1^{n_1} \ldots Q_N^{n_N}, \quad n = (n_1, \ldots, n_N), \tag{6.10b}$$

where R_n and I_n are real polynominals or power series in q_α and p_α, $\alpha = 1, \ldots, N$.

Our main goal is to find a method for the computation of the function $A^{(l)}(q, p)$ for which, in accordance with (4.1) and (6.2),

$$\mathrm{Tr}(\underset{\sim}{A}^{(l)}\rho) = \int_{\mathbb{R}^N} A^{(l)}(q, p)\rho^{(l)}(q, p)\, \mathrm{d}q\, \mathrm{d}p, \tag{6.11a}$$

$$\underset{\sim}{A}^{(l)}\rho = \tfrac{1}{2}\{A^{(l)}, \rho\}, \quad \rho \in \mathscr{L}(\Gamma^N_{(l)}), \tag{6.11b}$$

$$\rho^{(l)}(q, p) = \langle \bar{\xi}^{(l)}_{q,p} \mid \rho \bar{\xi}^{(l)}_{q,p}\rangle_\Gamma, \quad \bar{\xi}^{(l)}_{q,p} = W_{(l)}\, \xi^{(l)}_{q,p}. \tag{6.11c}$$

Since by the $L^2(\Gamma^N)$ counterpart of (1.7.2) and (1.7.3) we have

$$\mathbb{P}_{(l)} = \int_{\Gamma^N} |\bar{\xi}^{(l)}_{q,p}\rangle\, \mathrm{d}q\, \mathrm{d}p\, \langle \bar{\xi}^{(l)}_{q,p}|, \tag{6.12}$$

we immediately deduce that for any trace-class operator A in $L^2(\Gamma^N)$ we can write

$$\mathrm{Tr}(AP_{(l)}) = \int_{\Gamma^N} \langle \bar{\xi}^{(l)}_{q,p} | A\bar{\xi}^{(l)}_{q,p}\rangle\, \mathrm{d}q\, \mathrm{d}p. \tag{6.13}$$

On the other hand, by (6.10a) and (6.11b)

$$\begin{aligned}(\underset{\sim}{A}^{(l)}\rho^{(l)})(q, p) &\overset{\mathrm{def}}{=} \langle \bar{\xi}^{(l)}_{q,p} \mid (\underset{\sim}{A}^{(l)}\rho^{(l)})\bar{\xi}^{(l)}_{q,p}\rangle_\Gamma \\ &= \tfrac{1}{2}\sum_n (R_n(q, p)\, \{Q^n, \rho\}^{(l)}_{q,p} + iI_n(q, p)\, [Q^n, \rho]^{(l)}_{q,p}),\end{aligned} \tag{6.14}$$

where we have set, by definition,

$$\{Q^n, \rho\}^{(l)}_{q,p} = \langle \bar{\xi}^{(l)}_{q,p} \mid (Q^n\rho^{(l)} + \rho^{(l)}Q^n)\bar{\xi}^{(l)}_{q,p}\rangle_\Gamma, \tag{6.15a}$$

$$[Q^n, \rho]^{(l)}_{q,p} = \langle \bar{\xi}^{(l)}_{q,p} \mid (Q^n\rho^{(l)} - \rho^{(l)}Q^n)\bar{\xi}^{(l)}_{q,p}\rangle_\Gamma. \tag{6.15b}$$

Using (6.5) and (6.8), and taking into account that

$$\psi(q, p) = \langle \xi^{(l)}_{q,p} | \psi\rangle_\Gamma, \quad \psi \in L^2(\Gamma^N_{(l)}), \tag{6.16}$$

we deduce from the canonical expansions

$$\langle \xi^{(l)}_{q,p} \mid \rho^{(l)} \xi^{(l)}_{q',p'} \rangle_\Gamma = \sum_{i=1}^{\infty} \lambda_i \bar{e}_i(q,p) \bar{e}_i^*(q',p'), \tag{6.17}$$

which follow from (3.4) and (6.4) when $\rho \in \mathscr{L}(\Gamma^N_{(l)})$, that

$$\tfrac{1}{2} \{Q_\alpha, \rho\}^{(l)}_{q,p} = \left(q_\alpha + l^2 \frac{\partial}{\partial q_\alpha}\right) \rho^{(l)}(q,p), \tag{6.18a}$$

$$[Q_\alpha, \rho]^{(l)}_{q,p} = i\hbar \frac{\partial}{\partial p_\alpha} \rho^{(l)}(q,p). \tag{6.18b}$$

Hence the task of reducing (6.14) to the form of the integrand in (6.11a) becomes a matter of repeated use of the recursive relations

$$\{Q^n, \rho\}^{(l)}_{q,p} = \frac{i\hbar}{2} \frac{\partial}{\partial p_\alpha} [Q^{n-1_\alpha}, \rho]^{(l)}_{q,p} + \left(q_\alpha + l^2 \frac{\partial}{\partial q_\alpha}\right) \{Q^{n-1_\alpha}, \rho\}^{(l)}_{q,p}, \tag{6.19a}$$

$$[Q^n, \rho]^{(l)}_{q,p} = \frac{i\hbar}{2} \frac{\partial}{\partial p_\alpha} \{Q^{n-1_\alpha}, \rho\}^{(l)}_{q,p} + \left(q_\alpha + l^2 \frac{\partial}{\partial q_\alpha}\right) [Q^{n-1_\alpha}, \rho]^{(l)}_{q,p}, \tag{6.19b}$$

in which $n-1_\alpha$ denotes the n-tuple $(n_1, \ldots, n_\alpha - 1, \ldots, n_N)$ if n is as in (6.10b). These relations follow from (2.12) upon writing $\bar{e}_i = W_{(l)} e_i$, and then applying (6.4). After a finite number of successive applications, these relations obviously lead from (6.13) and (6.14) to (6.11a).

As a first example, let us consider the free Hamiltonian in $L^2(\Gamma^N_{(l)})$:

$$H^{(l)}_0 = \sum_\alpha P^{(l)2}_\alpha / 2m_\alpha. \tag{6.20}$$

Using (6.18) and (6.19) to compute (6.14) for $\underset{\sim}{H}^{(l)}_0 = \frac{1}{2} \{H^{(l)}_0, \cdot\}$ we get

$$(\underset{\sim}{H}^{(l)}_0 \rho^{(l)})(q,p) = \sum_\alpha \left\{ \frac{p_\alpha^2}{2m_\alpha} + \left(\frac{\hbar}{2l}\right)^2 \frac{1}{m_\alpha} \left(\frac{1}{2} + p_\alpha \frac{\partial}{\partial p_\alpha}\right) + \right.$$
$$\left. + \frac{\hbar^2}{8} \frac{1}{m_\alpha} \left(\frac{\hbar^2}{4l^4} \frac{\partial^2}{\partial p_\alpha^2} - \frac{\partial^2}{\partial q_\alpha^2}\right) \right\} \rho^{(l)}(q,p). \tag{6.21}$$

We see that, to the first order in $\hbar$, $\underset{\sim}{H}^{(l)}_0$ and $\underset{\sim}{H}^{\mathrm{cl}}_0$ coincide. Furthermore, upon integrating (6.21) over Γ^N to obtain $\overline{H}^{(l)}_0(q,p)$ according to (6.11a), we see that the last expression in round brackets in (6.21) contributes zero since $\rho(q,p)$ and its derivatives vanish at infinity, whereas integration by parts yields

$$\int_{\Gamma^N} p_\alpha \frac{\partial}{\partial p_\alpha} \rho^{(l)}(q,p)\, dq\, dp = -\int_{\Gamma^N} \rho^{(l)}(q,p)\, dq\, dp = -1.$$

Consequently, we get

$$H_0^{(l)}(q,p) = \sum_\alpha \left[\frac{p_\alpha^2}{2m_\alpha} - \left(\frac{\hbar}{2l}\right)^2 \frac{1}{m_\alpha} \right], \tag{6.22}$$

so that the quantum free Hamiltonian density in $\Gamma_{(l)}^N$ coincides with the classical free Hamiltonian density in Γ^N modulo an l-dependent constant, which vanishes in the limit $l \to \infty$ of infinitely precise stochastic momentum values.

As a second example, let us consider the interaction Hamiltonian in $L^2(\Gamma_{(l)}^N)$,

$$H_I^{(l)} = W_{(l)} \hat{H}_I W_{(l)}^{-1} = V(Q^{(l)}), \tag{6.23}$$

for $V(x)$ in (4.17). Then, in accordance with (6.14),

$$(\underset{\sim}{H}_I^{(l)} \rho^{(l)})(q,p) = \sum_{|n|=0}^{\infty} \frac{1}{n!} \frac{\partial^n V(q)}{\partial q^n} \operatorname{Re} \langle \rho^{(l)} \xi_{q,p}^{(l)} \mid (Q-q)^n \xi_{q,p}^{(l)} \rangle_\Gamma. \tag{6.24}$$

We can now use (6.14) and (6.19) on the terms of the above series, but it is more straightforward to employ (6.4) and (6.7) to develop recursive relations for these terms themselves. We then arrive at

$$\begin{aligned} &\operatorname{Re} \langle \rho^{(l)} \xi_{q,p}^{(l)} \mid (Q-q)^n \xi_{q,p}^{(l)} \rangle_\Gamma \\ &= (n_\alpha - 1) l^2 \operatorname{Re} \langle \rho^{(l)} \xi_{q,p}^{(l)} \mid (Q-q)^{n-2_\alpha} \xi_{q,p}^{(l)} \rangle_\Gamma + \\ &+ l^2 \frac{\partial}{\partial q_\alpha} \operatorname{Re} \langle \rho^{(l)} \xi_{q,p}^{(l)} \mid (Q-q)^{n-1_\alpha} \xi_{q,p}^{(l)} \rangle_\Gamma + \\ &+ \frac{\hbar}{2} \frac{\partial}{\partial p_\alpha} \operatorname{Im} \langle \rho^{(l)} \xi_{q,p}^{(l)} \mid (Q-q)^{n-1_\alpha} \xi_{q,p}^{(l)} \rangle, \end{aligned} \tag{6.25}$$

and the adjoint relation resulting from the substitutions Re → Im, Im → Re and $\hbar \to -\hbar$ in (6.25). Using these formulae in (6.24), we obtain this expression

$$(\underset{\sim}{H}_I^{(l)} \rho^{(l)})(q,p) = \left(V(q) + l^2 \frac{\partial V(q)}{\partial q_\alpha} \frac{\partial}{\partial q_\alpha} + \ldots \right) \rho^{(l)}(q,p), \tag{6.26}$$

in which we have written down all terms with $|n| \leqslant 1$. Consequently, upon integrating over Γ^N to obtain $H_I^{(l)}(q,p)$ according to (6.11a), we arrive at

$$H_I^{(l)}(q,p) = V(q) - l^2 \sum_\alpha \frac{\partial^2 V(q)}{\partial q_\alpha^2} + \ldots, \tag{6.27}$$

so that, up to terms containing second or higher powers of l, $H_I^{(l)}(q, p)$ and $H_I^{\mathrm{cl}}(q, p)$ coincide. Hence, as opposed to the free quantum Hamiltonian density in (6.22), the interacting one becomes exactly equal to the classical expression in the sharp-point limit $l \to +0$.

The same pattern is repeated with the Liouville operators in (6.3) for $\gamma = |\xi^{(l)}\rangle\langle\xi^{(l)}|$, where by (5.17)

$$(L_0^{(l)}\rho^{(l)})(q, p) = \hbar^{-1}\langle\bar{\xi}_{q,p}^{(l)} \mid [H_0^{(l)}, \rho^{(l)}]\,\bar{\xi}_{q,p}^{(l)}\rangle_\Gamma, \tag{6.28}$$

$$(L_I^{(l)}\rho^{(l)})(q, p) = \hbar^{-1}\langle\bar{\xi}_{q,p}^{(l)} \mid [H_I^{(l)}, \rho^{(l)}]\,\bar{\xi}_{q,p}^{(l)}\rangle_\Gamma. \tag{6.29}$$

Inserting (6.4) into (6.28), and using the easily verifiable equalities

$$l^2 \frac{\partial}{\partial q_\alpha} \rho^{(l)}(q, p) = \mathrm{Re}\,\langle\rho^{(l)}\,\xi_{q,p}^{(l)} \mid (\hat{Q}_\alpha - q_\alpha)\,\xi_{q,p}^{(l)}\rangle, \tag{6.30a}$$

$$\hbar l^2 \frac{\partial^2}{\partial q_\alpha\,\partial p_\alpha} \rho(q, p) = \mathrm{Im}\,\langle\rho^{(l)}\,\xi_{q,p}^{(l)} \mid (\hat{Q}_\alpha - q_\alpha)^2\,\xi_{q,p}^{(l)}\rangle, \tag{6.30b}$$

we arrive at the following free quantum Liouville operator in $\mathscr{P}(\Gamma_{(l)}^N)$:

$$L_0^{(l)} = -\sum_\alpha \left[\frac{p_\alpha}{m_\alpha}\,\frac{\partial}{\partial q_\alpha} + \left(\frac{\hbar}{2l}\right)^2 \frac{1}{m_\alpha}\,\frac{\partial^2}{\partial q_\alpha\,\partial p_\alpha}\right]. \tag{6.31}$$

Thus, as in the case of (6.2), the quantum Liouville operator in $\mathscr{P}(\Gamma_{(l)}^N)$ coincides with the classical one up to an l-dependent term that vanishes in the limit $l \to +\infty$. On the other hand, by (4.17), (6.23) and (6.29)

$$(L_I^{(l)}\rho^{(l)})(q, p) = -\frac{2}{\hbar}\sum_{|n|=1}^{\infty} \frac{1}{n!}\,\frac{\partial^n V(q)}{\partial q^n}\;\mathrm{Im}\langle\rho^{(l)}\,\bar{\xi}_{q,p}^{(l)} \mid (Q - q)^n\,\bar{\xi}_{q,p}^{(l)}\rangle_\Gamma, \tag{6.32}$$

where the inner product can be computed with the help of (6.25) and its adjoint relation. We get

$$L_I^{(l)} = \sum_\alpha \frac{\partial V(q)}{\partial q_\alpha}\,\frac{\partial}{\partial p_\alpha} + l^2 \sum_{\alpha,\beta} \frac{\partial^2 V(q)}{\partial q_\alpha\,\partial q_\beta}\left(\frac{\partial^2}{\partial q_\alpha\,\partial p_\beta} + \frac{\partial^2}{\partial p_\alpha\,\partial q_\beta}\right) + \ldots, \tag{6.33}$$

so that in this case the quantum Liouville operator equals the classical one modulo terms containing second or higher powers of l, and exact equality is achieved in the sharp-point limit $l \to +0$.

In conclusion, we can state that quantum statistical mechanics in the space $\mathscr{P}(\Gamma_{(l)}^N)$ of stochastic phase-space distribution functions can be approximated by the corresponding classical theory, modulo l-dependent terms which obey the complementarity principle *vis-à-vis* the sharp position and sharp momentum limits $l \to +0$ and $l \to +\infty$, respectively, in the following sense: the free parts of quantum quantities (Hamiltonian densities, Liouville operators), coincide with their classical counterparts in the limit $l \to +\infty$, whereas the interacting parts diverge from their classical counterparts in the same limit – with these roles being reversed in the limit $l \to +0$.

* 3.7. Unified Formulations of Classical and Quantum Boltzmann Equations and Scattering Cross-Sections

In the preceding three sections we have established a point of fundamental epistemic interest, namely that the concept of stochastic phase space constitutes the basis for a formalism that unifies classical and quantum nonrelativistic mechanics (ordinary as well as statistical) into an integral whole, in which the remaining distinctions are in technical detail (different choices of Liouville operators, Hamiltonians, etc.), rather than in fundamental structural features beyond those imposed by the uncertainty principle. Thus, Born's (1955c, 1956) fundamental thesis is validated: the consistent introduction of the stochastic values in physics brings about an all-embracing conceptual unification that extends all the way from nonrelativistic classical mechanics to the far reaches of quantum relativity, which will be studied in Part II.

The unified formalism of this chapter can be exploited, however, also at the more pragmatic level of model building. In this context it is obviously best suited to those phenomena that can be described with good accuracy by classical models, which are then to be improved quantum mechanically via the introduction of quantum 'corrections'. A good example of such a phenomenon is Brownian motion, whose classical treatment is included in most standard textbooks on statistical mechanics (see, e.g., Balescu, 1975). Quantum corrections to this model can be then computed by using the formalism of Section 3.6, as actually demonstrated for corrections of the first order in $\hbar$ by the techniques used[10] in a paper by McKenna and Frisch (1966).

Other good examples are supplied by transport equations in gases, and in particular by the equation

$$\partial_t \rho_t(\mathbf{q}_1, \mathbf{p}_1) + \frac{\mathbf{p}_1}{m} \cdot \nabla_{\mathbf{q}_1} \rho_t(\mathbf{q}_1, \mathbf{p}_1) = J_t(\mathbf{q}_1, \mathbf{p}_1), \tag{7.1}$$

derived by Boltzmann in 1872 in classical statistical mechanics. In (7.1) $\rho_t(\mathbf{q}_1, \mathbf{p}_1)$ is the one-particle Γ-distribution function of a low-density gas, and $J_t(\mathbf{q}_1, \mathbf{p}_1)$ is Boltzmann's collision term

$$J_t(\mathbf{q}_1, \mathbf{p}_1) = \int_{\mathbb{R}^3} d\mathbf{p}_2 \int d\omega \, \frac{|\mathbf{p}_1 - \mathbf{p}_2|}{m} \, \sigma(\omega) \times$$

$$\times \left[\rho_t(\mathbf{q}_1, \mathbf{p}_1^*)\rho_t(\mathbf{q}_1, \mathbf{p}_2^*) - \rho_t(\mathbf{q}_1, \mathbf{p}_1)\rho_t(\mathbf{q}_1, \mathbf{p}_2)\right] \tag{7.2}$$

for the collision of two particles of incoming momenta $\mathbf{p}_1$ and $\mathbf{p}_2$ in a neighborhood of $\mathbf{q}_1$, with $\mathbf{p}_1^*$ and $\mathbf{p}_2^*$ denoting the momenta after collision, and $\sigma(\omega)$ the classical differential cross-section for the direction $\omega = (\mathbf{p}_2^* - \mathbf{p}_1^*) \, |\mathbf{p}_2^* - \mathbf{p}_1^*|^{-1}$ relative to $\mathbf{p}_0 = \mathbf{p}_2 - \mathbf{p}_1$. In early attempts (Nordheim, 1928; Uhling and Uhlenbeck, 1933) at a quantum version of the Boltzmann equation (7.1), the Γ-distribution function was replaced by a momentum distribution function, but later attempts (Mori and Ono, 1952; Schönberg, 1953; Ono, 1954; Ross and Kirkwood, 1954; Saenz, 1957) were aimed at letting the Wigner transform (1.8) for a single particle take over the role of the classical Γ-distribution function $\rho_t(\mathbf{q}_1, \mathbf{p}_1)$. However, the Wigner function is not a probability distribution. Hence, more recently, purely operator versions of this equation have been proposed (Snider and Sanctuary, 1971; Eu, 1975). However, considering the very direct physical meaning (Balescu, 1975, Section 11.4) of the classical Boltzmann equation (7.1) and its collision term (7.2), it is very difficult to see how any equation not based on a quantum Γ-distribution function that is a *bona fide* probability density could be undisputedly taken to be the quantum version of (7.1).

When using the stochastic phase-space formalism, not only do we have quantum Γ-distribution functions that are probability densities, but it turns out (Prugovečki, 1978c) that some of these one-particle densities, such as $\rho_t^{(l)}(\mathbf{q}_1, \mathbf{p}_1)$, satisfy, for certain values of l, Equation (7.1) with a collision term that has the same form (7.2) in the classical and quantum case. Hence the only difference occurs in the differential cross-section $\sigma(\omega)$, which is based, respectively, on classical or quantum quantities in the master Liouville space which incorporates both cases in accordance with Section 3.6. The only surprise[11] is that in the quantum case $\sigma(\omega)$ appearing in (7.2) is not the conventional differential cross-section (for a particle of reduced mass $m/2$)

$$\hat{\sigma}(\mathbf{p}_0 \to \omega) = (2m\pi^2)^2 \, | \langle p_0' \omega \, | \, H_I \Omega_+ \, | \, \mathbf{p}_0 \rangle |^2, \tag{7.3}$$

but rather, as discussed at the end of this section, the quantum analogue (7.38) of the classical differential cross-section that is implicit in (7.2).

To derive the Boltzmann equation (7.1) simultaneously in the classical and quantum context, let us consider system of N_0 identical spinless particle, and let us describe the (classical and quantum) states of ensembles of such systems by density operators $\rho(1 \ldots N_0)$ belonging to the master Liouville space $\mathscr{L}(\Gamma^N)$ introduced in Section 3.5. Regarded as a Hilbert space identifiable with $L^2(\Gamma^N \times \Gamma^N)$, this space can be decomposed into a tensor product

$$\mathscr{L}(\Gamma^N) = \mathscr{L}(\Gamma^3_{(1)}) \otimes \ldots \otimes \mathscr{L}(\Gamma^3_{(N_0)}), \quad N = 3N_0, \tag{7.4}$$

of one-particle Liouville space $\mathscr{L}(\Gamma^3_{(\beta)})$, $\beta = 1, \ldots, N_0$. Thus reduced n-particle density operators

$$\rho(1 \ldots n) = \frac{N_0!}{(N_0 - n)!} \, \mathrm{Tr}^{(n+1, \ldots, N_0)} \rho(1 \ldots N_0) \tag{7.5}$$

belonging to a reduced master Liouville space $\mathscr{L}(\Gamma^3_{(1)}) \otimes \ldots \otimes \mathscr{L}(\Gamma^3_{(n)})$ can be introduced by taking the above partial traces as follows:

$$\langle \mathbf{q}_1, \mathbf{p}_1, \ldots, \mathbf{q}_n, \mathbf{p}_n | \operatorname{Tr}^{(n+1,\ldots,N_0)} \rho(1 \ldots N_0) | \mathbf{q}'_1, \mathbf{p}'_1, \ldots, \mathbf{q}'_n, \mathbf{p}'_n \rangle$$

$$= \int_{\mathbb{R}^{3(N_0-n)}} d\mathbf{q}_{n+1}\, d\mathbf{p}_{n+1} \ldots d\mathbf{q}_{N_0}\, d\mathbf{p}_{N_0} \times$$

$$\times \langle \mathbf{q}_1, \mathbf{p}_1, \ldots, \mathbf{q}_n, \mathbf{p}_n, \ldots, \mathbf{q}_{N_0}, \mathbf{p}_{N_0} | \rho(1 \ldots N_0) | \mathbf{q}'_1, \mathbf{p}'_1, \ldots,$$

$$\ldots, \mathbf{q}'_n, \mathbf{p}'_n, \ldots, \mathbf{q}_{N_0}, \mathbf{p}_{N_0} \rangle, \tag{7.6}$$

We observe that by (6.7b) $\operatorname{Tr}^{(1 \ldots N_0)}$ coincides with the ordinary trace of any $\bar{A} \in \mathscr{L}(\Gamma^N)$, and that upon the identification of $\mathscr{L}(\Gamma^3_{(1)}) \otimes \ldots \otimes \mathscr{L}(\Gamma^3_{(n)})$ with $L^2(\Gamma^n \times \Gamma^n)$, the partial trace in (7.6) indeed yields an element of this reduced master Liouville space, which due to the symmetry of $\rho(1 \ldots N_0)$ under any permutation of all particles is an n-particle density operator for $1 \leqslant n < N_0$.

We shall assume now that the N_0 particles in this system interact via two-body forces given by a potential $V(\mathbf{q})$, so that the Hamilton operators in (4.21) and (4.23) are of the form

$$H_0^{(\omega)} = \sum_{\beta=1}^{N_0} H_\beta, \qquad H_I^{(\omega)} = \sum_{\beta'>\beta=1}^{N_0} H_{\beta\beta'}, \tag{7.7}$$

for the classical case (when we formally set $\omega = \mathrm{cl}$) as well as for the quantum case. In fact, specializing (4.23) to two-body forces, we immediately see that

$$H_\beta = -i\hbar \frac{\mathbf{p}_\beta}{m} \cdot \nabla_{\mathbf{q}_\beta} + \left\{ \frac{\mathbf{p}_\beta^2}{2m} - \frac{\hbar^2}{2m} \Delta_{\mathbf{q}_\beta} \right\}, \tag{7.8}$$

$$H_{12} = i\hbar \nabla_{\mathbf{q}} V(\mathbf{q}) \cdot \nabla_{\mathbf{p}} + \{V(\mathbf{q}) - \mathbf{q} \cdot \nabla_{\mathbf{q}} V(\mathbf{q}) + \ldots\},$$

$$\mathbf{q} = \mathbf{q}_2 - \mathbf{q}_1, \quad \mathbf{p} = \frac{1}{2}(\mathbf{p}_2 - \mathbf{p}_1), \tag{7.9}$$

in the quantum case, whereas in the classical case the terms in the brackets are not present. In either case, we can introduce in accordance with (5.18) the super-operators

$$\underline{H}_\beta = \hbar^{-1} [H_\beta, \cdot], \qquad \underline{H}_{\beta\beta'} = \hbar^{-1} [H_{\beta\beta'}, \cdot], \tag{7.10}$$

which by (5.18) govern the time evolution of the density operators $\rho_t(1 \ldots N_0)$ in $\mathscr{L}(\Gamma^N)$,.

$$i\partial_t \rho_t(1 \ldots N_0) = \left(\sum_{\beta=1}^{N_0} \underline{H}_\beta + \sum_{\beta'>\beta=1}^{N_0} \underline{H}_{\beta\beta'} \right) \rho_t(1 \ldots N_0), \tag{7.11}$$

and therefore by (7.5), also that of each of the reduced density operators $\rho_t(1 \ldots n)$. The self-adjointness of $\rho(1 \ldots N_0)$ implies that

$$\mathrm{Tr}^{(n+1, \ldots, N_0)} [\underline{H}_\beta \rho(1 \ldots N_0)] = \mathrm{Tr}^{(n+1, \ldots, N_0)} [\underline{H}_{\beta\beta'} \rho(1 \ldots N_0)] = 0,$$

$$\beta' > \beta = n+1, \ldots, N_0, \tag{7.12}$$

so that a BBGKY hierarchy (Balescu, 1975) is obtained for $\rho_t(1 \ldots n)$ by taking all the partial traces for $n = N_0 - 1, N_0 - 2, \ldots, 1$ on both sides of (7.11). The first two members of this hierarchy obviously are:

$$i\partial_t \rho_t(1) = \underline{H}_1 \rho_t(1) + \mathrm{Tr}^{(2)} [\underline{H}_{12} \rho_t(12)], \tag{7.13}$$

$$i\partial_t \rho_t(12) = (\underline{H}_1 + \underline{H}_2 + \underline{H}_{12}) \rho_t(12) + \mathrm{Tr}^{(3)} [(\underline{H}_{13} + \underline{H}_{23}) \rho_t(123)]. \tag{7.14}$$

It can be argued (Hess, 1967; Snider and Sanctuary, 1971) that in a low-density gas of N_0 molecules interacting via short-range forces, the last term in (7.14) can be dropped on grounds that the likelihood of three-body collision in the neighborhood of one and the same point $\mathbf{q}_1$ is negligible. Furthermore, using basic properties of the scattering wave superoperators (Prugovečki, 1981a, p. 487)

$$\underline{\Omega}_{12}^{\pm} = \lim_{t \to \mp\infty} \exp[i(\underline{H}_1 + \underline{H}_2 + \underline{H}_{12})t] \exp[-i(\underline{H}_1 + \underline{H}_2)t] \tag{7.15}$$

we can relate the density operator $\rho_t(12)$ for an ensemble of interacting two-particle systems, representing two colliding molecules in the gas, to the incoming asymptotic state of such an ensemble represented by a freely evolving density operator $\rho_t^{\mathrm{in}}(12)$:

$$\rho_t(12) = \underline{\Omega}_{12}^{+} \rho_t^{\mathrm{in}}(12), \quad \rho_t^{\mathrm{in}}(12) = \exp[-i(\underline{H}_1 + \underline{H}_2)t] \rho_0^{\mathrm{in}}(12). \tag{7.16}$$

Hence, (2.9) assumes the form

$$i\partial_t \rho_t(12) = (\underline{H}_1 + \underline{H}_2 + \underline{H}_{12}) \underline{\Omega}_{12}^{+} \rho_t^{\mathrm{in}}(12). \tag{7.17}$$

Boltzmann's famous *Stosszahlansatz* (Balescu, 1975, p. 393) can be expressed in the form [12]

$$\rho_t^{\mathrm{in}}(12) = \rho_t(1) \otimes \rho_t(2), \tag{7.18}$$

so that when (7.17) is used in (7.13) we get

$$(\partial_t + i\underline{H}_1) \rho_t(1) = -\mathrm{Tr}^{(2)} \{ i\underline{H}_{12} \underline{\Omega}_{12}^{+} [\rho_t(1) \otimes \rho_t(2)] \}. \tag{7.19}$$

Equation (7.19) has been derived by a number of authors (Snider and Sanctuary, 1971; Tip, 1971; Eu, 1975) as a kind of density-operator counterpart in the quantum realm of the Boltzmann equation (7.1). In the reduced master Liouville spaces $\mathscr{L}(\Gamma^3_{(1)})$ and $\mathscr{L}(\Gamma^3_{(1)}) \otimes \mathscr{L}(\Gamma^3_{(2)})$ it is, however, equally valid in the classical as well as in the quantum realm – the difference between the two cases lying only in the choice of H_1 in (7.8) and H_{12} in (7.9), and in the initial conditions that are allowed in the quantum case (namely, $\rho_0(\beta) \in \mathscr{L}(\Gamma^3_{(\beta)\gamma})$, $\beta = 1, 2$, for some generalized resolution generator γ). The transition to Γ-distribution functions can be made according to the general methods of Section 3.6. Hence, setting

$$\rho_t(\mathbf{q}_\beta, \mathbf{p}_\beta) = \langle \mathbf{q}_\beta, \mathbf{p}_\beta | \rho_t(\beta) | \mathbf{q}_\beta, \mathbf{p}_\beta \rangle, \qquad \beta = 1, 2, \tag{7.20}$$

$$(L_1 \rho_t)(\mathbf{q}_1, \mathbf{p}_1) = -i \langle \mathbf{q}_1, \mathbf{p}_1 | \underline{H}_1 \rho_t(1) | \mathbf{q}_1, \mathbf{p}_1 \rangle, \tag{7.21}$$

and restricting ourselves in the quantum case to $\gamma = |\xi^{(l)}\rangle \langle \xi^{(l)}|$, we get from (7.19) by specializing (6.31) to the case of a single particle:

$$\left(\partial_t + \frac{\mathbf{p}_1}{m} \cdot \nabla_{\mathbf{q}_1} + \frac{\hbar^2}{4ml^2} \nabla_{\mathbf{q}_1} \cdot \nabla_{\mathbf{p}_1} \right) \rho_t(\mathbf{q}_1, \mathbf{p}_1) = I_t(\mathbf{q}_1, \mathbf{p}_1), \tag{7.22}$$

$$I_t(\mathbf{q}_1, \mathbf{p}_1) = \int_{\Gamma^3} (L_{12} \underline{\Omega}^+_{12} \rho_t(1) \otimes \rho_t(2)) (\mathbf{q}_1, \mathbf{p}_1; \mathbf{q}_2, \mathbf{p}_2) \, d\mathbf{q}_2 \, d\mathbf{p}_2 . \tag{7.23}$$

The l-dependent term in (7.22) is absent in the classical case. We shall also remove it in the quantum case by choosing l so large that $ml^2 \gg \hbar T$ (where T is the duration of the collision during which ρ_t changes appreciably), and so that $\rho_t(\mathbf{q}, \mathbf{p})$ does not vary significantly over phase-space cells of volume equal to $\hbar$. Then, upon comparing (7.22) with (7.1), we see that we are left only with the task of investigating the circumstances in which (7.23) assumes the form (7.2).

Let us introduce the centre-of-mass coordinates

$$\mathbf{q}_{\text{c.m.}} = \tfrac{1}{2}(\mathbf{q}_1 + \mathbf{q}_2), \qquad \mathbf{p}_{\text{c.m.}} = \mathbf{p}_1 + \mathbf{p}_2, \tag{7.24}$$

which together with the relative-motion coordinates in (7.9) can be used to re-express the diagonal elements of $\rho_t(1) \otimes \rho_t(2)$, by setting

$$\rho_t(\mathbf{q}, \mathbf{p}; \mathbf{q}_{\text{c.m.}}, \mathbf{p}_{\text{c.m.}}) = \rho_t(\mathbf{q}_1, \mathbf{p}_1) \rho_t(\mathbf{q}_2, \mathbf{p}_2). \tag{7.25}$$

We observe that if we similarly decompose L_{12} in (7.23),

$$L_{12} = L_{\text{c.m.}} \otimes L_I, \; (L_{12} \rho_t)(\zeta_1, \zeta_2) = \hbar^{-1} \langle \zeta_1, \zeta_2 | \underline{H}_{12} \rho_t | \zeta_1, \zeta_2 \rangle, \tag{7.26}$$

then by (7.9) $L_{\text{c.m.}}$ is just the identity operator. On the other hand, in accordance with (6.33)

$$L_I = \nabla_{\mathbf{q}} V(\mathbf{q}) \cdot \nabla_{\mathbf{p}} + \left\{ l^2 \sum_{i,j=1}^{3} \frac{\partial^2 V(q)}{\partial q_i \, \partial q_j} \left(\frac{\partial^2}{\partial q_i \, \partial p_j} + \frac{\partial^2}{\partial q_j \, \partial p_i} \right) + \ldots \right\} \tag{7.27}$$

in the quantum case, whereas in the classical case the terms in curly brackets do not occur. The free Lagrangian operator in (6.31) for the case of two particles can be similarly decomposed into a center-of-mass part and relative motion part, with the latter corresponding to a particle of reduced mass $m/2$,

$$L_0 = -\frac{2\mathbf{p}}{m}\nabla_{\mathbf{q}} - \left\{\frac{\hbar^2}{2ml^2}\nabla_{\mathbf{q}}\cdot\nabla_{\mathbf{p}}\right\}, \tag{7.28}$$

where again the term in curly brackets does not occur in the classical case. The eigenfunction expansions $f_{\mathbf{w},\mathbf{k}}(\mathbf{q},\mathbf{p})$ of L_0 depend on six parameters $\{\mathbf{w},\mathbf{k}\}$, and for $\mathbf{w}=\mathbf{0}$ they are $\mathbf{q}$-independent and equal to (cf. (7.53)–(7.57)):

$$f^{\text{cl}}_{\mathbf{0},\mathbf{k}}(\mathbf{p}) = \delta^3(\mathbf{p}-\mathbf{k}), \tag{7.29}$$

$$f^{(l)}_{\mathbf{0},\mathbf{k}}(\mathbf{p}) = \left(\frac{2l^2}{\pi\hbar^2}\right)^{3/3} \exp\left[-\frac{2l^2}{\hbar^2}(\mathbf{p}-\mathbf{k})^2\right]. \tag{7.30}$$

Thus, in the classical case we can formally insert after $\underline{\Omega}^+_{12}$ in (7.23):

$$\int_{\mathbb{R}^3} f^{\text{cl}}_{\mathbf{0},\mathbf{w}}(\mathbf{p}) f_{\mathbf{0},\mathbf{w}}(\mathbf{p}')\, d\mathbf{w} = \delta^3(\mathbf{p}-\mathbf{p}'). \tag{7.31}$$

We then obtain for (7.23) the expression

$$J_t(\mathbf{q}_1,\mathbf{p}_1) = \int d\mathbf{p}_2\, d\mathbf{k} \int d\mathbf{q} (L_I \overset{+}{f}_{\mathbf{0},\mathbf{k}})(\mathbf{q};\mathbf{p}) \int d\mathbf{p}'\, f_{\mathbf{0},\mathbf{k}}(\mathbf{p}')\rho_t(\mathbf{0},\mathbf{p}';\mathbf{q}_{\text{c.m.}},\mathbf{p}_{\text{c.m.}}) \tag{7.32}$$

if we use Boltzmann's assumption that the distribution function in (7.25) changes little over the region of integration in $\mathbf{q}$. Due to the presence of L_I in (7.32), and to its form (7.27), this region coincides with the support of $V(\mathbf{q})$, i.e. with the region of effective interaction. On the other hand, in the quantum case we have the relation

$$\int_{\mathbb{R}^3} f^{(l)}_{\mathbf{0},\mathbf{k}}(\mathbf{p}) f^{(l)}_{\mathbf{0},\mathbf{k}}(\mathbf{p}')\, d\mathbf{k} = \left(\frac{2l^2}{\hbar^2}\right)^{3/2} \exp\left[-\frac{l^2}{\hbar^2}(\mathbf{p}-\mathbf{p}')^2\right] \tag{7.33}$$

which coincides with (7.31) only in the limit $l \to +\infty$. However, under our earlier assumption that $ml^2 >> \hbar T$, the procedure leading to (7.32) can be repeated to a good approximation. In both cases, $\overset{+}{f}_{\mathbf{0},\mathbf{k}}$ denotes the retarded eigenfunction expansions $\underline{\Omega}^+ f_{\mathbf{0},\mathbf{k}}$ (Prugovečki, 1978b) of $L_0 + L_I$ with zero eigennumber, for which, therefore, by (7.28):

$$L_I \overset{+}{f}_{\mathbf{0},\mathbf{k}}(\mathbf{q},\mathbf{p}) = \left(\frac{2\mathbf{p}}{m}\cdot\nabla_{\mathbf{q}} + \left\{\frac{\hbar^2}{2ml^2}\nabla_{\mathbf{q}}\cdot\nabla_{\mathbf{p}}\right\}\right)\overset{+}{f}_{\mathbf{0},\mathbf{k}}(\mathbf{q},\mathbf{p}). \tag{7.34}$$

The term in curly brackets appears only in the quantum case, but under our assumption $ml^2 >> \hbar T$ it can be ignored. Upon inserting the resulting expression for $L_I f_{0,\mathbf{k}}^+$ into (7.32), and introducing the new variables of integration

$$q_{||} = |\mathbf{p}|^{-1}\, \mathbf{q}\cdot\mathbf{p}, \qquad \mathbf{q}_\perp = \mathbf{q} - q_{||}\, |\mathbf{p}|^{-1}\, \mathbf{p}, \tag{7.35}$$

we can perform the integration in $q_{||}$ by first integrating along a cylinder centered at the origin, and with axis parallel to $\mathbf{p}$. This standard procedure (Balescu, 1975, Section 20.3) leads to

$$J_t(\mathbf{q}_1, \mathbf{p}_1) = \int d\mathbf{p}_2\, d\mathbf{q}_\perp\, \frac{2\,|\mathbf{p}|}{m} \int d\mathbf{p}'\, \rho_t(\mathbf{0}, \mathbf{p}'; \mathbf{q}_{\mathrm{c.m.}}, \mathbf{p}_{\mathrm{c.m.}}) \times$$

$$\times \int d\mathbf{k}\, f_{0,\mathbf{k}}(\mathbf{p}')\, [f_{0,\mathbf{k}}^+(+\infty, \mathbf{q}_\perp; \mathbf{p}) - f_{0,\mathbf{k}}^+(-\infty, \mathbf{q}_\perp; \mathbf{p})]. \tag{7.36}$$

In the classical case we can follow $f_{0,\mathbf{k}}^+(\mathbf{q}, \mathbf{p})$ from $t \to -\infty$ to $t \to +\infty$ along any trajectory $\mathbf{q}(t)$ of a reduced particle of mass $m/2$ (Balescu, 1975, p. 387), with the result that

$$f_{0,\mathbf{k}}^{+\mathrm{cl}}(-\infty, \mathbf{q}_\perp; \mathbf{p}) = f_{0,\mathbf{k}}^{\mathrm{cl}}(\mathbf{p}), \qquad f_{0,\mathbf{k}}^{+\mathrm{cl}}(+\infty, \mathbf{q}_\perp; \mathbf{p}) = f_{0,\mathbf{k}}^{\mathrm{cl}}(\mathbf{p}^*), \tag{7.37}$$

where $\mathbf{p}^*$ denotes the incoming momentum for a trajectory of outgoing momentum $\mathbf{p}$ that asymptotically approaches $\mathbf{q}_\perp$. Using (7.36) and (7.37), we then easily arrive at (7.2).

In the quantum case the notion of trajectory is nonexistent, so one has to adopt a different strategy. It turns out that by working in the master Liouville space $\mathscr{L}(\Gamma^3)$ for the particle of reduced mass $m/2$, both the classical and quantum cases can be treated simultaneously. The outcome is an altogether different derivation of the formula (7.2) from (7.36). In the classical case this derivation does not involve the concept of trajectory, and it is equally well valid in the quantum case. The only difference between the two cases is that in the following universal formula for a particle of reduced mass $m/2$,

$$\sigma(\omega) = -(2\pi)^3\, i \int_0^\infty dp\, p^2 \int d\mathbf{k}\, f_{0,\mathbf{k}}(p\omega)\, \frac{m}{2\,|\mathbf{k}|}\, \langle \mathbf{k} |\, \underline{H_I \Omega_+}\, | \mathbf{p} \rangle, \tag{7.38}$$

one has to choose the appropriate (i.e.) classical or quantum values for $f_{0,\mathbf{k}}$ and for the quantities entering the formal inner product in (7.38). Since this derivation (Prugovečki, 1978c, Section 3) of (7.2) is somewhat lengthy, and it first requires the systematic development (Prugovečki, 1978b) of classical and quantum scattering theory in master Liouville spaces, we shall limit ourselves to a discussion of the main physical ideas involved in the derivation of (7.38), and of its physical and mathematical meaning and significance.

The derivation[13] of (7.38) can proceed by starting with the conventional operational definition (Messiah, 1962) of the scattering differential cross-section $\sigma(\mathbf{p}_0 \to \omega)$ for an incoming beam of particles of sharp momentum $\mathbf{p}_0$ into the direction ω: $\sigma(\mathbf{p}_0 \to \omega)$, namely the number of particles scattered per unit time and unit solid angle in the direction ω divided by the incident flux j_{inc} (i.e. the number of incident particles per unit time and unit beam cross-section). Since the particles in the incoming beam are supposed not to interact with each other, such a beam can be considered to be an ensemble of one-particle systems. Implicit in the above operational definition is also a limiting process, in which one starts with a finite beam, that can be viewed to occupy a cylinder $\mathscr{C}$ of height $\hbar$ and base b, in which the particles are spatially distributed with uniform randomness; then one lets $\hbar \to \infty$ and $b \to \infty$, i.e. in practical terms one chooses the dimensions of the beam to be much larger than the size of the target particles and the range of interaction. In mathematical terms, such a finite beam is therefore described by an incoming density operator ρ_t^{in} which evolves freely in accordance with the von Neumann equation (5.17):

$$\rho_t^{\text{in}} = \exp(-i\underline{H}_0 t)\rho_{\text{in}}, \quad \underline{H}_0 = \underline{H} - \underline{H}_I. \tag{7.39}$$

Ideally, such a ρ_t^{in} represents an ensemble of particles for which the probability density of finding a particle at the (sharp) point $\mathbf{q}$ equals $(\hbar b)^{-1}\chi_{\mathscr{C}}(\mathbf{q})$ (i.e. it is nonzero and uniform within the cylinder $\mathscr{C}$, and zero outside it), and whose momentum probability density should be a δ-like function:

$$\delta^3_{\hbar,b}(\mathbf{p} - \mathbf{p}_0) \to \delta^3(\mathbf{p} - \mathbf{p}_0), \quad \hbar, b \to \infty. \tag{7.40}$$

Thus, a good choice for $\rho_{\text{in}} \in \mathscr{L}(\Gamma^3)$ in the classical case would be

$$\langle \mathbf{q}, \mathbf{p} | \rho_{\text{in}}^{\text{cl}} | \mathbf{q}, \mathbf{p} \rangle = (\hbar b)^{-1} \chi_{\mathscr{C}}(\mathbf{q}) \delta^3_{\hbar,b}(\mathbf{p} - \mathbf{p}_0), \tag{7.41}$$

whereas in the quantum case it should equal $\rho_{\text{in}} \in \mathscr{L}(\Gamma^3_{(l)})$ with a configuration space representative $\hat{\rho}_{\text{in}} = W_l^{-1}\rho_{\text{in}} \in \mathscr{L}(\mathbb{R}^3)$ such that

$$\langle \mathbf{x}' | \hat{\rho}_{\text{in}} | \mathbf{x}'' \rangle = (\hbar b)^{-1} \exp\left[\frac{i}{\hbar} \mathbf{p}_0 \cdot (\mathbf{x}' - \mathbf{x}'')\right] \chi_{\mathscr{C}}(\mathbf{x}')\chi_{\mathscr{C}}(\mathbf{x}''). \tag{7.42}$$

The number of particles scattered in a narrow cone C_ω of (sharp or spread-out stochastic) momentum values around the direction ω will be equal to the total number of particles in the beam times the probability

$$\lim_{t \to +\infty} P_{\rho_t}(\mathbb{R}^3 \times C_\omega) = P_{\rho_{\text{out}}}(\mathbb{R}^3 \times C_\omega) = \int_{\mathbb{R}^3} d\mathbf{q} \int_{C_\omega} \langle \mathbf{q}, \mathbf{p} | \rho_{\text{out}} | \mathbf{q}, \mathbf{p} \rangle \, d\mathbf{p} \tag{7.43}$$

of finding each particle with (stochastic) momentum within C_ω long after the scattering has taken place. In (7.43) ρ_t is the density operator describing the actual state of the ensemble, whose asymptotic states are described by ρ_t^{in} and ρ_t^{out}, so that (Prugovečki, 1981a, p. 486)

$$\lim_{t\to\mp\infty} \operatorname{Tr}[(\rho_t - \rho_t^{\text{ex}})^2] = 0, \quad \text{ex} = \text{in, out}. \tag{7.44}$$

These asymptotic states are related at $t = 0$ (and at all other times) by the scattering superoperators $\underline{S}$:

$$\rho_{\text{out}} = \underline{S}\rho_{\text{in}} = S\rho_{\text{in}}S^*, \quad \underline{S} = \underline{\Omega}_-\underline{\Omega}_+^*. \tag{7.45}$$

The mean incident flux of the still finite beam obviously equals the number of particles per unit beam volume times their mean velocity $|\mathbf{p}_0|/m$. Hence, if $|C_\omega|$ denotes the solid angle cut out by the cone C_ω of momentum values, then by (7.43)

$$\sigma(\mathbf{p}_0 \to \omega) = \lim_{b,\hbar\to\infty} \frac{\hbar bm}{|\mathbf{p}_0|} \lim_{|C_\omega|\to 0} \frac{1}{\Delta t} \frac{1}{|C_\omega|} \int_{C_\omega} \rho_{\text{out}}(\mathbf{p})\, d\mathbf{p}, \tag{7.46}$$

where $\Delta t = \hbar m |\mathbf{p}_0|^{-1}$ is the length of time between the arrival of the first and the last particle in the beam within interaction range, and

$$\rho_{\text{out}}(\mathbf{p}) = \int_{\mathbb{R}^3} \langle \mathbf{q}, \mathbf{p}|\, \rho_{\text{out}}\, |\mathbf{q}, \mathbf{p}\rangle\, d\mathbf{q}. \tag{7.47}$$

It should be noted that by (7.41) and (7.42) ρ_{in} (and therefore also ρ_{out}) depends on b and $\hbar$, so that in

$$\sigma(\omega) = \sigma(\mathbf{p}_0 \to \omega) = \lim_{b,\hbar\to\infty} b \int_0^\infty \rho_{\text{out}}(p\omega)p^2\, dp \tag{7.48}$$

the limit in $\hbar$ cannot be bypassed. Furthermore, the density $\rho_{\text{out}}(\mathbf{p})$ is in the classical case (7.41) at sharp momentum values $\mathbf{p}$, whereas in the quantum case (7.42) it is at the stochastic values $\underset{\sim}{\mathbf{p}} = (\mathbf{p}, \overline{\chi}_{\mathbf{p}}^{(l)})$ with the confidence function $\overline{\chi}_{\mathbf{p}}^{(l)}$ in (1.6.27b). However, as a special case of the marginality property (3.14b) we get from (7.47) in case of $\rho_{\text{out}} = \rho_{\text{out}}^{(l)} = W_{(l)}\hat{\rho}_{\text{out}}$

$$\rho_{\text{out}}^{(l)}(\mathbf{p}) = \int_{\mathbb{R}^3} \overline{\chi}_{\mathbf{p}}^{(l)}(\mathbf{k})\langle\mathbf{k}|\, \hat{\rho}_{\text{out}}\, |\mathbf{k}\rangle\, d\mathbf{k} \longrightarrow \langle\mathbf{p}|\, \hat{\rho}_{\text{out}}\, |\mathbf{p}\rangle, \quad l \to +\infty \tag{7.49}$$

In view of (7.49), it might be thought that if we take the limit $l \to +\infty$ in (7.48) as well, we should eventually arrive at the conventional differential cross-section formula (7.3). This, however, is not the case, since all rigorous derivations of (7.3) from the transition probabilities for scattering of particles in an initially finite beam require,[14] prior to taking the limits in (7.48), an averaging of $\rho_{\text{out}}(\mathbf{p})$ over an array of randomly distributed scattering centers in a planar target perpendicular to the incident beam. Thus $\sigma(\omega)$ in (7.3) can be appropriately called a *planar-target* scattering differential cross-section. If, however, this averaging is not performed – i.e. if we are dealing with a single target particle – then we get (Progovečki, 1978c) in both the classical and quantum case the formula (7.38) for $\sigma(\omega)$ that enters (7.2). Hence this $\sigma(\omega)$ can be appropriately called a *point-target* scattering differential cross-section. The more detailed form of the formal inner product in (7.38) is:

$$\langle \mathbf{w} | \, \underline{H}_I \underline{\Omega}_+ \, | \mathbf{p}_0 \rangle = \int_{\Gamma^3} \Phi^*_{0,\mathbf{w}}(\mathbf{q}, \mathbf{p}) \underline{H}_I \Phi^+_{0,\mathbf{p}_0}(\mathbf{q}, \mathbf{p}) \, d\mathbf{q} \, d\mathbf{p}. \tag{7.50}$$

This should be compared with the detailed form of the formal inner product in (7.3):

$$\langle \mathbf{w} | \, H_I \Omega_+ \, | \mathbf{k} \rangle = \int_{\mathbb{R}^3} \hat{\Phi}^*_{\mathbf{w}}(\mathbf{x}) \hat{H}_I \hat{\Phi}^+_{\mathbf{k}}(\mathbf{x}) \, d\mathbf{x}. \tag{7.51}$$

In (7.51) $\{\Phi_\mathbf{p} \mid \mathbf{p} \in \mathbb{R}^3\}$ constitutes an eigenfunction expansion for $\hat{H}_0$ (i.e. in the configuration representations for $\hat{H}_0$ in (4.12) these are plane waves), and

$$\hat{\Phi}^+_\mathbf{k} = \hat{\Omega}_+ \hat{\Phi}_\mathbf{k}, \qquad \hat{\Phi}_\mathbf{k}(\mathbf{x}) = (2\pi\hbar)^{-3/2} \exp\left(\frac{i}{\hbar} \mathbf{k} \cdot \mathbf{x}\right), \tag{7.52}$$

provide an eigenfunction expansion for the continuous part of the spectrum of $H = H_0 + H_I$ (Prugovečki, 1981a, pp. 491–510). In $L^2(\Gamma^3_{(l)})$ the eigenfunction expansion for $H_0^{(l)} = W_{(l)} \hat{H}_0 W^{-1}_{(l)}$ is given by

$$\Phi^{(l)}_\mathbf{k}(\mathbf{q}, \mathbf{p}) = \exp\left(-\frac{i}{\hbar} \mathbf{q} \cdot \mathbf{p}\right) \tilde{\xi}^{(l)*}_{\mathbf{q},\mathbf{p}}(\mathbf{k}), \tag{7.53}$$

with $\tilde{\xi}^{(l)}_{\mathbf{q},\mathbf{p}}(\mathbf{k})$ in (1.6.30) (the exponential factor in (7.53) being due to the choice of gauge in (6.4)), so that (7.51) can be rewritten as

$$\langle \mathbf{w} | \, H_I \Omega_+ \, | \mathbf{k} \rangle = \int_{\Gamma^3} \Phi^{(l)*}_\mathbf{w}(\mathbf{q}, \mathbf{p}) H_I^{(l)} \Phi^{(l)+}_\mathbf{k}(\mathbf{q}, \mathbf{p}) \, d\mathbf{q} \, d\mathbf{p}. \tag{7.54}$$

The eigenfunction expansions for the corresponding superoperators $\underline{H}_0^{(l)}$ are:

$$\Phi^{(l)}_{\mathbf{w},\mathbf{k}}(\mathbf{q}, \mathbf{p}; \mathbf{q}', \mathbf{p}') = \left(\frac{\hbar}{2}\right)^{3/2} \Phi^{(l)}_{(\mathbf{k}+\hbar\mathbf{w})/2}(\mathbf{q}, \mathbf{p}) \Phi^{(l)*}_{(\mathbf{k}-\hbar\mathbf{w})/2}(\mathbf{q}', \mathbf{p}'). \tag{7.55}$$

Hence they satisfy the eigenvalue equation

$$\underline{H}_0^{(l)} \Phi^{(l)}_{\mathbf{w},\mathbf{k}} = \frac{2\mathbf{w}\cdot\mathbf{k}}{m} \Phi^{(l)}_{\mathbf{w},\mathbf{k}}, \tag{7.56}$$

with eigennumber adjusted to the classical part of L_0 in (7.28), so that the eigenfunctions of $L_0^{(l)}$ are

$$f^{(l)}_{\mathbf{w},\mathbf{k}}(\mathbf{q},\mathbf{p}) = (2\pi\hbar)^{3/2}\, \Phi^{(l)}_{\mathbf{w},\mathbf{k}}(\mathbf{q},\mathbf{p};\mathbf{q},\mathbf{p}), \tag{7.57}$$

and give rise to (7.30). On the other hand, the classical counterpart of (7.57) is

$$f^{\mathrm{cl}}_{\mathbf{w},\mathbf{k}}(\mathbf{q},\mathbf{p}) = (2\pi)^{-3/2} \exp(i\mathbf{w}\cdot\mathbf{q})\, \delta^3(\mathbf{p}-\mathbf{k}), \tag{7.58}$$

and it gives rise to (7.30). Hence, in the classical case (7.38) reduces to the formula

$$\sigma^{\mathrm{cl}}(\omega) = m \int_0^\infty \mathrm{d}p\, p \int_{\mathbb{R}^3} \mathrm{d}\mathbf{q}\, \nabla_{\mathbf{q}} V(\mathbf{q}) \cdot \nabla_{\mathbf{p}} f^{\mathrm{cl}}_{\mathbf{0},\mathbf{p}_0}(\mathbf{q}, p\omega) \tag{7.59}$$

first derived by Miles and Dahler (1970, Eq. (42)) for the classical differential cross-section, which equals the one usually derived by the more direct trajectory method (Balescu, 1975, Eq. (11.4.19)).

In the quantum case, a formula similar to (7.59)

$$\sigma^{(l)}(\omega) \approx m \int_0^\infty \mathrm{d}p'\, p' \int \mathrm{d}\mathbf{k}\, f^{(l)}_{\mathbf{0},\mathbf{k}}(p'\omega)\, \frac{p'}{|\mathbf{k}|} \int \mathrm{d}\mathbf{q} (L_I^{(l)} f^{(l)+}_{\mathbf{0},\mathbf{p}})(\mathbf{q},\mathbf{k}) \tag{7.60}$$

follows from (7.38) only in an approximative sense, and it leads to the collision term (7.2) only at values of l that are small enough to allow us to neglect the term in curly brackets in (7.27) when computing $f^{(l)+}_{\mathbf{0},\mathbf{p}}$ by the Green's operator method. However, earlier in this section we had to choose l sufficiently large to neglect the term in curly brackets in (7.34). Thus, the Boltzmann equation (7.1) with the collision term (7.2) is approximately valid in the quantum case for a limited range of values of l. In that range the accuracy of localization of the two-body collisions that give rise to the quantum Boltzmann equation for $\Gamma^3_{(l)}$ distribution functions $\rho_t^{(l)}(\mathbf{q},\mathbf{p})$ should be very good by macroscopic standards, and yet at the same time it should be related only to regions which are much larger than the range of interaction in the collision process.

Notes

[1] Spin as well as other internal degrees of freedom could have been easily included into all subsequent considerations, without changing the validity of the results discussed in this chapter. However, for the sake of notational simplicity we refrain from dealing explicitly with these degrees of freedom.

2 A mathematically rigorous derivation of all properties of Liouville spaces (i.e. of Hilbert–Schmidt classes) that are of physical significance can be found in Section 8 of Chapter IV by Prugovečki (1981a).

3 That is, a map preserving convex combinations $\lambda_1 \rho_1 + \lambda_2 \rho_2$, $\lambda_1, \lambda_2 \geq 0$, $\lambda_1 + \lambda_2 = 1$, rather than all possible linear combinations. Indeed, the set $\mathscr{R}(\mathscr{H})$ of density operators is convex, but it is not linear.

4 By Cohen (1966), Agarwal and Wolf (1970), Srinivas and Wolf (1975) and many others. A unified and mathematically rigorous treatment has been presented by Guz (1983a).

5 In discussing precisely this point, Wigner (1963, p. 14) has remarked: "For some observables, in fact for the majority of them (such as xyp_z), nobody seriously believes that a measuring apparatus exists. It can even be shown that no observable which does not commute with the additive conserved quantities (such as linear or angular momentum or electric charge) can be measured precisely, and in order to increase the accuracy of the measurement one has to use a very large measuring apparatus."

6 Stochastic spin can be introduced in a consistent manner (Prugovečki, 1977b; Busch, 1982), and POV measures for the measurement of noncommuting spin components can be constructed (Schroeck, 1982a).

7 The general aspects of the theory of Weyl transforms has been presented by Pool (1966). Its intimate relation to Wigner transforms has been studied by Balazs (1980). A development of the theory of stochastic phase spaces from the point of view of Weyl transforms has been given by Schroeck (1978, 1981, 1982b).

8 It should be evident by now that in this monograph we subscribe to the orthodox point of view with regard to the relation of wave functions to quantum systems (Heisenberg, 1930; Bohr, 1961), according to which wave functions describe the states of single quantum systems, and not merely that of ensembles. This point of view is reinforced by the concept of proper wave function.

9 In fact, as illustrated by the quotes at the beginning of chapters, it has been forcefully argued by Born (1955c, 1956, 1962) that the physically correct description of classical systems of particles should be in terms of the probability distributions (4.28), rather than in the deterministic terms of trajectories, since due to the impossibility of ever constructing perfectly accurate instruments it is the former rather than the latter that reflects empirical realities. It should be noted, however, that (4.28) corresponds to a perfectly sharp preparatory measurement. In a completely realistic set-up, the preparatory measurement provides only stochastically spread-out values, and the time evolution of the classical system can then exhibit features that are conventionally thought of as purely quantum mechanical in nature (Born, 1955c).

10 McKenna and Frisch (1966) employed Husimi (1940) transforms, which essentially equal $\rho^{(l)}(q, p)$ for $l^2 = 1/2$ in any system of units with $\hbar = 1$. The first-order corrections to the friction coefficient in the Fokker–Planck equation for Brownian motion which they obtained are not found in the Wigner transform approach (Hynes and Deutsch, 1969). This indicates that if l would be treated as a parameter, those corrections would disappear in the limit $l \longrightarrow +\infty$ (Prugovečki, 1978a, p. 119).

11 Actually, Ross and Kirkwood (1954, p. 1100) have already noted that something is amiss with Equation (7.3), when it is used beyond the first Born approximation in the context of the Uhling–Uhlenbeck (1933) equation, where "the usual formulation of the quantum-mechanical cross-sections may be inadequate to describe the transition probabilities due to scattering".

12 See, e.g., Snider and Sanctuary (1971), or Tip (1971). The right-hand side of (7.18) also presupposes the Maxwell–Boltzmann statistics for the ensemble even in the quantum case, but it can be argued (Uhlenbeck, 1973) that in the realm of low densities, where the Boltzmann equation provides a good approximation of the actual dynamics contained in (7.11) or the BBGKY hierarchy, the effects of the Bose–Einstein or Fermi–Dirac statistics can be ignored.

[13] We present the derivation given by Prugovečki (1978c). A different approach to this derivation has been presented by Turner and Snider (1980), who label $\sigma(\omega)$ in (7.38) a 'generalized differential cross-section'. It will be clear, however, from the subsequent discussion that (7.38), or its equivalent forms (7.46) and (7.59)–(7.60), cannot be regarded as a generalization of (7.3), since (7.3) does not emerge as a special case of (7.38), but rather as operationally distinct from (7.38) – as indicated by the suggested terminology of 'point-target' versus 'planar-target' differential cross-section for σ and $\hat{\sigma}$, respectively.

[14] See, e.g., Messiah (1961, Ch. X, Secs. 5–6), Taylor (1972, Sec. 3–e), Newton (1979), Prugovečki (1981a, Ch. 5, Sec. 1.8). It is sometimes maintained that this averaging reflects the fact that one is dealing with an incident beam of particles, rather than with a single particle. In view of the fact that (7.42) already describes a beam, and that the same procedure in the classical case leads via (7.41) to the cconventional differential cross-section without additional averagings, this claim cannot be truly sustained (Prugovečki, 1978c, p. 246; 1979, p. 586).

Part II

Quantum Spacetime

"If one disregards quantum structure, one can justify the introduction of the [metric tensor] g_{ik} 'operationally' by pointing to the fact that we can hardly doubt the physical reality of the elementary light cone which belongs to a point. In doing so one implicitly makes use of the existence of an arbitrarily sharp optical signal. Such a signal, however, as regards quantum facts, involves infinitely high frequencies and energies, and therefore a complete destruction of the field to be determined. That kind of a physical justification for the introduction of the g_{ik} falls by the wayside, unless one limits oneself to the 'macroscopic'."

EINSTEIN (1949, p. 676)

"Since all knowledge (or almost all) is doubtful, the concept 'uncertain knowledge' must be admitted ... The supposed absolute concept 'knowledge' should be replaced by the concept 'knowledge with degree of certainty *p*', where *p* will be measured by mathematical probability when this can be ascertained."

RUSSELL (1948, p. 498)

Chapter 4

Reciprocity Theory and the Geometrization of Stochastic Quantum Mechanics

"The question whether the structure of this [spacetime] continuum is Euclidean, or in accordance with Riemann's general scheme, or otherwise, is . . . properly speaking a physical question which must be answered by experience, and not a question of a mere convention to be selected on practical grounds."

EINSTEIN (1923; 1953, p. 193)

"Ordinary relativity is based on the invariance of the 4-dimensional distance, or its square $R = x_k x^k$. Can only really define distance of two particles in sub-atomic dimensions independently of their velocity? This seems to me not evident at all."

BORN (1949a, p. 208)

In Part I of this monograph we have been striving to solve the localizability problem in relativistic quantum mechanics by viewing a quantum particle as a kind of probability 'cloud' against a classical spacetime background. Our main aim was to arrive at a mathematical framework *capable* of a consistent physical interpretation due to the presence of *bona fide* probability densities and currents that transform in a covariant manner under the change of inertial frames. The epistemologically deeper question as to the existence of such frames, and, furthermore, as to the ultimate validity at the nuclear or even the subnuclear level of the pure classical description of spacetime in terms of a continuum – such as some differentiable manifold with a metric structure – had not been considered, or even posed.

And yet, as seen from the quote heading Part II of this monograph, Einstein had himself pointed out that the pseudo-Riemannian (or for that matter purely continuum) description of spacetime adopted in classical relativity theory cannot, strictly speaking, be reconciled with quantum mechanics in general, and the uncertainty principle in particular. In fact, from the very early days of quantum mechanics, there have been speculations as to a fundamentally atomistic structure of spacetime (Latzin, 1927; Pokrowski, 1928; Schames, 1933). The introduction by Heisenberg (1927) of the uncertainty principle promptly spurred papers on spacetime quantization related to elementary space and time uncertainties (Ruark, 1928; Flint and Richardson, 1928; Fürth, 1929; Landau and Peierls, 1931; Glaser and Sitte, 1934; Flint, 1937). The subsequent introduction of a fundamental length l_0 in quantum theory was again pioneered by Heisenberg (1938, 1942) himself, and as we shall see when we discuss reciprocity theory later in this chapter, Born (1938,

1949) based his idea of a metric operator on the existence of such an l_0. Other researchers (March, 1934, 1937; Markov, 1940) realized that this idea could be related to a statistical concept of a metric a few years prior to Menger's (1942) first formulation of axioms for the statistical metric spaces discussed in Section 1.1.

By the late forties the possible connection of a quantum structure of spacetime with the problem of divergences in quantum field theory was clearly recognized, and a notion of discrete quantized spacetime was proposed (Snyder, 1947) and pursued by a host of researchers in an effort to provide a cure for these divergences (Yang, 1947; Flint, 1948; Schild, 1948; Hill, 1950; Das, 1960; Gol'fand, 1960, 1963; Kadyshevskiĭ, 1962, 1963). However, the chimera born out of the numerical successes of renormalization theory in quantum electrodynamics apparently kept away all pragmatically minded theoreticians from pursuing this line of research, albeit, as we shall discuss in the next chapter, the renormalization program has left just about all truly fundamental difficulties facing quantum field theory totally unresolved. On the other hand, the fortuitous fact that gravitational fields have resisted (Deser, 1980) inncorporation into this program even at the most formal of computational levels has kept alive even in recent times the interest in advancing new schemes for the quantization of spacetime (Penrose and MacCallum, 1973; Papp, 1974, 1975; Taylor, 1978, 1978) – or in the alternative formulations of stochastic spacetimes (Frederick, 1966; Namsrai, 1980, 1981; Roy, 1980, 1981; Halbwachs *et al*., 1982).

Judged against the extensive background[1] of very diverse and sometimes elaborate schemes for the quantization of spacetime, the approach advocated in this chapter offers a new *mathematically consistent* framework, whose foundations have been laid in Part I. This framework shall be now combined in a *physically consistent* manner with the basic premise underlying Einstein's (1905, 1916) approach to the mathematical description of spacetime – namely, the view that this description should not be chosen on *a priori* grounds (be they motivated by tradition, metaphysics, convenience, aesthetics, utilitarianism, etc.) but should reflect *operationally defined* spatio-temporal relationships between test particles. Thus, the mathematical framework developed in Part I, and especially in Chapter 2, can now be applied to the problem of reconsidering those operational procedures from a quantum rather than classical perspective. The purely logical outcome of this analysis is a quantum description of spacetime by means of stochastic geometries formulated in terms of families of propagators describing quantum test particles in free fall. In the absence of gravitational fields, these propagators are supplied by the results of Sections 2.6, 2.7 and 2.9. The extrapolation to nonvanishing gravity requires, however, a purely quantum formulation of the strong equivalence principle (Carmeli, 1977, p. 132) as to the existence of local Lorentz (classical) reference frames, thereby enabling the introduction of local Minkowski coordinates in general relativity. We find the key to that extrapolation in the relativistic canonical commutation relations (RCCRs). This in turn suggests a very general formulation of relativistic stochastic quantum mechanics which can be appropriately described (Prugovečki, 1982c) by the term[2] 'geometrization'.

This geometrization of relativistic stochastic quantum mechanics is capable of supplying a general concept of quantum spacetime that is self-consistent as well as consistent with the basic tenets of both quantum mechanics and general relativity. However, the development of a general theory of elementary particles within this framework requires new physical principles capable of reducing the very wide class of proper wave functions, arrived at in Chapter 2 on general grounds of relativistic covariance and probability current conservation, to a narrower family that should describe only the extended elementary particles that actually occur in Nature. It turns out that Born's (1949) reciprocity principle, as well as other key ingredients of the reciprocity theory developed by M. Born and his coworkers from 1938 to 1949, can be incorporated in a most natural manner into the present quantum spacetime framework. The outcome is a new version of the reciprocity theory that is capable of producing experimentally verifiable predictions even at its most elementary level – as discussed in the last two sections of this chapter.

4.1. The Operational Foundations of Special and General Relativity

The perusal of Einstein's (1905, 1916) seminal papers on special and general relativity reveals that Einstein's key epistemological idea – the idea which constituted the cornerstone of the entire ensuing mathematical framework, such as Lorentz transformations, pseudo-Riemannian structure, etc. – resided in the premise that spatio-temporal relationships do not have an *a priori* mathematical meaning (as was assumed in the neo-Kantian outlook that had been prevailing until Einstein's times), but rather that such a meaning has to *result* from empirically verifiable properties of operational definitions[3] of such relationships. For example, although the principle of general covariance is stated in textbooks on general relativity (Adler *et al.*, 1975, p. 117; Weinberg, 1973, p. 92, etc.) in various forms, that are not exactly mathematically equivalent (Trautman, 1965) but primarily invoke tensorial features of physical laws, in the original form stated by Einstein (1916) its connotation is straightforward and very clear: physical reality does not reside in the coordinates themselves, but rather in the events which those coordinates label! Thus, Einstein (1918) totally concurred with Kretschmann's (1917) comment that *any* physical law dealing with coordinate-dependent quantities can be written in covariant form. Hence, it might appear that all the textbook versions of the covariance principle (such as that of the statement that the equations of physics should have tensorial form) are mere mathematical tautologies. That is indeed so, as long as it is overlooked that the Minkowski coordinates in special relativity are not mere labels, but do have a direct operational meaning. Since in general relativity the *strong principle of equivalence* (Carmeli, 1977, p. 132; Dicke, 1964, p. 4) essentially states that *any* spacetime event O can serve as the origin of a *local Lorentz frame*, which therefore *operationally defines local inertial coordinates* x^μ, the nontautological nature of the principle of general covariance emerges only in combination with the strong principle of equivalence in the following sense: By the strong principle of equivalence, one can always transform an arbitrarily chosen system of coordinate labels $\bar{x}^\mu$ of events in a neighborhood of O into local Minkowski inertial coordinates x^μ with respect to O, adopted as the origin of the local frame and labeled therefore by $x^\mu = 0$:

$$\bar{x}^\mu \mapsto x^\mu = x^\mu(\bar{x}^\nu), \qquad \mu, \nu = 0, \ldots, 3, \tag{1.1}$$

$$\bar{g}^{\mu\nu}(\bar{x}^\alpha) \mapsto g^{\mu\nu}(x^\alpha) = \frac{\partial x^\mu}{\partial \bar{x}^\kappa} \frac{\partial x^\nu}{\partial \bar{x}^\lambda} \bar{g}^{\kappa\lambda}(\bar{x}^\alpha), \tag{1.2}$$

$$g^{\mu\nu}(0) = \eta^{\mu\nu} = \mathrm{diag}(1, -1, -1, -1). \tag{1.3}$$

In a neighborhood of O, the coordinates x^μ are operationally defined. Consequently, the transformation laws of any physical quantity expressed in terms of x^μ become subject to empirical verification as we change local Minkowski coordinates by (operationally) changing local Lorentz frames. Thus, if a *formally covariant* formulation of a physical law holds in one local Lorentz frame at O, but not in some other such frame, that formulation is not *operationally covariant*. It therefore violates the principle of covariance interpreted operationally rather than mathematically, and should therefore be discarded.

Although perhaps never explicitly stated in the above form by Einstein, there is little doubt that the above analysis reflects the true (and physically nontrivial) connotation of the principle of general covariance that emerges from Einstein's original writings on this subject. It is also clear from these early writings (Einstein *et al.*, 1923) that Einstein's operational formulation of relativistic spacetime rested on five primitive concepts: test particle, light signal, event, rigid rod and standard clock. Of these, the last one is redundant "since a light signal, which is reflected back and forth between the ends of a rigid rod, constitutes an ideal clock, provided that the postulate of the constancy of the light-velocity in vacuum does not lead to contradictions" (Einstein, 1949, p. 55). On the other hand, the concept of rigid rod violates the principle of special relativity since, if such an object existed in Nature, one could instantaneously transmit a signal from one end to the other end of the rod by a purely kinematical operation – such as imparting an impulse to one of the two ends and observing the instantaneous recoil at the other end.

Einstein was very much aware of this contradiction, but it appears that even as late as 1951 he maintained that at least the use of 'infinitesimal' rigid rods was justified in establishing the spacetime geometry in an existing physical situation – such as on a rotating disk (Stachel, 1980, p. 8). However, rigid rods can be altogether eliminated from classical relativity theory through the introduction of the concept of the geodesic light clock (Marzke and Wheeler, 1964; Harvey, 1976), defined by a light signal that bounces between two test particles, one in free fall and therefore following a worldline which is a timelike geodesic, and the second moving along a 'parallel' worldline constructed by Schild's ladder method (Misner *et al.*, 1973, p. 397). This operational construction presupposes, of course, the existence of a pseudo-Riemannian geometry from the outset. On the other hand, one can eliminate this assumption by introducing the concept of 'test particle in free fall' as primitive, and then postulating a number of features about the behavior of test particles and light signals (Castagnino, 1971; Ehlers *et al.*, 1972). With such ideal clocks in existence, one can then *operationally* define the (infinitesimal) spacetime distance

$$ds^2 = g_{\mu\nu}(\bar{x}^\lambda)\, d\bar{x}^\mu\, d\bar{x}^\nu \tag{1.4}$$

between two lightlike separated neighboring events O and O' (labeled by some general coordinates $\bar{x}^\mu$ and $\bar{x}^\mu + d\bar{x}^\mu$, respectively), by placing ideal clocks at O and O' and measuring the duration required of a light signal to reach from O to O', or vice versa – depending on whether O' lies in the future or past light cone of O. Similar commonsensical operations can be used to define the spacetime distance of timelike or spacelike separated events, as well as to operationally define other basic geometric relationships (Marzke and Wheeler, 1964). However, a totally rigorous derivation of the pseudo-Riemannian nature of the ensuing geometries requires a number of axioms concerning the manner in which all the primitive concepts (namely the *events, light signals, test particles*, and *freely falling particles*) are interrelated (Kronheimer and Penrose, 1967; Ehlers *et al.*, 1972; Ehlers and Schild, 1973; Woodhouse, 1973; Ehlers, 1973, Section 2).

It is with regard to these basic postulates that fundamental physical objections can be raised. First of all, as Margenau (1949, p. 258) pointed out in a critical essay dealing with Einstein's conception of physical reality, the existence of test particles that can be regarded as material points is unwarranted. Second, and even more important, the postulated behavior of test particles lies totally beyond the realm of quantum laws of motion. In other words, is it sensible to postulate the existence of point test particles that behave in a totally deterministic manner, which is totally unaffected by the uncertainty principle?

Einstein's (1949, p. 676) answer to this kind of question was that all operationally based justification for a pseudo-Riemannian description of spacetime "falls by the wayside, unless one limits himself to the 'macroscopic'. The application of the formal basis of general relativity to the 'microscopic' can, therefore, be based only upon the fact that the metric tensor is the simplest covariant structure which can come under consideration." Naturally, using (formal) covariance to justify the retention of a geometric description that gives meaning to the formal concept of covariance in the first place is rather circular reasoning, and Einstein hastens to admit in the same paragraph that "such argumentation, however, carries no weight with anyone who doubts that we have to adhere to the continuum at all". But then he rhetorically asks: "All honor to his doubt – but where else is there a passable road?"

For anyone not as totally dedicated to strict determinacy in physics as Einstein had been, alternative passable roads lie in an obvious direction: stochastic geometries, that embody the uncertainty principle in their framework as intimately as conventional geometries, from Euclidean to Riemannian, do embody a deterministic outlook when they are applied to the physical world around us. That this is a correct general direction in which to search if one believes in the basic tenets of quantum theory is quite obvious as soon as one regards all test particles in quantum terms. The light signals assume then the appearance of beams of photons, whose collisions with a quantum test particle produce *quantum events*, and the spatio-temporal relationships of such quantum events can be described only in stochastic terms if the uncertainty principle is universally valid. Indeed, in order to execute the necessary compensations (Heisenberg, 1930; Bohr, 1961) for the recoil momentum imparted by a photon of 3-momentum **k** to a quantum test particle, one has to send a photon (or some other kind of particle) of momentum $-\mathbf{k}$ on a collision course with the test particle. To make the compensation *exact*, one would have to arrange that this second photon meets the test particle at the exact time and place of collision between the first photon and the test particle. Hence, the setup presupposes an exact knowledge of simultaneous position and momentum for both photons – contrary to the uncertainty principle.

Any of the various alternatives for *gedanken* experiments (Heisenberg, 1930; Bohr, 1961) lead to the same conclusion: a strictly deterministic operational definition of spatio-temporal relationships is impossible if the uncertainty principle is generally correct. Of course, that leaves the possibility of an asymptotically deterministic description corresponding to the infinite-mass limit $m \to \infty$ for the test particles, since for given uncertainties Δx^j in position, the uncertainties in velocity

$$\Delta v^j \approx \hbar/2m \quad \Delta x^j, \quad j = 1, 2, 3, \tag{1.5}$$

could be reduced indefinitely by working with sufficiently massive test particles. However, this procedure presupposes the factual possibility of finding in Nature, or at least constructing in the laboratory, arbitrarily massive test bodies of arbitrarily small spatial extensions in their own rest frames. Not only is such an assumption unwarranted,[4] but it stands in direct contradiction to any cosmological model which envisages a Universe with a finite number of particles (Eddington, 1946; Steigman, 1976), and in which therefore the limit $m \to \infty$ would be operationally meaningless.

Setting aside such cosmological arguments, which at the present stage of scientific knowledge might be deemed esoteric and not properly founded (Alfvén, 1977), we are still left with the fact that in any actual measurement the test bodies have a finite mass, and therefore, by the uncertainty principle the correct description of the outcome of a measurement procedure can be only in stochastic terms.[5] The question is, therefore, how to devise stochastic geometries which in the macroscopic limit, as well as in the infinite mass limit $m \to \infty$ for test particles, merge in the mean with their pseudo-Riemannian counterparts that describe classical models of spacetime. It is to this fundamental problem that we turn to next.

4.2. Coordinate Fluctuation Amplitudes for Coherent Flows of Quantum Test Particles

The analysis of the nontautological aspects of the principle of general covariance which we had presented in Section 4.1 indicated that the following paraphrasing of the customary textbook formulations of this principle reflects more accurately its physically nontrivial aspects: all spatio-temporal relationships are operationally defined by coincidences between test particles in free fall and light signals; these events can be labeled by coordinates, but all physical laws should be expressible in a manner in which the only dependence on coordinates is via the events themselves. We shall refer to this paraphrasing as the *principle of operational covariance*. It can be most easily implemented if we adopt a geometrodynamic (Wheeler, 1962, 1979; Misner *et al.*, 1973) point of view, in which spacetime is viewed dynamically, i.e. as an evolution of 3-geometries in the following sense: In principle we can execute a foliation of spacetime into synchronous slices (Lifshitz and Khalatnikov, 1963; Landau and Lifshitz, 1971). In pseudo-Riemannian spacetime these slices would be spacelike hypersurfaces σ, orthogonal to timelike geodesics resulting during the introduction of Gaussian normal coordinates (Adler *et al.*, 1975, p. 63) – which in the flat case can be chosen to be the Minkowski coordinates (σ being then spacelike hyperplanes). The earlier-mentioned evolving 3-geometries are those of the hypersurfaces σ, and can be most easily visualized by setting at the points of an initial-data hypersurface σ_0 test particles of a given species A (e.g. electrons, protons, etc.) which are locally at rest relative to each other. If we assume that all these test particles do not interact and that they are allowed to fall freely, we obtain a classical coherent flow (cf. Section 2.3) if we imagine that these test particles behave classically, i.e. follow strictly deterministic worldlines, namely the earlier-mentioned timelike geodesics. If we parametrize these worldlines by the proper time τ of the test particles that describe them, then each locus of points corresponding to $\tau = \text{const}$ will yield the hypersurface σ of operational simultaneity, thus presenting an operationally defined image of evolving 3-geometries.[6]

Of course, the execution of any operational procedure involving the exchange of light signals (or any other kind of signals) between the test particles in the coherent flow has to incorporate recoil-compensating procedures meant to insure that no test particle is deflected from its natural free-fall worldline. It is at this stage that, as described in the preceding section, the uncertainty principle draws a sharp line between the classical and the quantum analysis of one and the same set of operational procedures (Salecker and Wigner, 1958).

Classically, the compensations can be made exact, and each test particle will remain at rest in relation to other test particles in its immediate vicinity, i.e. its 4-velocity u^μ will always point in the direction of the normal of σ at any point $\mathbf{x} \in \sigma$ where its worldline impinges a $x^0 = \tau = \text{const}$ hypersurface σ. On the other hand, quantum mechanically, its 4-momentum and therefore its 4-velocity u^μ cannot be controlled with exactitude, but only stochastically: under repeated measurements, one can made sure only in the average that its 3-velocity relative to particles in its neighborhood remains zero. However, when subsequent readings are compared, fluctuations in positions as well as in velocity are observed. In other words, whereas classically the Gaussian normal coordinates x^j, as well as u^μ, would remain constant for all values of x^0, quantum mechanical fluctuations are to be observed around the x^j and the u^μ values. These fluctuations can be described mathematically by a *coordinate probability density* $G_A(x, u; q, v)$ for obtaining readings q^μ and v^μ instead of x^μ and u^μ. Since in quantum mechanics all probability densities are derivable from probability amplitudes, we postulate that

$$G_A(x, u; q, v) = N_A K_A^*(x, u; q, v) \cdot K_A(x, u; q, v). \tag{2.1}$$

where K_A is a *coordinate fluctuation (probability) amplitude*, assuming for each $\zeta = (x, u)$ and $\zeta' = (q, v)$ values $K_A(x, u; q, v) \in \mathcal{H}_{\text{int}}$ in some Hilbert space $\mathcal{H}_{\text{int}}$ for internal degrees of freedom, such as spin, isospin, etc., and N_A is a normalization constant.

Thus, we observe that the uncertainty principle alone suggests that in describing quantum fluctuations in the values of coordinates labelling *quantum* test particles, the 4-momentum degrees of freedom have to be taken into account.[7] Of course, by itself, this principle provides no clue as to the form of the coordinate probability amplitudes K_A, or as to their basic properties. On the other hand, since any K_A obviously describes the propagation of an A-type quantum test particle of mean 4-momentum $p^\mu = mu^\mu$ and mean Gaussian normal coordinate x^j along the x^0 = const hypersurfaces σ, we expect that K_A will possess the basic properties of propagators. Extrapolating from (2.9.4), we should therefore require that

$$K_A(x, u; q, v) = K_A^*(q, v; x, u)$$

$$= \int_{\Sigma_A^\pm} K_A(x, u; q', v') \cdot K_A(q', v'; q, v) \, d\Sigma(q', v') \tag{2.2}$$

where $\Sigma_A^\pm$ will now be hypersurfaces in general relativistic phase spaces (Ehlers, 1971, p. 27), or more exactly in:

$$\mathcal{M}_m^\pm = \{(q, v) \mid q \in \mathcal{M}(1, 3), p = mv \in \mathcal{V}_m^\pm\}. \tag{2.3}$$

Thus $\mathcal{M}_m^\pm$ is a subset of the cotangent bundle over the pseudo-Riemannian manifold $\mathcal{M}(1, 3)$ obtained by assigning to each $q \in \mathcal{M}(1, 3)$ the tangent space at q as a fibre.

We can operationally define spacetime as the family of all conceivable coherent flows of test particles. If we envisage these test particles to be pointlike, and their behavior to be governed by classical laws, then due to the weak equivalence principle (namely the hypothesis that gravitational mass is exactly equal to the inertial mass) we can eliminate from explicit consideration all species of test particles except one. Indeed, for identical initial conditions (same spacetime location along σ_0 and same

4-velocity), the weak equivalence principle ensures that the worldlines of any two test particles in free fall will be identical.

Quantum mechanically, however, due to the evanescence of free wave packets, such a simplification is not altogether feasible. Indeed, already in nonrelativistic quantum mechanics the wave functions of free particles of different masses evanesce at different rates from a bounded region of space even when the original configuration space functions are identical. The present mathematical description of quantum spacetime consists not only of a pseudo-Riemannian manifold $\mathscr{M}(1, 3)$, whose points q represent the mean spacetime locations of the quantum test particles in coherent flows in relation to each other, but also of the propagators $K_A(x, u; q, v)$ for these test particles. These propagators will in general be different for different species $A, B, C, \ldots$ of test particles. Hence, until some additional physical principle (such as the existence of a quantum metric operator, postulated in Section 4.6) is introduced in the quantum spacetime framework, the simultaneous consideration of all conceivable coherent flows and their coordinate fluctuation amplitudes has to be retained.

An interesting aspect of quantum spacetime as a conceptual framework that combines the geometrodynamic point of view with operational principles by retaining at the quantum level Einstein's notion of 'event' is that it avoids the epistemological pitfalls of the many-world hypothesis required in quantum geometrodynamics (Wheeler, 1962). Indeed, if one regards the concept of 3-geometry ${}^{(3)}\mathscr{G}$ as more fundamental than that of test particle and event, one is forced to introduce at the quantum level the probability amplitude function $\psi({}^{(3)}\mathscr{G})$ in the superspace ${}^{(3)}\mathscr{S}$ of all 3-geometries (Wheeler, 1967, 1979). The physical interpretation of $\psi({}^{(3)}\mathscr{G})$ entails a sample space of infinitely many worlds (Everett, 1957; De Witt and Graham, 1973), out of which the act of quantum measurement singles out one or more elements. The metaphysical aspect of this assumption, which aprioristically has to deny the verifiability of the 'existence' or 'nonexistence' of all the other 'worlds' in order to remain within the realm of mundane as well as scientific common sense, is clearly at odds with Occam's razor underlying any philosophy of scientific rationalism. In fact, a simultaneous and infinitely precise measurement of the entire (classical) 3-geometry of the Universe would have to be envisaged, in order to have an omniscient 'observer' decide in exactly which world he is actually operating. Thus, in such an approach, all measurement – theoretical difficulties of conventional relativistic quantum mechanics are compounded with those of a superhuman act of 'observation', if any kind of ultimate 'physical' meaning is to be imparted to $\psi({}^{(3)}\mathscr{G})$.

By contrast, in quantum spacetime it is not some imaginary classical geometry that 'fluctuates' as a whole, but rather the operationally defined infinitesimal Minkowski coordinates of a quantum test particle in relation to other quantum test particles in its neighborhood. These fluctuations manifest themselves, in principle, if a series of measurements of spacetime distances are performed in quick succession on two or more neighboring quantum test particles in free fall, with the proviso that optimally accurate compensatory mechanisms are employed during each of these observations. Clearly, in addition to avoiding metaphysical assumptions, such an outlook is bound to supply new methods for the 'covariant quantization of gravity'.

4.3. Quantum Lorentz Frames and Minkowski Quantum Spacetime

We have explained in the preceding section how any spacetime geometry (classical or quantum, in special or in general relativity) can be conceptualized in purely operational terms as given by a totality of coherent flows of test particles between whose spatio-temporal relationships are operationally defined[8] by means of light (i.e. photon) signals. Each coherent flow defines a foliation of the spacetime into (for that flow) synchronous hypersurfaces. The hypersurfaces possess geometries whose evolution in the proper time of particles in that flow specifies the geometrodynamics of that spacetime. Whether the spacetime will be that of special or general relativity, or whether it will be classical or quantum, depends exclusively on the (true or assumed) physical nature and kinematical properties of the test particles and of the signals (Salecker and Wigner, 1958).

In classical Minkowski spacetime it is implicitly assumed that the test particles in each coherent flow are pointlike and that they behave deterministically – their worldlines being straight and parallel. Operationally defined Minkowski coordinates can be then introduced by setting up an inertial Lorentz frame, envisaged by Einstein as "three rigid material lines perpendicular to one another, and issuing from a point" (Einstein *et al.*, 1923, p. 43). Obviously, however, the atomic structure of any such 'material lines' imposes a lower limit to the length of intervals that can be marked on them, so that such a setup is suitable only in the measurement of macroscopic distances. Even more to the point, the concept of rigid rods is essentially contradictory to the basic postulates of special relativity, and as we have seen in Section 4.1, it has been eliminated from all modern operational formulations of classical spacetime.

The elimination of rigid material lines from the definition of classical inertial Lorentz frames can be carried out by selecting seven test particles O, $A_{\pm 1}$, $A_{\pm 2}$, $A_{\pm 3}$ in any (classical) coherent flow, so that they are at rest in relation to each other. The choice should be such that O marks the origin of the frame, and $\overline{OA_{\pm j}}$, j = 1, 2, 3, its three coordinate axes. This can be operationally achieved if we assume that all the chosen particles carry synchronized standard clocks. Indeed, we can then operationally set up the $\overline{OA_j}$ lines at right angles to each other by measuring the duration of light signals sent from O to each A_j, so as to ascertain that the distances O–$A_{\pm j}$, j = 1, 2, 3, are all equal, and then verifying in the same manner that the distances A_i–A_j for $|i| \neq |j|$ are also equal.

In the above operational alignment of the Cartesian coordinate axes $\overline{OA_j}, |j|$ = 1, 2, 3, one has to make sure that the aligning procedure itself does not disturb the kinematical relationship of relative rest of the particles O, $A_{\pm 1}$, $A_{\pm 2}$, $A_{\pm 3}$, and that the equidistances established between some of these particles are not disturbed by the recoil momenta imparted by the light signals. Hence, the sending of each light signal has to be accompanied by an *exactly* compensating countersignal. It is at this stage, however, that we again encounter the limitations imposed by the uncertainty principle, which stipulates that the compensation cannot be deterministically exact, but will display under repeated and operationally identical verifications fluctuations around a stochastic mean. Thus, if we regard our particles O, $A_{\pm 1}$, $A_{\pm 2}$, $A_{\pm 3}$ as quantum rather than classical, in the *quantum Lorentz frame* which they constitute the alignment of the axes can be only stochastic (rather than deterministic), with the three axes being verifiably at right angles only in the mean.

On the other hand, at the quantum level an artificial and problematic (Schlegel, 1980) assumption in the operational definition of classical spacetimes can be eliminated, and thus a conceptual simplification can be achieved: we no longer have to assume that test particles 'carry a standard clock', since due to the fundamental quantum relation

$$mc^2 = h\nu = \frac{2\pi\hbar}{T} \tag{3.1}$$

each quantum test particle of rest mass m already *is* a natural clock[9] of proper period T. In fact, in contemporary experimental praxis it has been already realized that "the best physical definition of length and time intervals is provided by a particular light or radio source which acts as standard of both length and time" (Arzeliés, 1966, p. 21). Hence, if in typical experimental arrangements for accurate measurements of distance, such as those based on the interference phenomenon between the two components of a split beam emanating from a monochromatic light source (e.g. a laser), we replace the light source by a source of identical particles accurately boosted to the same velocity, we arrive, at least in principle, at a method for measuring distances[10] by means of some given species of quantum test particles.

The fluctuations in mean coordinate values x^ν and mean 4-velocity components u^ν measured in this manner with respect to a quantum Lorentz frame $\mathscr{L} = \{O, A_j\}$ will be described by coordinate probability densities (2.1), which in case the test particles have spin zero and no internal degrees of freedom should assume the form:

$$G_A^{\mathscr{L}}(x, u; q, v) = N_A \; |K_A^{\mathscr{L}}(x, u; q, v)|^2 . \tag{3.2}$$

Viewed from another quantum Lorentz frame $\mathscr{L}' = \{O', A_j'\}$ in which the mean coordinates of the same event are

$$x' = a + \Lambda x, \qquad u' = \Lambda u, \tag{3.3}$$

the coordinate fluctuation amplitudes $K_A^{\mathscr{L}'}$ should be modulo gauge transformations of the general form (2.10.41),

$$K_A^{\mathscr{L}'}(x', u'; q', v') = K_A^{\mathscr{L}}(x, u; q, v), \tag{3.4}$$

provided that the element of measure with respect to which we interpret $G_A^{\mathcal{L}}$ and $G_A^{\mathcal{L}'}$ to obtain coordinate probabilities has the covariant form (Ehlers, 1971)

$$d\Sigma(q, v) = 2\epsilon(v^0)v^\mu \, d\sigma_\mu(q) \, d\Omega(v). \tag{3.5}$$

Following Born's (1949) suggestion, we shall introduce a *universal length* $\hat{l}_0$, so that we can work in dimensionless coordinates q^μ, v^μ and p^μ, related to the corresponding values $\hat{q}^\mu$, $\hat{v}^\mu$ and $\hat{p}^\mu$ in some conventional system of units (such as c.g.s.) as follows:

$$q^\mu = \hat{q}^\mu/\hat{l}_0, \qquad v^\mu = \hat{p}^\mu/\hat{m}\hat{c} = p^\mu/m, \tag{3.6a}$$

$$p^\mu = \hat{p}^\mu/\hat{m}_0\hat{c}, \qquad \hat{m}_0 = \hat{\hbar}/\hat{c}\hat{l}_0, \, m = \hat{m}/\hat{m}_0. \tag{3.6b}$$

Consequently, in (3.5) we have to set

$$d\Omega(v) = \delta(v^2 - 1) \, d^4 v, \quad v \in \mathcal{V}^+ = \{u \mid u_\mu u^\mu = 1\}. \tag{3.7}$$

Then the mean square deviation in the spacetime distance

$$(x - y)^2 = g_{\mu\nu}(x^\mu - y^\mu)(x^\nu - y^\nu) \tag{3.8}$$

of two stochastic events with mean coordinates x^μ and y^μ in $\mathcal{L}$ shall be

$$\int_{\Sigma'} d\Sigma(q', v') \int_{\Sigma''} d\Sigma(q'', v'') \, (q' - q'')^2 G_A^{\mathcal{L}}(x, u; q', v') G_A^{\mathcal{L}}(y, v; q'', v''), \tag{3.9}$$

where $\Sigma' = \sigma' \times \mathcal{V}^+$ and $\Sigma'' = \sigma'' \times \mathcal{V}^+$ are determined by the hyperplanes $q'^0 =$ const and $q''^0 =$ const, with σ' and σ'' containing x and y, respectively. Thus, although the mean spacetime distance itself does not depend on the 4-velocities of the test particles marking those events, the coordinate fluctuations themselves do so – as anticipated by Born (1949a).

To arrive at some specific forms for the coordinate fluctuation amplitudes K_A, we have to impose, in addition to the covariance condition (3.4), the requirement that the nonrelativistic contraction of each K_A should lead to nonrelativistic propagators that satisfy, in the configuration and momentum representations, marginality conditions of the type (1.5.1), so that conventional nonrelativistic quantum mechanics can be recovered in the nonrelativistic limit. By Theorem 3 in Section 3.3, this immediately leads to nonrelativistic propagators of type (1.9.21), and therefore to relativistic ones of type (2.6.21) (in the gauge with $\omega \equiv 0$), where by (2.8.9) and (2.8.10) η has to be real. Thus, in the dimensionless coordinates (3.6) we should have

$$K_A(q, v; q', v') = \int_{\mathcal{V}^\pm} \exp[im_A(q' - q) \cdot u] \eta_A(v \cdot u) \eta_A(v' \cdot u) \, d\Omega(u), \tag{3.10}$$

where K_A and η_A are, respectively, related to K and η in (2.6.21) as follows:

$$K_A(q, v; q', v') = (\hat{l}_0 \hat{m}_A \hat{c})^3 K(\hat{q}, \hat{p}; \hat{q}', \hat{p}'), \tag{3.11}$$

$$\eta_A(u \cdot v) = \hat{l}_0^{3/2} (\hat{m}_A \hat{c})^{5/2} \eta(\hat{m}_A^2 \hat{c}^2 u \cdot v). \tag{3.12}$$

The factors in (3.11) and (3.12) have been chosen in such a manner as to make both K_A and η_A dimensionless, and at the same time retain the basic relations

$$K_A(q, v; q', v') = K_A^*(q', v'; q, v)$$

$$= \int_{\Sigma^\pm} K_A(q, v; q'', v'') K_A(q'', v''; q', v')\, d\Sigma(q'', v'') \tag{3.13}$$

for propagators when the integration is carried out with respect to the dimensionless measure (3.5) along any $\Sigma^\pm = \sigma \times \mathscr{V}^\pm$. Hence, we shall have for G_A in (3.2)

$$\int_{\Sigma^\pm} G_A(x, u; q, v)\, d\Sigma(q, v) = 1 \tag{3.14}$$

if in terms of $m_A = \hat{m}_A / \hat{m}_0$ we have set

$$G_A(x, u; q, v) = \left(\frac{2\pi}{m_A}\right)^3 |K_A(x, u; q, v)|^2. \tag{3.15}$$

Performing the substitutions (3.11) and (3.12) for the propagator in (2.9.19), we get

$$K_A(q, v; q', v') = \frac{(2m_A^3)^{1/2}}{\pi \omega_A [K_2(4\omega_A^{-2})]^{1/2}} \frac{K_1(\mathbf{f}_A(q, v; q', v'))}{\mathbf{f}_A(q, v; q', v')}, \tag{3.16a}$$

$$\mathbf{f}_A(q, v; q', v') = \omega_A^{-1} \left\{ \left[\frac{i}{l_A} (q - q') + 2 \frac{v + v'}{\omega_A} \right]^2 \right\}^{1/2}, \qquad \omega_A^{-1} = l_A m_A, \tag{3.16b}$$

where $l_A = \hat{l}_A / \hat{l}_0$ is the proper radius of a quantum test particle of species A expressed in units based on the universal length $\hat{l}_0$. We note that by (2.9.7) and (3.12), the propagator K_A in (3.16a) corresponds to the choice of

$$\eta_A'(u \cdot v) = N_{\hat{m}_A, \hat{l}_A} \exp\left(-\frac{2u \cdot v}{\omega_A^2}\right) \tag{3.16c}$$

for proper wave function. This propagator obeys, as all propagators (3.10) do, the Klein–Gordon equation (2.3.11), which in dimensionless coordinates assumes the form

$$(\Box_q + m_A^2)\phi(q, p) = 0. \tag{3.17}$$

Landé (1939) and Born (1939) have proposed an equation which is reciprocally related[11] to the Klein–Gordon equation for $\hat{m}_A = \hat{m}_0$, and which in dimensionless coordinates assumes the form

$$(\Box_p - l_A^2)\phi(q, p) = 0, \quad p = m_0 v, \; l_A = l_0. \tag{3.18}$$

It is interesting to note that if we request that (3.10) should also satisfy the Born–Landé equation (3.18) then we immediately find that K_A will obey that equation for all $(q', v') \in \mathcal{M}^{\pm} = M(1, 3) \times \mathcal{V}^{\pm}$ if and only if η_A equals the solution η_0 of

$$\left(\frac{d^2}{dv_0{}^2} - 1\right) \eta_0 (v_0) = 0. \tag{3.19}$$

The integrating condition (3.14) singles out

$$\eta_0(u \cdot v) = C_0 \exp(-u \cdot v) \tag{3.20}$$

as the only acceptable solution of (3.19) for an appropriate choice of the normalization constant C_0. When it is inserted into (3.10), η_0 in (3.20) leads to a propagator K_0 that equals K_A in (3.16) at $\omega_A = 1$. Thus, for any given choice $\hat{l}_0$ of fundamental length, the Born–Landé equation leads to a unique choice of propagator for particles of proper radius $l_0 = 1$.

When it is viewed in the role of coordinate fluctuation amplitudes, each one of the propagators K_A in (3.10) is related to coherent flows of a single species of (spinless and structureless) test particles. However, quantum spacetime has been operationally defined in terms of a totality of coherent flows, corresponding to all species of test particles. Since it is always operationally feasible to measure spatio-temporal relationships between any two distinct species A and B of test particles, *inter*-flow coordinate fluctuation amplitudes $K_{A,B}$ have to be supplied in addition to those of *intra*-flow, such as K_A in (3.10).

Clearly, each K_A should be equal to the special case of $K_{A,B}$ corresponding to $B = A$. In addition, if we apply the principle of nuclear democracy to coordinate probability densities, we infer that in

$$G_{A,B}(q, v; q', v') = N_{A,B} \, |K_{A,B}(q, v; q', v')|^2 \tag{3.21}$$

the roles of A and B as 'test particle' and 'system' should be interchangeable. This suggests that as an extrapolation of (3.13) we have to impose on $K_{A,B}$ the consistency conditions

$$K_{A,B}(q, v; q', v') = K^*_{B,A}(q', v'; q, v), \tag{3.22a}$$

$$K_A(q', v'; q'', v'') = \int_{\Sigma^\pm} K_{A,B}(q', v'; q, v) K_{B,A}(q, v; q'', v'')\, \mathrm{d}\Sigma(q, v), \tag{3.22b}$$

and that in (3.22b) A and B should be interchangeable. All these conditions are satisfied by

$$K_{A,B}(q, v; q', v') = \int_{\mathcal{V}^\pm} \exp[iu \cdot (m_B q' - m_A q)]\, \eta_A(u \cdot v) \eta_B(u \cdot v')\, \mathrm{d}\Omega(u). \tag{3.23}$$

Indeed, (3.10) is the special case of (3.23) for $A = B$, and (3.22a) in an immediate consequence of (3.23), whereas (3.22b) follows from (2.4.24) upon making the transition to the dimensionless coordinates (3.6). Thus, for coherent flows of spinless and structureless test particles, the Minkowski quantum spacetime is completely described by the flat pseudo-Riemannian manifold $M(1, 3)$ of *mean* spacetime locations, and by the coordinate fluctuation amplitudes (3.23).

4.4. General Quantum Spacetimes and Stochastic Geometries

The mathematical description of quantum spacetimes that are operationally defined by means of measurement procedures of a spatio-temporal nature consisting of exchanges of light signals and of kinematical operations executed on coherent flows of quantum test particles, has two components: the geometrodynamic component, assuming the form of evolving *mean* 3-geometries, which should display a pseudo-Riemannian structure (or, more generally, some pseudo-Finslerian structure); and a stochastic component, assuming the form of intraflow and interflow coordinate fluctuation (probability) amplitudes K_A and $K_{A,B}$, respectively. These amplitudes provide the probability densities $G_{A,B}$ in (3.21) ($A = B$ in case of K_A) for obtaining the readings (q', v') for a test-particle in a B-flow which *in the mean* would follow the same worldline as the A-flow test particle located at the stochastic position $\underset{\sim}{q} \in \underset{\sim}{\sigma}$ and having the stochastic 4-velocity $\underset{\sim}{v}$. Consequently, except if future research reveals that the coherent flows of a particular species of particle are in some sense more fundamental than those of all other species, the present quantum spacetime framework has to deal not with only one, but with an infinity of stochastic metric spaces of the kind described in Section 1.1, and possessing an indefinite (mean) metric $g_{\mu\nu}$. In semiclassical models, obtained by superimposing on a classical spacetime background a family of coordinate fluctuation amplitudes $K_{A,B}$ (as was the case with Minkowski quantum spacetime in the last section), the mean metric tensor $g_{\mu\nu}$ depends only on the mean coordinates q^ν. However, in general we can envisage also a dependence on ν^ν, as suggested by reciprocity theory (Born, 1949; Caianiello, 1980, 1981; Prugovečki, 1981g, h). The possibility of curved phase space (Caianiello and Vilasi, 1981; Brooke and Prugovečki, 1982; Caianiello *et al.*, 1982) with metric

$$ds^2 = g_{\mu\nu}(\zeta)\, d\zeta^\mu\, d\zeta^\nu, \quad \mu, \nu = 0, \ldots, 7, \tag{4.1}$$

cannot be dismissed out of hand, and should remain the subject of further research.

Let us limit ourselves for the time being to semiclassical models. The question arises already at this level as to how to construct the coordinate fluctuation amplitudes $K_{A,B}$. Our main guidelines have to be provided, as in the classical case, by the strong equivalence principle. However, since we are not dealing with clasical point test particles, the textbook arguments (Misner *et al.*, 1973; Adler *et al.*, 1975) based on the concept of sharp (classical) locality are not applicable. It is evident, however, that when the Riemann curvature tensor approaches zero, then the coordinate fluctuation amplitudes $K_{A,B}$ should approach for spinless test particles the values (3.23) provided in the flat case.

In this context, one has to recall that in general relativity most coordinate systems supply only a means of labeling events, with the coordinates themselves having no *direct* operational meaning. To arrive at an operational interpretation of a pseudo-Riemannian manifold describing a classical spacetime, one has to adopt Gaussian normal coordinates q^ν. These coordinates assume in the neighborhood of any event marking the origin of a local Lorentz frame the operational meaning discussed in Section 4.1. The time-ordered family $\mathscr{S}^\uparrow$ of reference hypersurfaces consists in that case of the synchronous surfaces q^0 = const. The task of computing the coordinate fluctuation amplitudes $K_{A,B}$ reduces to determining $K_{A,B}(\zeta; \zeta')$ for ζ and ζ' on the initial data hypersurface $\hat{\Sigma}$ with $q^0 = 0$, and then introducing a time-evolution law that would take us from $\hat{\Sigma}$ to all the other hypersurfaces in the reference family $\mathscr{S}^\uparrow$. Once the task of computing the coordinate fluctuation amplitudes has been completed in Gaussian normal coordinates q^μ, the transition to any other system of coordinates x^μ becomes one of replacing q^μ with $q^\mu(x^\nu)$ in the resulting expressions.[12]

Since in the flat case the evolution law has to reduce to the one in Minkowski spacetime, the conventional approach (Parker, 1977) to its derivation in the spin-zero case consists of replacing in the Klein–Gordon equation (2.1.1) partial derivatives by the covariant derivatives of the 4-vector $\phi^\nu = \partial^\nu \phi$, i.e. by

$$\nabla_\mu \phi^{,\nu} = \partial_\mu \phi^{,\nu} + \Gamma^\nu_{\mu\lambda} \phi^{,\lambda}, \tag{4.2}$$

where $\Gamma^\nu_{\mu\lambda}$ is the Christoffel symbol corresponding to the Riemannian connection on the chosen classical spacetime. It is then discovered (Parker, 1973; Hu *et al.*,

1973; Parker and Fulling, 1974) that the time-evolution governed by the ensuing Klein–Gordon equation with minimal coupling to the external gravitational field,

$$\left(\nabla_\mu \nabla^\mu + \frac{\hat{m}^2 \hat{c}^2}{\hat{\hbar}^2}\right) \phi = 0, \tag{4.3}$$

– to which a conformal coupling term is sometimes added (Parker, 1973; Avis *et al.*, 1978; Bunch and Parker, 1979) – does not leave (with or without that term) the space of positive-energy solutions invariant in nonstatic universes. This mathematical fact is then interpreted (Parker, 1973; Raine and Winlove, 1975; Unruh, 1976) as a manifestation of spontaneous particle production due to the expansion of the Universe. It should be recalled, however, that this *mathematical* feature (namely the instability in time of the set of positive-energy solutions describing particles) also occurs in the Klein–Gordon equation (2.10.1) with minimal coupling to an external electromagnetic field (cf. Section 2.10). Consequently, this mathematical phenomenon might very well be due to the intrinsic inconsistencies of all 'local' models in relativistic quantum mechanics, rather than to an objective *physical* phenomenon.

In any event, in the present framework the procedure used in Section 2.10 in the electromagnetic case proves equally effective in the gravitational case (Prugovečki, 1981e). Indeed, in Gaussian normal coordinates (expressed in dimensionless manner, in accordance with (3.6)) we have

$$ds^2 = (dx^0)^2 - h_{ij}(x)\, dx^i\, dx^j, \qquad i, j = 1, 2, 3. \tag{4.4}$$

We can then use the well-known representation (Abraham and Marsden, 1978)

$$\Delta_\sigma = h^{-1/2} \frac{\partial}{\partial x^i}\left(h^{1/2} h^{ij} \frac{\partial}{\partial x^j}\right), \qquad h = \det h^{ij} = -g \tag{4.5}$$

of the Laplace–Beltrami operator Δ_σ on the spacelike hypersurface σ in the chosen classical spacetime to rewrite (4.3) in the form

$$(h^{-1/2} \partial_0 (h^{1/2} \partial_0) - h^{-1/2} \partial_i (h^{1/2} h^{ij} \partial_j) + m_A^2)\phi = 0. \tag{4.6}$$

Then the mathematical source of the aforementioned problems, which supposedly reflect spontaneous particle production (not related to the phenomenon of pair creation and annihilation!) becomes rather obvious: the operator Δ_σ is not related in nonstatic cases to a symmetric operator on the Hilbert space $L^2(\sigma^\pm)$ with inner product of form (2.1.28), i.e. in this respect the status of (4.6) is quite similar to that of (2.10.1).

On the other hand, we can consider instead of $L^2(\sigma^\pm)$ the Hilbert spaces $L^2(\Sigma^\pm)$ with inner products

$$\langle \phi_1 | \phi_2 \rangle_{\Sigma^\pm} = \int_{\Sigma^\pm} \phi_1^*(\zeta) \phi_2(\zeta)\, d\Sigma(\zeta), \tag{4.7}$$

where $d\Sigma(\zeta)$ is given by (3.5). Therefore in Gaussian normal coordinates $d\Sigma$ assumes the form

$$d\Sigma(\zeta) = h(q)\,\mathbf{dq}\,\mathbf{dv}, \quad \zeta = (q, v) \in \Sigma_A^{\pm}, \tag{4.8}$$

where $\mathbf{v} = (v^1, v^2, v^3)$, so that the usual 3-velocity equals $\gamma_v^{-1}\mathbf{v}$. In the Hilbert spaces $L^2(\Sigma_A^{\pm})$ the operators

$$\Delta_\Sigma = h^{-1/2}\partial_i(h^{1/2}h^{ij}\partial_j), \quad \partial_i = \partial/\partial q^i, \tag{4.9}$$

are symmetric (and, in fact, self-adjoint), as it becomes totally evident when we combine (4.7)–(4.9) into

$$\langle\phi_1|\Delta_\Sigma\phi_2\rangle_{\Sigma^\pm} = \int_{\Sigma^\pm} \phi_1^*(\zeta)h^{1/2}(q)\partial_i(h^{1/2}h^{ij}\partial_j)\phi_2(\zeta)\,\mathbf{dq}\,\mathbf{dv}$$

$$= -\int_{\Sigma^\pm} h^{ij}\partial_i(h^{1/2}\phi_1)^*\partial_j(h^{1/2}\phi_2)\,\mathbf{dq}\,\mathbf{dv}. \tag{4.10}$$

Furthermore, setting above $\phi_1 = \phi_2$, we see that $-\Delta_\Sigma \geqslant 0$ due to the (1,1,1) signature of the metric tensor h^{ij}, which insures the nonnegativity of the last integrand at each $\zeta = (q, v) \in \Sigma^\pm$. Hence, the evolution described by the equation

$$i\frac{\partial}{\partial q^0}\phi(q, v) = (-\Delta_\sigma + m_A^2)^{1/2}\phi(q, v) \tag{4.11}$$

is governed by the unitary operators

$$U(\sigma'', \sigma') = T\exp\left[-i\int_{\sigma'}^{\sigma''}(-\Delta_\sigma + m_A^2)^{1/2}\,dq^0\right] \tag{4.12}$$

from $L^2(\Sigma')$ to $L^2(\Sigma'')$, and consequently it conserves the total probability. In fact, if we write

$$P_\mu = i\nabla_\mu, \quad \mu = 0, \ldots, 3, \tag{4.13}$$

we see that (4.11) formally enjoys *vis-à-vis* the equation

$$(P_\mu P^\mu - m_A^2)\phi(q, v) = 0 \tag{4.14}$$

the same status that Equations (2.10.6)–(2.10.8) or (2.10.39) had enjoyed *vis-à-vis* (2.10.1). Thus $K_{A,B}(\zeta; \zeta')$ can be characterized as solutions of (4.11) in $\zeta = (q, v)$, and of the corresponding equation for B-particles in $\zeta' = (q', v')$.

As an example, let us consider the Robertson–Walker model with spatially flat σ-surfaces (Adler *et al*., 1975), for which (4.4) assumes the form

$$ds^2 = (dq^0)^2 - R^2(q^0)\delta_{ij}\, dq^i\, dq^j, \tag{4.15}$$

where δ_{ij}, $i, j = 1, 2, 3$, is the usual Kronecker symbol. Upon introducing in a neighborhood of the initial-data hypersurface $\hat{\Sigma}$, described in the above normal coordinates by $q^0 = \tilde{q}^0$, the rescaled coordinates

$$\hat{q}^0 = q^0, \quad \hat{q}^i = R(\tilde{q}^0)q^i, \quad \hat{u}^0 = u^0, \quad \hat{u}^i = R(\tilde{q}^0)u^i, \quad i = 1, 2, 3, \tag{4.16}$$

the metric along $\hat{\Sigma}$ assumes the Minkowski form in these new coordinates, i.e. the new mean stochastic coordinates along $\hat{\Sigma}$ correspond to local stochastic inertial frames with (stochastic) origins on $\hat{\sigma}$. Consequently, by the strong equivalence principle, $K_{A,B}$ should assume in these coordinates the form (3.23) for $\zeta, \zeta' \in \hat{\Sigma}$. Since $\hat{v} \cdot \hat{u}$ and $\hat{v}' \cdot \hat{u}$ in the resulting expression are invariants, whereas $\hat{u}_\nu \hat{q}^\nu = u_\nu q^\nu$ along $\hat{\Sigma}$, we get upon reverting to the original Gaussian normal coordinates,

$$K_{A,B}(q, v; q', v') = \int \exp[iu_\nu(m_B q'^\nu - m_A q^\nu)]\, \eta_A(v \cdot u)\eta_B(v' \cdot u)\, d\Omega(u), \tag{4.17}$$

where the integration is performed over all coordinate labels u^ν such that

$$u_\nu u^\nu = (u^0)^2 - R^2(q^0)\mathbf{u}^2 = 1. \tag{4.18}$$

Thus (4.17) provides the coordinate fluctuation amplitudes along $\hat{\Sigma}$ for the normal Gaussian coordinates that appear in (4.15). In turn, the amplitudes supply the coordinate probability densities in (3.21) with respect to the measure

$$d\Sigma(q, v) = R^3(q^0)\, d\mathbf{q}\, d\mathbf{v}, \quad (q = q^0, \mathbf{q}),\ v = (v^0, \mathbf{v}), \tag{4.19}$$

on $\hat{\Sigma}$. As mentioned earlier, the transition to an arbitrary system of coordinates along $\hat{\Sigma}$ is made by performing the appropriate substitutions in (4.17), and by adopting the covariant form (3.5) for $d\Sigma$.

In the Gaussian normal coordinates appearing in (4.19) we have

$$h_{ij}(q) = R^4(q^0)h^{ij}(q) = R^2(q^0)\delta_{ij}, \quad h(q) = R^6(q^0), \tag{4.20}$$

so that the evolution equation (4.11) assumes the form

$$i\frac{\partial}{\partial q^0}\phi(q, v) = (-R^{-2}(q^0)\Delta_{\mathbf{q}} + m_A^2)^{1/2}\phi(q, v), \tag{4.21}$$

where $\Delta_{\mathbf{q}}$ is the ordinary Laplacian in the variables $\mathbf{q}$. Hence, as a solution of (4.21) with the initial conditions (4.17), $K_{A,B}$ assumes at $\zeta = (q, v) \in \Sigma$ the form:

$$K_{A,B}(\zeta; \zeta') = \int \exp\{iu^\mu g_{\mu\nu}(q)[m_B q'^\nu - m_A q^\nu]\}\, \eta_A(v \cdot u)\eta_B(v' \cdot u)\, d\Omega(u). \tag{4.22}$$

If we recall, however, that the general relativistic phase spaces (2.3) lie within the cotangent buddle over the (mean) spacetime manifold $\mathscr{M}(1, 3)$, then we immediately see that $K_{A,B}$ can be rewritten for arbitrary $\zeta, \zeta' \in \mathscr{M}_1^{\pm}$ as

$$K_{A,B}(\zeta;\zeta') = \int \exp[i(m_B q' \cdot u - m_A q \cdot u)] \eta_A(v \cdot u)] \eta_B(v' \cdot u) \, d\Omega(u). \quad (4.23)$$

In (4.23) we have set for $u \in \mathscr{V} \subset \mathscr{T}_q$ and $u' \in \mathscr{V} \subset \mathscr{T}_{q'}$, respectively (under the understanding that q and q' are represented in normal coordinates),

$$u \cdot q = g_{\mu\nu}(q) u^\mu q^\nu, \; u \cdot q' = g_{\mu\nu}(q') u^\mu q'^\nu, \quad (4.24)$$

albeit q and q' can no longer be identified with elements of the respective tangent spaces $\mathscr{T}_q$ and $\mathscr{T}_{q'}$, as was the case in Minkowski spacetime. Naturally, implicit in this manner of writing $K_{A,B}$ is an identification of the tangent spaces at different point $q \in \mathscr{M}(1, 3)$ reflecting an alignment of fibres in the cotangent bundle over the pseudo-Riemannian manifold of mean spacetime values. This involves (Atiyah, 1979) a choice of gauge, and (4.23) reflects a specific choice that corresponds to that adopted in (2.6.21) and (2.7.2) in case of Minkowski spacetime.

Clearly, the issue of gauge is as intimately related now to the choice of representation of RCCRs as it was in the Minkowski case. In the general relativistic case the formal appearance of the RCCRs (2.7.17) for spinless particles can be retained, so that in the dimensionless coordinates (3.6)

$$[Q^\mu, Q^\nu] = [P^\mu, P^\nu] = 0, \quad (4.25a)$$

$$[Q^\mu, P^\nu] = -ig^{\mu\nu}(q). \quad (4.25b)$$

The imposition of (4.25) when the above commutators are applied to $K_{A,B}$ can be viewed as an extension of the strong equivalence principle to the quantum domain: albeit the Poincaré group loses its globally operational meaning, on curved spacetimes its local significance is preserved in a stochastic sense by the existence for any quantum event q of stochastic mean coordinates q^ν in which $g^{\mu\nu}$ assumes the Minkowski values $\eta^{\mu\nu}$ at q, and in a neighborhood of which the coordinate fluctuation amplitudes shall therefore retain their Minkowski spacetime values.

Thus, the task of constructing semiclassical quantum spacetime models becomes one of finding coordinate fluctuation amplitudes $K_{A,B}$ that are functions of the mean-spacetime coordinates q^ν in the pseudo-Riemannian manifold $\mathscr{M}(1,3)$ supplied by the corresponding classical model, and of the 4-momentum components p^ν, $p \in \mathscr{V}_m \subset \mathscr{T}_q$. For spin-zero particles, $K_{A,B}$ comply with the RCCRs (4.25), and once given along an initial data hypersurface $\hat{\Sigma}^{\pm}$, they can be in principle computed along any other hypersurface $\Sigma^{\pm}$ in the reference family $\mathscr{S}^{\uparrow}$ to which $\hat{\Sigma}^{\pm}$ belongs by applying the evolution operators in (4.12). For particles with spin, and for internal structure, the procedure has to be modified in the manner discussed in the next two sections.

In the above-described semiclassical models, the structure of the pseudo-Riemannian manifold $\mathscr{M}(1, 3)$ representing the mean location of quantum particles in free fall is not affected by the presence of those particles, or of quantum matter in general. To take into account such effects, we have to impose Einstein's equations

$$R_{\alpha\beta} - \frac{1}{2} g_{\alpha\beta} R = \kappa \langle T_{\alpha\beta} \rangle, \tag{4.26}$$

where $\langle T_{\alpha\beta} \rangle$ is the expectation value of the quantum stress-energy tensor $T_{\alpha\beta}(q, v)$ averaged over all 4-velocities v. Examples of such operators $T_{\alpha\beta}$ are provided by the field theoretical models considered in the next section. Naturally, in such fully quantized models the structure of $\mathscr{M}(1, 3)$ will generally depend on the initial quantum state, and therefore on the relative abundance of various species of quantum particles at each mean spacetime point q, as well as on their 4-velocity probability distributions.

It is noteworthy that in quantum spacetime there is no physical or mathematical imperative for an extra 'quantization of the gravitational field', albeit, on a formal level, such quantization can be carried out by any of the various methods devised thus far – which lead in the conventional approach to a variety of difficulties (Isham, 1975; van Nieuwenhuizen, 1977) only partly resolved by standard re-normalization procedures (De Witt, 1975; Davies, 1980). Indeed, in the geometrized version of quantum theory advocated in Section 4.6, spatio-temporal properties are amalgamated with the internal structure properties of elementary particles. Consequently, gravitation as a geometric property of spacetime is treated on par with internal gauge freedom, and its quantization thus becomes an integral part of the 'quantization of gauge fields'. As illustrated in Section 5.5 by the abelian case of the electromagnetic field, quantum gauge fields are a necessary consequence of the operationally based concept of quantum spacetime: since quantum events are the 'seats' of quantum test particles playing the role of markers for such events, the presence of any gauge field quanta, such as photons, gluons, or gravitons, manifests itself only by its effect on a quantum test particle, which is necessarily of nonzero mass. Thus 'quantized fields' are not so much the product of a 'second quantization', as they are the natural manifestation of a consistently quantum point of view, which juxtaposes fields on quantum spacetime next to the concept of 'classical field' viewed as a field on classical spacetime.

4.5. Reciprocity Theory and Born's Quantum Metric Operator

In the preceding sections we have shown that a probabilistically consistent description of quantum spacetimes operationally defined in terms of spatio-temporal measurements performed on quantum test particles can be provided in terms of stochastic geometries specified by mean phase-space coordinates ζ^{ν} and coordinate fluctuation amplitudes $K_{A,B}$. For spin-zero particles these amplitudes assume the general form (3.23), and since $u, v, v' \in \mathscr{V}^{\pm}$, they can be written in the alternative form

$$K_{B,A}(q, v; q', v') = \int_{\mathscr{V}^{\pm}} \exp[iu \cdot (m_A q' - m_B q)\hat{\eta}_B(m_B(u - v))\hat{\eta}_A(u - v')\, d\Omega(u), \tag{5.1}$$

$$\hat{\eta}_A(u - v) = \eta_A(u \cdot v), \qquad \hat{\eta}_B(m_B(u - v)) = \eta_B(u \cdot v). \tag{5.2}$$

This form stresses the formal analogy with the nonrelativistic free propagators (1.9.21), once the latter are expressed in accordance with (1.4.6b) and (1.2.11) as:

$$K(\mathbf{q}, \mathbf{p}, t; \mathbf{q}', \mathbf{p}', t') = \int_{\mathbb{R}^3} \exp\left\{\frac{i}{\hbar}\left[-\frac{\mathbf{k}^2}{2m}(t - t') + \mathbf{k} \cdot (\mathbf{q} - \mathbf{q}')\right]\right\} \times$$

$$\times\, \tilde{\xi}(\mathbf{k} - \mathbf{p})\tilde{\xi}(\mathbf{k} - \mathbf{p}')\, d\mathbf{k}. \tag{5.3}$$

The actual realization of a quantum spacetime model requires an appropriate choice of the proper wave functions $\eta_A, \eta_B, \ldots$ that correspond to those elementary particles[13] that actually occur in Nature. Born's (1938, 1949) reciprocity theory, with its emphasis on the symmetric roles played by spacetime coordinates and 4-momentum components, appears to be the most natural vehicle for arriving at such proper wave functions of existing elementary particles.

Born's reciprocity principle is based on an observation that is as deep as it is simple, namely that in special relativity the spacetime 'distance'

$$q^2 = q_\mu q^\mu = (q^0)^2 - \mathbf{q}^2 \tag{5.4}$$

and the 4-momentum 'magnitude'

$$p^2 = p_\mu p^\mu = (p^0)^2 - \mathbf{p}^2 \tag{5.5}$$

are formally totally analogous, but that in elementary particle physics the latter is taken to assume only discrete values (namely the square-masses of actually existing particles), whereas the former is thought to assume a continuum of values. Noting the symmetry of many well-established physical laws and basic physical quantities (e.g. the Hamilton equations, the canonical commutation relations, the angular momentum, etc.) under the *reciprocity transformation*

$$q^\mu \mapsto p^\mu, \quad p^\mu \mapsto -q^\mu, \tag{5.6}$$

in the nonrelativistic realm (i.e. for $\mu = 1, 2, 3$), Born (1949b) conjectured that such a reciprocity symmetry should extend to the relativistic domain. Consequently, if the rest masses of the elementary particles assume only discrete values, so also should their 'sizes'. This has led Born to postulate the existence of a so-called 'metric operator', whose eigenvalues would then yield mass formulae that would predict the masses of all elementary particles occurring in Nature.

Although the reciprocity theory developed by Born and his collaborators (Born and Fuchs, 1940; Born *et al.*, 1949) met with some partial numerical successes, its mathematical embodiment led to unresolved fundamental difficulties.[14] However, the principle of reciprocity as well as Born's metric operator were incorporated by Yukawa (1950a, 1953) in his nonlocal quantum theory. This in turn led Markov (1956), Katayama and Yukawa (1968), as well as Takabayashi (1967), to a relativistic harmonic oscillator model, which may subsequent researchers (Fujimura *et al.*, 1970, 1971; Takabayashi, 1970, 1971; Feynman *et al.*, 1971; Kim and Noz, 1973, 1975, 1977; etc. – cf, Takabayashi, 1979, for a review) chose to interpret in the context of the quark–antiquark structural model for mesons. However, as a model for extended elementary particles the relativistic harmonic oscillator model also belongs to a large class of models (Barut and Böhm, 1965; Dothan *et al.*, 1968; Mukunda *et al.*, 1965; Malkin and Man'ko, 1965; Böhm, 1966, 1968; van Dam and Biedenharn, 1976; Mukunda *et al.*, 1980; van Dam *et al.*, 1981) for particles with internal structure, which are based on mass-spectrum generating groups and are not necessarily interpreted in the context of the conjecture of permanently confined hadronic constituents. It is in this context that we shall discuss how Born's metric operator has been incorporated (Prugovečki, 1981g, h) into the quantum spacetime framework – albeit an alternative treatment based on the constituent idea could be equally well adapted to this framework.[15]

In adapting Born's quantum metric operator

$$D^2 = \hat{Q}_\mu \hat{Q}^\mu + \hat{P}_\mu \hat{P}^\mu \tag{5.7}$$

to the quantum spacetime framework, we shall interpret $\hat{Q}^\mu$ and $\hat{P}^\mu$ as *internal* spacetime and 4-momentum operators for extended particles with mean spacetime Minkowski coordinates q^μ and mean 4-momentum components p^μ, so that

$$\hat{Q}^\mu = Q^\mu - q^\mu, \qquad \hat{P}^\mu = P^\mu - p^\mu, \tag{5.8}$$

where Q^μ and P^μ are (external) realizations of the RCCRs (4.25) in Minkowski quantum spacetime. The general form of these realizations in arbitrary units and arbitrary gauges is given in (2.7.25). However, since the metric operator D^2 in (5.7) is going to give rise to exciton states – in the sense of Katayama and Yukawa (1968), and Katayama *et al.* (1968) – let us replace the dimensionless coordinates (3.6) by the new coordinates

$$\mathring{q}^\mu = q^\mu / l_A, \qquad \mathring{p}^\mu = p^\mu / m_A, \tag{5.9}$$

where l_A and m_A are the proper radius and rest mass respectively, for ground exciton states. The species of particles representing some such ground state will therefore provide the units for measuring spacetime distances as well as 4-momentum, and therefore such particles will be referred to as *standards*. In the coordinates (5.9) the RCCRs assume the form

$$[Q^\mu, Q^\nu] = [P^\mu, P^\nu] = 0, \tag{5.10a}$$

$$[Q^\mu, P^\nu] = -i\omega_A g^{\mu\nu}, \quad \omega_A = (l_A m_A)^{-1} = \frac{\hat{\hbar}}{\hat{l}_A \hat{m}_A \hat{c}}, \tag{5.10b}$$

and their realizations (2.7.25) assume the form

$$Q^\mu = \mathring{q}^\mu - \frac{\partial\omega}{\partial\mathring{p}_\mu} - i\omega_A \frac{\partial}{\partial\mathring{p}_\mu}, \qquad P^\mu = \frac{\partial\omega}{\partial\mathring{q}_\mu} + i\omega_A \frac{\partial}{\partial\mathring{q}_\mu}. \tag{5.11}$$

The corresponding internal operators in (5.8) will therefore provide realizations of the *internal RCCRs*:

$$[\hat{Q}^\mu, \hat{Q}^\nu] = [\hat{P}^\mu, \hat{P}^\nu] = 0, \qquad [\hat{Q}^\mu, \hat{P}^\nu] = i\omega_A g^{\mu\nu}. \tag{5.12}$$

For example, in the gauge with $\omega = 0$ that corresponds to (5.1), these operators equal

$$\hat{Q}^\mu = -i\omega_A \frac{\partial}{\partial\mathring{p}_\mu}, \qquad \hat{P}^\mu = i\omega_A \frac{\partial}{\partial\mathring{q}_\mu} - \mathring{p}^\mu. \tag{5.13}$$

The interflow coordinate fluctuation amplitudes in (5.1) can be obtained from the *exciton transition amplitudes*

$$K_{\cdot,A}(\mathring{q}, \mathring{p}; \mathring{q}', \mathring{p}') = \int_{\mathcal{V}_A^\pm} \exp\left[\frac{i}{\omega_A}\left(\mathring{q}' - \frac{m}{m_A}\mathring{q}\right)\cdot\mathring{k}\right] \times$$

$$\times\, \hat{\eta}\left(\frac{m}{m_A}\mathring{k} - \mathring{p}\right) \hat{\eta}_A\left(\frac{\mathring{k} - \mathring{p}'}{m_A}\right) \frac{d\Omega(\mathring{k})}{m_A^2} \tag{5.14}$$

by setting $m = m_B$ and by restricting $\mathring{p}$ to the mass-shells $\mathcal{V}_B^\pm$ of B particles or antiparticles. However, *prior* to the imposition of any such restrictions, (5.14) can be viewed as a transition amplitude in the exciton space of internal degrees of freedom that relates to kinematical operations *within* the hadron. It is at this *pregeometry* stage that we can impose Born's eigenvalue condition

$$D^2 K_{\cdot,A} = \lambda_{\cdot,A} K_{\cdot,A} \tag{5.15}$$

that restricts the range of available exciton states to the eigenfunctions of the metric operator in (5.7).

To find the solutions (5.14) that satisfy (5.15), we note that the action of $\hat{Q}^\mu$ and $\hat{P}^\mu$ in (5.13) can be transferred to the would-be proper wave functions

$$\mathring{\eta}(\mathring{w}) = \hat{\eta}\left(\frac{m}{m_A}\,\mathring{k} - \mathring{p}\right), \quad \mathring{w} = \frac{m}{m_A}\,\mathring{k} - \mathring{p}, \tag{5.16}$$

on which they effectively act as

$$\mathring{Q}^\mu \mathring{\eta} = i\omega_A \frac{\partial}{\partial w_\mu}\,\mathring{\eta}, \qquad \mathring{P}^\mu \mathring{\eta} = \mathring{w}^\mu \mathring{\eta}. \tag{5.17}$$

Hence (5.15) is equivalent to the eigenvalue equation

$$\left(-\omega_A^2 \frac{\partial^2}{\partial \mathring{w}_\mu\, \partial \mathring{w}^\mu} + \mathring{w}_\mu \mathring{w}^\mu\right) \mathring{\eta} = \lambda_{\cdot,A} \mathring{\eta}, \tag{5.18}$$

which in turn is formally equivalent to the relativistic, harmonic oscillator equation – as it is easily seen by formally taking the 4-dimensional Fourier transforms

$$\eta(\mathring{x}) = (2\pi)^{-2} \int \exp(-i\mathring{x} \cdot \mathring{w})\eta(\mathring{w})\, \mathrm{d}^4 \mathring{w}. \tag{5.19}$$

However, in the relativistic harmonic oscillator quark model, $\mathring{\eta}(\mathring{w})$, and therefore also $\eta(\mathring{x})$, are assumed to be square integrable.[16] This leads to a spectrum of eigenvalues which is not bounded from below, and from which the negative eigenvalues can be eliminated only by imposing an *ad hoc* subsidiary condition (Takabayashi, 1979, Eq. (1.19)). In contradistinction, the existence of the integral in (5.14) does not entail the square integrability of $\hat{\eta}$ over $\mathbb{R}^4$ with respect to $\mathrm{d}^4 k$. This leads to different boundary conditions at infinity for the sought-after solutions of (5.18), and therefore also to a spectrum of eigenvalues that is bounded from below without the need to impose the *ad hoc* subsidiary conditions required in the quark relativistic harmonic oscillator model.

To arrive at these eigenvalues, let us compute the eigenfunctions in (5.18) by means of the raising and lowering exciton operators

$$\bar{a}^\mu = 2^{-1/2}(\mathring{Q}^\mu + i\mathring{P}^\mu), \qquad a^\mu = 2^{-1/2}(\mathring{Q}^\mu - i\mathring{P}^\mu), \tag{5.20}$$

which, according to (5.17), satisfy the commutation relations

$$[\bar{a}^\mu, a^\nu] = \omega_A g^{\mu\nu}, \qquad [\bar{a}^\mu, \bar{a}^\nu] = [a^\mu, a^\nu] = 0. \tag{5.21}$$

The ground-exciton state $\mathring{\eta}_A$, therefore, has to be the solution of the equations

$$a^\mu \mathring{\eta}_A = 2^{-1/2} i \left(\omega_A \frac{\partial}{\partial \mathring{w}_\mu} - \mathring{w}^\mu\right) \mathring{\eta}_A = 0, \tag{5.22}$$

specifying that there are no eigenstates with any lower eigenvalues. The unique solution of (5.22) is

$$\mathring{\eta}_A(\mathring{w}) = \mathring{Z}_A^{-1/2} \exp\left(\frac{\mathring{w}_\mu \mathring{w}^\mu}{2\omega_A}\right), \tag{5.23}$$

where the normalization constant $\mathring{Z}_A$ has to be chosen in such a manner that in case of

$$K_{A,A}(\mathring{q}, \mathring{p}; \mathring{q}', \mathring{p}') = \mathring{Z}_A^{-1} \int_{\mathscr{V}^\pm} \exp\left[\frac{i}{\omega_A}(\mathring{q}' - \mathring{q}) \cdot \mathring{k} + \frac{(\mathring{k} - \mathring{p})^2 + (\mathring{k}' - \mathring{p}')^2}{2\omega_A}\right] \mathrm{d}\Omega(\mathring{k}) \tag{5.24}$$

the reproducibility property in (2.2) is satisfied in terms of the present dimensionless variables $\mathring{q}$, $\mathring{p}$, $\mathring{q}'$ and $\mathring{p}'$. In view of the considerations of Section 2.4, this is equivalent to requiring that

$$\mathring{\eta}_A(\mathring{p} \cdot \mathring{k}) = \mathring{Z}_A^{-1/2} \exp(-\omega_A^{-1} \mathring{p} \cdot \mathring{k}) \tag{5.25a}$$

should satisfy (2.4.10) in the dimensionless coordinates $\mathring{k}$ and $\mathring{p}$, and with $\hbar$ replaced by ω_A, so that, by (2.9.12) and (2.9.13),

$$\mathring{Z}_A = \frac{(2\pi\omega_A)^4}{2} \exp\left(\frac{2m_A^2}{\omega_A}\right) K_2\left(\frac{2}{\omega_A}\right). \tag{5.25b}$$

Comparing (3.20) with (5.25a), we see that $K_{A,A}$ in (5.24) will satisfy the Born–Landé equation (3.18) if $\hat{l}_A = \hat{l}_0$ and $\hat{m}_A = \hat{m}_0$.

Let us express the action of D^2 on $K_{\cdot,A}$ in terms of that of

$$\mathring{D}^2 = \mathring{Q}_\mu \mathring{Q}^\mu + \mathring{P}_\mu \mathring{P}^\mu = 2\bar{a}_\mu a^\mu - 4\omega_A \tag{5.26}$$

on $\mathring{\eta}$. If we assume that

$$\mathring{\phi}_{\bar{n}} = (\bar{a}^0)^{n_0} (\bar{a}^1)^{n_1} (\bar{a}^2)^{n_2} (\bar{a}^3)^{n_3} \phi_{\bar{0}}, \quad \bar{n} = (n_0, n_1, n_2, n_3), \tag{5.27}$$

is an eigenvector of $\bar{a}^\mu a^\mu$ (with μ fixed) so that

$$-\bar{a}^\mu a^\mu \mathring{\phi}_{\bar{n}} = n^\mu \omega_A \mathring{\phi}_{\bar{n}} = g^{\mu\nu} n_\nu \omega_A \mathring{\phi}_{\bar{n}}, \tag{5.28}$$

then by (5.21) we obtain

$$-(\bar{a}^\mu a^\mu)\bar{a}^\mu \mathring{\phi}_{\bar{n}} = (n^\mu + g^{\mu\mu})\bar{a}^\mu \mathring{\phi}_{\bar{n}}. \tag{5.29}$$

Since (5.28) is satisfied with $n_\mu = 0$, $\mu = 0, \ldots, 3$, we deduce by induction that (5.27) satisfies (5.28) for all values $n_\mu = 0, 1, 2, \ldots$. Consequently,

$$-\mathring{D}^2 \mathring{\phi}_{\bar{n}} = 2\omega_A(2 + n_0 + n_1 + n_2 + n_3)\mathring{\phi}_{\bar{n}}, \tag{5.30}$$

and we see that $-D^2$ is positive definite in the space of all exciton transition amplitudes $K_{\cdot,A}$ given by (5.14). Its role is, of course, to single out from this amorphous 'sea' of exciton transition amplitudes those that lead to actual coordinate fluctuation amplitudes. Hence this role is analogous to that played by the Hamiltonian of a nonrelativistic system of bound particles when it singles out from the space of all wave functions for those particles the bound state wave functions that represent actual stationary states of the system.

However, not all eigenfunctions of $\mathring{D}^2$ that are superpositions of those in (5.30) yield actual coordinate fluctuation amplitudes $K_{B,A}$. Indeed, any such amplitudes have to correspond to test particles that are elementary particles and therefore exhibit a definite spin. Consequently we have to single out those solutions of (5.13) which correspond to irreducible representation of SO(3). Using the eigenfunctions of the isotropic harmonic oscillator in three dimensions (Bates, 1961, Sec. 4.3), we arrive at the solutions

$$\mathring{\eta}_{J,n_0,n}(\mathring{w}, s) = H_{n_0}(w^0)F_{J,n}(\mathring{\rho}) \exp(\mathring{w}^2/2\omega_A) Y_J^s\left(\frac{\mathring{\mathbf{w}}}{|\mathring{\mathbf{w}}|}\right), \tag{5.31a}$$

$$F_{J,n}(\mathring{\rho}) = \mathring{\rho}^{(J+1)/2} L^{J+1/2}_{n+J+1/2}(\mathring{\rho}), \quad \mathring{\rho} = \frac{|\mathring{\mathbf{w}}|^2}{\omega_A^2}, \tag{5.31b}$$

for n_0, $n = 0, 1, 2, \ldots$, and $J = 0, 1, 2, \ldots$, where H_{n_0} and L^α_β denote the Hermite polynomials and the associated Laguerre functions, respectively. In (5.31a) $s = -J, -J+1, \ldots, +J$ refers to the spin components in some (stochastic) direction, such as that specified by any coordinate axis of some stochastic inertial frame. If that direction is the same at all $\zeta = (q, p)$, then (Brooke and Prugovečki, 1983a)

$$K_{B,A}(\mathring{\zeta}, s; \mathring{\zeta}') = \int_{\mathscr{V}_\pm} \exp\left[\frac{i}{\omega_A}\left(\mathring{q}' - \frac{m_B}{m_A}\mathring{q}\right)\cdot \mathring{k}\right] \times$$
$$\times\, \eta_{J,n_0,n}(\mathring{p}\cdot\mathring{k})\eta_{0,0,0}(\mathring{p}'\cdot\mathring{k}) Y_J^s\left(\frac{-R\mathbf{p}}{|\mathbf{p}|}\right) d\Omega(\mathring{k}), \tag{5.32}$$

$$\eta_{J,n_0,n}(\mathring{p}\cdot\mathring{k}) = N_{J,n_0,n} H_{n_0}\left(\frac{\mathring{k}\cdot\mathring{w}}{m_A}\right)\times$$
$$\times\, F_{J,n}\left(\omega_A^{-2}\left[\left(\frac{\mathring{k}\cdot\mathring{w}}{m_A}\right)^2 - \mathring{w}^2\right]\right) \exp\left(\frac{\mathring{w}^2}{2\omega_A}\right), \tag{5.33a}$$

$$R = \Lambda^{-1}_{\Lambda^{-1}_{\mathring{v}}\mathring{u}}\Lambda^{-1}_{\mathring{v}}\Lambda_{\mathring{u}}, \qquad \mathring{u} = \mathring{\mathbf{p}}/m_B, \qquad \mathring{v} = \mathring{\mathbf{k}}/m_A, \tag{5.33b}$$

where $N_{J,n_0,n}$ are normalization constants whose values are decided by the consistency conditions (3.22b) – which now involve also summation over the spin component values $s = -J, \ldots, +J$.

As a measure of l_B^2, where l_B is the *proper radius* of an exciton state describing an extended elementary particle, we can adopt the expectation value

$$l_A^2 \langle \mathring{q}^2 \rangle_B = \sum_{s=-J}^{+J} \int \mathring{q}_\mu \mathring{q}^\mu \, |K_{B,A}(\mathring{q}, \mathring{p}, s; 0, \mathring{p}')|^2 \, d\Sigma(\mathring{q}, \mathring{p})|_{\mathring{\mathbf{p}}'=0} \tag{5.34}$$

multiplied by a constant adjusted so as to obtain for $B = A$ the value l_A that occurs in (5.24) as part of the value of ω_A in (5.10b). This proportionality constant is then uniquely determined if the rest mass m_A of the standard is known. However, to determine the rest masses m_B of the remaining excitons we require an additional equation, such as the equation

$$\left[P_\mu P^\mu + \kappa_A D^2 + \left(\frac{l_B}{l_A} \right)^2 \right] K_{B,A} = 0 \tag{5.35}$$

of the type first suggested by Yukawa (1953), in which κ_A is a proportionality constant to be determined from experiments. Since $K_{B,A}$ is a solution of the Klein–Gordon equation

$$\left[P_\mu P^\mu - \left(\frac{m_B}{m_A} \right)^2 \right] K_{B,A} = 0, \tag{5.36}$$

the imposition of (5.35) in conjunction with Born's eigenvalue equation (5.15) leads to the following reciprocally symmetric formula:

$$\left(\frac{m_B}{m_A} \right)^2 + \left(\frac{l_B}{l_A} \right)^2 + \kappa_A \lambda_{B,A} = 0. \tag{5.37}$$

On the other hand, the eigenvalues $\lambda_{B,A}$ of D^2 that correspond to (5.32) are (Prugovečki, 1981h; Brooke and Prugovečki, 1983a):

$$\lambda_{B,A} = -2\omega_A (2 + n_0 + 2n + J_B). \tag{5.38}$$

Consequently, we arrive at the mass formula

$$m_B^2 = [2\kappa_A \omega_A (J_B + n_0 + 2n + 2) - (l_B/l_A)^2] m_A^2, \tag{5.39}$$

which for fixed values of n, $n_0 = 0, 1, 2, \ldots$, predicts the well-established linear Regge trajectories (Chiu, 1972; Pasupathy, 1976; Collins, 1977) of hadronic square-masses versus hadronic spins. The analysis of up-to-date experimental data by Brooke and Guz (1982), and by Holstein and Schroeck (1983), exhibit very good agreement between experimentally obtained masses and spins and the predictions of the mass formula (5.39) interpreted in the light of the considerations of the next section, that take into account isospin degrees of freedom.

* 4.6. Unitary Internal Symmetries and Generalized Representations of Relativistic Canonical Commutation Relations

The realizations of the external RCCRs (5.10) as well as of the internal RCCRs (5.12) with which we have been dealing in the preceding section have led us, via Born's metric operator (5.7) and via the ensuing eigenvalue condition (5.15), to spin internal degrees of freedom that are embodied in the resulting coordinate fluctuation amplitudes (5.32) and the mass formulae (5.39). Indeed, each realization of the external RCCRs leads (Brooke and Prugovečki, 1983b) in the case of Minkowski quantum spacetime to an irreducible representation of the Poincaré group, whose generators P_μ and $M_{\mu\nu}$ are linked to that realization by means of the relation

$$M_{\mu\nu} = Q_\mu P_\nu - Q_\nu P_\mu + S_{\mu\nu}, \tag{6.1}$$

where the operators $S_{\mu\nu}$ characterize the spin of the excitons corresponding to that representation. However, according to the contemporary classification scheme of elementary particles (Kelly *et al.*, 1982), mesons and baryons display also internal unitary symmetry degrees of freedom. To take those into account, we have to consider realizations (Brooke and Prugovečki, 1982)

$$Q_\mu/\omega_A = -i\left[\frac{\partial}{\partial p^\mu} + \Phi_\mu(q, p)\right], \qquad P_\mu/\omega_A = i\left[\frac{\partial}{\partial q^\mu} + \Psi_\mu(q, p)\right] \tag{6.2}$$

of the external RCCRs (5.10) – as well as of the internal ones (5.12) for $\hat{Q}_\mu$ and $\hat{P}_\mu$ related to Q_μ and P_μ by (5.8) – in which Φ_μ and Ψ_μ assume matrix values corresponding to some representation of an internal symmetry group G^{int}, such as SU(N). Thus, the present operators shall act in spaces that are tensor products of the n-dimensional spaces for SU(N) multiplets and of the Hilbert spaces considered in preceding sections – such as $L^2(\Sigma_A)$ for the operators Q_μ and P_μ of spinless particles that represent ground exciton states.

The necessary and sufficient conditions for (6.2) to be realizations of the RCCRs in the absence of internal symmetry, when the gauge group at each $\zeta = (q, p)$ is U(1) and $\{\Phi_\mu, \Psi_\mu\}$ are complex-valued functions, is that

$$\frac{\partial \Phi_\mu}{\partial q^\nu} - \frac{\partial \Psi_\nu}{\partial p^\mu} = \frac{i}{\omega_A} g_{\mu\nu}, \qquad \frac{\partial \Phi_\nu}{\partial p^\mu} - \frac{\partial \Phi_\mu}{\partial p^\nu} = 0, \qquad \frac{\partial \Psi_\nu}{\partial q^\mu} - \frac{\partial \Phi_\mu}{\partial q^\nu} = 0. \tag{6.3}$$

If we introduce the differential 1-form

$$\lambda = \Phi_\mu \, dp^\mu + \Psi_\mu \, dq^\mu \tag{6.4}$$

then (6.3) stipulates that the exterior derivative

$$d\lambda = \frac{1}{2}\left(\frac{\partial\Phi_\nu}{\partial p^\mu} - \frac{\partial\Phi_\mu}{\partial p^\nu}\right) dp^\mu \wedge dp^\nu + \left(\frac{\partial\Psi_\nu}{\partial p^\mu} - \frac{\partial\Phi_\mu}{\partial q^\nu}\right) dp^\mu \wedge dq^\nu +$$

$$+ \frac{1}{2}\left(\frac{\partial\Psi_\nu}{\partial q^\mu} - \frac{\partial\Psi_\mu}{\partial q^\nu}\right) dq^\mu \wedge dq^\nu \tag{6.5}$$

should equal $(-i/\omega_A)g_{\mu\nu}\, dp^\mu \wedge dq^\nu$, i.e. that

$$d\left(\lambda - \frac{i}{\omega_A}\, g_{\mu\nu} q^\mu\, dp^\nu\right) = 0. \tag{6.6}$$

By Poincaré's lemma (Spivak, 1970) this will be the case if and only if there is (locally) a scalar function ω such that

$$\lambda - \frac{i}{\omega_A}\, g_{\mu\nu} q^\mu\, dp^\nu = i\, d\omega = i\left(\frac{\partial\omega}{\partial q^\mu}\, dq^\mu + \frac{\partial\omega}{\partial p^\mu}\, dp^\mu\right). \tag{6.7}$$

Hence (5.11) and (5.13) are indeed the most general solutions when the local gauge group is U(1), i.e. in the absence of isospin and of internal symmetries in general.

In the presence of nonzero isospin, i.e. in case of the nonabelian local gauge group U(1) × G^{int} (Taylor, 1976; Faddeev and Slavnov, 1980), the RCCRs (5.10) shall be satisfied if and only if

$$\frac{\partial\Phi_\mu}{\partial q^\nu} - \frac{\partial\Psi_\nu}{\partial p^\mu} + [\Psi_\nu, \Phi_\mu] = \frac{i}{\omega_A}\, g_{\mu\nu}, \tag{6.8a}$$

$$\frac{\partial\Phi_\nu}{\partial p^\mu} - \frac{\partial\Phi_\mu}{\partial p^\nu} + [\Phi_\mu, \Phi_\nu] = 0, \tag{6.8b}$$

$$\frac{\partial\Psi_\nu}{\partial q^\mu} - \frac{\partial\Psi_\mu}{\partial q^\nu} + [\Psi_\mu, \Psi_\nu] = 0. \tag{6.8c}$$

A geometrized form of the realizations of these RCCRs is arrived at if we introduce the covariant derivatives

$$\nabla_\mu = \frac{\partial}{\partial q^\mu} + \Psi_\mu, \qquad \nabla_{\bar{\mu}} = \frac{\partial}{\partial p^\mu} + \Phi_\mu, \tag{6.9}$$

i.e. treat (Φ_μ, Ψ_μ) as a connection on a fibre bundle (Atiyah, 1979). Indeed, in that case (6.2) assumes the form

$$Q_\mu = -i\omega_A \nabla_{\bar{\mu}}, \qquad P_\mu = i\omega_A \nabla_\mu. \tag{6.10}$$

If we now compute $D\lambda$, namely the exterior covariant derivative (Kobayashi and Nomizu, 1963) for λ in (6.4), we get:

$$D\lambda = d\lambda + \frac{1}{2}[\lambda, \lambda] = \frac{1}{2}\left(\frac{\partial \Phi_\nu}{\partial p^\mu} - \frac{\partial \Phi_\mu}{\partial p^\nu} + [\Phi_\mu, \Phi_\nu]\right) dp^\mu \wedge dp^\nu +$$

$$+ \left(\frac{\partial \Psi_\nu}{\partial p^\mu} - \frac{\partial \Phi_\mu}{\partial q^\nu} + [\Phi_\mu, \Psi_\nu]\right) dp^\mu \wedge dq^\nu +$$

$$+ \frac{1}{2}\left(\frac{\partial \Psi_\nu}{\partial q^\mu} - \frac{\partial \Psi_\mu}{\partial q^\nu} + [\Psi_\mu, \Psi_\nu]\right) dq^\mu \wedge dq^\nu. \tag{6.11}$$

Consequently, by (6.7) we conclude that the Yang–Mills (1954) type of equation

$$D\lambda = \frac{1}{i\omega_A} g_{\mu\nu}\, dp^\mu \wedge dq^\nu = \frac{1}{i\omega_A}\, dp_\mu \wedge dq^\mu, \tag{6.12}$$

is a necesary and sufficient condition for (6.10) to be realizations of RCCRs in (5.10).

In case that $G^{\text{int}} = \text{SU}(N)$, and T_a, $a = 1, \ldots, N^2 - 1$, are its infinitesimal generators, so that

$$[T_a, T_b] = t^c_{ab} T_c, \tag{6.13}$$

where t^c_{ab} are the structure constants of SU(N), then the set $\{I, T_a\}$ of infinitesimal generators of U(1) × SU(N) constitute a basis in its N^2-dimensional regular representation space. Writing

$$\Phi_\mu = i(\Phi^{\text{e.m.}}_\mu I + W^a_\mu T_a), \qquad \Psi_\mu = i(\Psi^{\text{e.m.}}_\mu I + W^a_\mu T_a), \tag{6.14}$$

we observe that in the absence of internal symmetries, i.e. for $N = 1$, according to (5.11)

$$\Phi^{\text{e.m.}} = \frac{1}{\omega_A} g_{\mu\nu} q^\nu + \frac{\partial \omega}{\partial p^\mu}, \qquad \Psi^{\text{e.m.}}_\mu = \frac{\partial \omega}{\partial q^\mu}, \tag{6.15}$$

so that (6.3) is equivalent to

$$F_{\hat{\mu}\hat{\nu}} = \partial_{\hat{\mu}} A_{\hat{\nu}} - \partial_{\hat{\nu}} A_{\hat{\mu}} = 0, \quad \hat{\mu} \in \{\mu, \bar{\mu}\}, \hat{\nu} \in \{\nu, \bar{\nu}\}, \tag{6.16}$$

$$A_{\hat{\mu}} = \Psi^{\text{e.m.}}_\mu, \quad \hat{\mu} = \mu, \qquad A_{\hat{\mu}} = \Phi^{\text{e.m.}}_\mu - \frac{1}{\omega_A} g_{\mu\nu} q^\nu, \quad \hat{\mu} = \bar{\mu}, \tag{6.17}$$

i.e. to a vanishing external electromagnetic field. On the other hand, when $N > 1$ the relations (6.8) yield,

$$\frac{\partial W_\nu^c}{\partial p^\mu} - \frac{\partial W_\mu^c}{\partial q^\nu} + it_{ab}^c W_\mu^a W_\nu^b = 0, \tag{6.18a}$$

$$\frac{\partial W_\nu^c}{\partial p^\mu} - \frac{\partial W_\mu^c}{\partial p^\nu} + it_{ab}^c W_\mu^a W_\nu^b = 0, \tag{6.18b}$$

$$\frac{\partial W_\nu^c}{\partial q^\mu} - \frac{\partial W_\mu^c}{\partial q^\nu} + it_{ab}^c W_\mu^a W_\nu^b = 0. \tag{6.18c}$$

In turn, the above relations (6.18) yield the generalization of (6.16) to internal gauge degrees of freedom:

$$F_{\hat{\mu}\hat{\nu}} = (\partial_{\hat{\mu}} A_{\hat{\nu}} - \partial_{\hat{\nu}} A_{\hat{\mu}})I + (\partial_{\hat{\mu}} W_{\hat{\nu}} - \partial_{\hat{\nu}} W_{\hat{\mu}} + i[W_{\hat{\mu}}, W_{\hat{\nu}}]) = 0. \tag{6.19}$$

We see that a geometrization of relativistic stochastic quantum theory has been achieved in the sense that the problem of constructing quantum spacetime models has been reduced to finding solutions of the RCCRs expressed in terms of covariant derivatives. In flat quantum spacetimes, these covariant derivatives are related to connections which describe how the internal states of test particles change under parallel transport from $\underset{\sim}{\zeta}'$ to $\underset{\sim}{\zeta}''$ in combination with the (external) phase of their wave functions (cf. (2.7.25c)). In the generic curved case however, a Levi–Civita term $\Gamma_{\hat{\mu}}$ has to be added in order to arrive at the full connection, which therefore consists of a metric and an internal gauge part relating to the parallel transport of external and internal tensor quantities, respectively.

It has to be realized, however, that in the generic case the operators P_μ lose all *direct* physical significance (namely that of sharp global momentum operators), since the momentum of a test particle in free fall is no longer a conserved quantity. Thus, P_μ undergo the same kind of loss of status that the positive operators Q^j had undergone during the transition from the nonrelativistic to the relativistic regime – similar to the loss of status of position as a canonical variable (Currie *et al.*, 1963; Kerner, 1964) in classical relativistic mechanics, indicating some of the intrinsic deficiencies of the concept of sharp localizability (Hill, 1967, pp. 1772–1773). Hence, in the ultimate analysis, all direct (i.e. operational) physical meaning in quantum spacetime is retained only by the coordinate fluctuation amplitudes $K_{B,A}$, which give rise to coordinate probability densities $G_{B,A}$. A heuristic[17] physical meaning (Prugovečki, 1982a) can be assigned, however, to the internal operators $\hat{Q}^\mu$ and $\hat{P}^\mu$ viewed as operators related to spacetime and 4-momentum oscillations *within* the extended particle. In this context, it is to be noted that even in case of Minkowski quantum spacetime, not all four of the $\hat{Q}^\mu$ operators can be chosen to be simultaneously self-adjoint in the Hilbert spaces $L^2(\Sigma_A)$. Rather, if σ in the reference hypersurface $\Sigma_A = \sigma \times \mathscr{V}_A$ corresponds to $q^0 = \text{const}$, then only

$\hat{Q}^j$, $j = 1, 2, 3$, are self-adjoint. To secure the self-adjointness of $\hat{Q}^0$ we have to adopt a timelike hyperplane σ, and to restrict ourselves to the subspace of TCP invariant wave functions (Prugovečki, 1982a). These features are, however, totally in keeping with the nonexistence of time operators in nonrelativistic quantum mechanics (Misra and Sudarshan, 1977; Misra *et al.*, 1979), and with the physical meaning of the time-energy uncertainty relations (Aharanov and Bohm, 1961; Wigner, 1972; Rayski and Rayski, 1977): the particular combination of position–momentum and time–energy uncertainty relations that will result from the observation process, depends on the manner in which that observation is performed, e.g. on the choice of the (stochastic mean) coordinate q^μ that is kept fixed while accumulating a sample of data. In other words, whether the registered coordinate fluctuations will be in spacelike or timelike directions depends on the choice of reference hypersurface, namely whether its spacelike component σ is spacelike or timelike, respectively.

On the more pragmatic side, the injection of reciprocity theory into the present geometrized version of relativistic quantum mechanics can also provide directly verifiable predictions. Thus, in analyzing the most up-to-date data (Kelly *et al.*, 1982) on baryons from the point of view of the mass formula (5.39), and with $G^{\text{int}} = SU(3)$, Brooke and Guz (1983a) could arrive at a very good fit for the predicted Regge trajectories by adopting the members of the positive-parity baryon octet $\{p, n, \Lambda, \Sigma^+, \Sigma^0, \Sigma^-, \Xi^0, \Xi^-\}$ for $J_A = 1/2$, and of the decuplet $\{\Delta^{++}, \Delta^+, \Delta^0, \Delta^-, \Sigma^{*+}, \Sigma^{*0}, \Sigma^{*-}, \Xi^{*0}, \Xi^{*-}, \Omega^-\}$ for $J_A = \frac{3}{2}$, as standards. In fact, the average masses of both the octet and the decuplet fit the formula as it stands. However, a separation of trajectories occurs, suggesting the exact SU(3) symmetry is broken into $SU(2)_I \times U(1)_Y$ in the presence of medium strong interactions.

Despite these initial successes in adapting Born's metric operator to the present framework for quantum spacetime, the idea of applying the outcome directly to hadrons has to be regarded primarily as an illustration of a methodological approach. Clearly, the ultimate strategy will have to contend with the question as to which are the true excitons: are they the elementary particles themselves, or are they the leptons and the quarks, as the at present prevalent point of view suggests, or do they represent an even more fundamental substratum of localized and yet undetected physical entities? This is an epistemic as much as an experimental question that has to be approached with an open mind, and only the elimination of competing models on basis of internal inconsistencies (such as the quark confinement issue) as well as disagreement with experiments can point future research in the right direction.

Notes

1 We have cited only a small sample of representative papers on the subject. Some further references can be found in a review article by Papp (1977), and in a monograph as well as in a review article by Blokhintsev (1973, 1975). A monograph by Vyaltsev (1965) on discrete

and cellular spacetimes contains references to most of the main papers on that subject, whereas one by Ingraham (1967) contains most references to his formulation of stochastic spacetime (Ingraham, 1962–1964, 1968, 1972).

[2] Not to be confused with the notion of 'geometric quantization' (Simms and Woodhouse, 1976; Śniatycki, 1979), albeit some indirect connections might exist (Brooke and Prugovečki, 1982).

[3] There is some (debatable) evidence (Sachs, 1979) that in later years, in his fruitless pursuit of a unified field theory, Einstein had retreated somewhat from his uncompromising advocacy of an operational definition of basic spacetime concepts (metric, spacetime intervals, etc.). This vacillation can be attributed, however, to a great extent to his basic reluctance to accept the uncertainty principle, and indeterminacy in general, as a basic feature of Nature rather than a temporary aberration in the historical development of physics, combined concomitantly with his failure to produce operationally-based proofs in support of this metaphysical belief during his protracted debates with Bohr (1961) and Born (1971) on the foundations of quantum mechanics. However, in his public reply to criticisms (Einstein, 1949) – from which the quote heading Part II is taken – he remained steadfast in his opposition to any neo-Kantian approach to the concept of spacetime. It is interesting, however, that historically (Moyer, 1979) there had been attempts by many physicists to disassociate the predictions of relativity theory from its fundamentally operational basis even after the spectacular confirmation of light deflection during the 1919 solar eclipse. It was only very gradually that Einstein's conceptualization achieved (almost) general acceptance.

[4] In fact, there have been speculations as to the existence of 'maximons' (Markov, 1966; Blokhintsev, 1975), i.e. particles with a limiting rest mass $m_{\max} = \hbar/cl_g$. For l_g equal to Plancks' length of 0.82×10^{-32} cm, this maximal rest mass would be $\approx 10^{-5}$ g.

[5] Those who tend to interpret Bohr's (1961, p. 39) dictum that "the account of ... the results of observations must be expressed in unambiguous language with suitable application of the terminology of classical physics" as an advocacy of a deterministic description of measurement results, should be reminded of Born's (1955, 1956) conclusion that "determinism is out of the question in the original sense of the word even in case of the simplest classical science, that of mechanics" (Born, 1962, p. 34).

[6] Except in the flat case, the worldlines of the test particles in the coherent flow will in general intersect, but the locus of these intersections is at most a manifold of lower dimensionality (Lifshitz and Khalatnikov, 1963), and therefore of Riemannian measure zero.

[7] The fact that a consistent quantum mechanical operational definition of spatio-temporal relationships has to incorporate momentum (or, equivalently, velocity) degrees of freedom along with the purely spatio-temporal ones has been first realized by M. Born (1939) in the course of developing reciprocity theory (cf. also the quotes heading Part II and the present chapter).

[8] In Einstein's (1923, p. 117) own words, "all our spacetime verifications invariably amount to a determination of spacetime coincidences". Those coincidences are "meetings of the material points of our measuring instruments with other material points".

[9] According to a 1971 article by de Broglie (1979, p. 7), the idea that each quantum particle is a natural clock was implicit in his 1923 formulation of wave-particle duality. This idea has been revived by Penrose (1968, p. 129) in the context of giving to spacetime distances an operational meaning which would not rely on rigid rods and standard clocks.

[10] Such a procedure would be very accurate, due to the very short wavelengths already present in beams of electrons used in electron microscopes. Presumably, however, the accurate control of velocities, and other technical aspects, make such devices for the measurement of distance feasible only to the technologies of the future.

[11] Whereas the Klein–Gordon equation is obtained by the substitution $\hat{k}^\nu \to i\hbar\partial/\partial\hat{q}_\nu$ in $\hat{k}^\nu\hat{k}_\nu = \hat{m}^2c^2$, the Born–Landé equation is obtained by the substitution $\hat{x}^\nu \to -i\hbar\partial/\partial\hat{p}_\nu$ in $\hat{x}^\nu\hat{x}_\nu = -\hat{l}^2$.

[12] The favored status in the quantization procedure of normal coordinates tied to free (or free-fall) test bodies, can be traced all the way from quantization in the nonrelativistic regime, where inertial Cartesian coordinates play a distinguished role (since the imposition of CCRs on operators associated with coordinate variables in certain canonical coordinate systems can lead to inconsistencies with that in Cartesian coordinates (Abraham and Marsden, 1978; Peierls, 1979)), to the special relativistic case where, for example, quantization in Rindler coordinates tied to uniformly accelerated test bodies is inconsistent (Fulling, 1973) *vis-à-vis* the conventional method of quantization in Minkowski coordinates tied to test particles moving by inertia.

[13] If the existence of quarks is accepted as established physical reality that represents an actual spatio-temporal structure within hadrons, then the proper wave functons of quarks would take over the role of hadronic proper wave functions in the subsequent considerations. Minority points of view, such as Santilli's (1978) 'eleton' conjecture, or Kurşunoglu's (1979) 'orbiton' idea, envisage instead already well-established elementary particles in the role of fundamental constituents of the Universe. Born's quantum metric idea can embrace either of these viewpoints, but in the absence of *direct* experimental confirmation of quark existence (cf. Section 1.12) it appears more prudent to leave the question of fundamental constituent open.

[14] According to one of Born's collaborators (H. S. Green), Pauli raised basic objections with regard to gauge and translational invariance as well as causality in reciprocity theory. Whereas the difficulties with the former could be dealt with, the problem "concerning causality appeared insuperable" (Green, 1982 – private communication). In contradistinction, the quantum spacetime approach is based on the consistent solution of the particle localizability problem, into which (as we have seen in Sections 1.3 and 2.3) gauge and translational invariance is embedded from the beginning, and classical causality is replaced by stochastic causality – as further discussed in Chapter 5.

[15] This could be achieved in a straightforward manner in the old (static) quark models (Kokkedee, 1969) by using a metric operator for the computation of the quark proper wave functions, and introducing, as in atomic models (cf. Section 1.12), a potential that binds three quarks into a nucleon and a quark–antiquark pair into a meson for the computation of hadronic wave functions. In QCD, in the absence of an adequate understanding of quark confinement, the corresponding bound-state problem can be approached via the concept of valons, i.e. valence-quark clusters (Hwa, 1980; Hwa and Zahir, 1981). The reservations expressed in Section 1.12 and based on Heisenberg's (1976) epistemic criticism of all constituent-based models for hadrons would, however, still remain in effect.

[16] According to Kim and Noz (1979), $|\eta(x)|^2$ is to be interpreted as a probability density over all the spacetime values of x. This leads to the kind of stochastic existence–nonexistence of one constituent in the causal and acausal past and future of the other that has been discussed in Section 1.3, and found unacceptable on physical and epistemic grounds if quarks and antiquarks are deemed to be objectively existing entities. On the other hand, according to Takabayashi (1979, p. 44) any 'imperfectness of the probabilistic interpretation [of relativistic harmonic oscillator quark models] does not imply the imperfectness of [the] theory since the constituents (quarks) are not directly observable particles". However, that still leaves open the 'observability' of quarks as localized objects in well-defined spatio-temporal relationships to each other within the hadron which, if actually feasible, should be describable in physically consistent probabilistic terms.

[17] This meaning would achieve operational status if Landé's (1939) original idea about the possibility of signals with speeds $> c$ *within* an extended particle could be validated. More recently, Santilli (1982), as well as De Sabbata and Gasperini (1981), have suggested that such signals might indeed propagate within hadronic matter.

Chapter 5

Field Theory on Quantum Spacetime

"Our present quantum theory is very good provided we do not try to push it too far – we do not try to apply it to particles with very high energies and we do not try to apply it to very small distances. When we try to push it in these directions, we get equations which do not have sensible solutions . . . It is because of these difficulties that I feel that the foundations of quantum mechanics have not yet been correctly established. Working with the present foundations, people have done an awful lot of work in making applications in which they find rules for discarding the infinities. But these rules, even though they may lead to results in agreement with observation, are artificial rules, and I just cannot accept that the present foundations are correct."

DIRAC (1978a, p. 20)

"If predictive power were indeed the only criterion of truth, Ptolemy's astronomy would be no worse than Newton's."

HEISENBERG (1971, p. 212)

"The renormalization idea would be sensible only if it was applied with finite renormalization factors, not with infinite ones."

DIRAC (1978b, p. 5)

Quantum field theory dates back to the same year when matrix and wave mechanics were both founded: indeed, in one of the pioneering papers on matrix mechanics by Born and Jordan (1925), the quantization of the electromagnetic field was considered for the first time. However, the first paper devoted entirely to this topic was that by Dirac (1927), who treated this subject in a systematic manner, and has been henceforth considered to be the founder of 'local' quantum field theory (LQFT). The publication of that paper preceded only slightly that of another key paper by Dirac (1928), namely the one in which he launched the equation bearing his name. In the beginning it appeared that the problem of sharp localizability of relativistic quantum particles of spin ½ had been thereby solved, but, as discussed in Sections 2.1 and 2.2, subsequent developments proved that hope to have been groundless. In fact, as was pointed out at the end of Section 2.2, Hegerfeldt's (1974) theorem eventually supplied the mathematical proof of the unsolvability of that problem for particles of any spin within the confines of the concept of sharp localizability. The descriptive term 'local' has nevertheless remained entrenched in this context, with the phraseology 'relativistic quantum field $\psi(x)$ at the spacetime point

x' being habitually used ever since Dirac's pioneering work, despite the mathematically proven nonexistence (Bogolubov *et al*., 1975, Sec. 10.4) of any such entities.

On the other hand, long before the problem of localizability of relativistic particles was resubmitted to careful scrutiny (cf. Section 2.2), it had been discovered (Heisenberg and Pauli, 1929) that the self-energy of the electron in quantum electrodynamics (QED) was given by a divergent integral. This discovery was soon followed by the discovery of divergences in just about all formal expressions for quantities of physical interest in QED as well as other LQFT models, indicating that the newly founded LQFT was riddled with mathematical inconsistencies. After almost two decades of fruitless attempts at dealing with these divergences – at least to the extent of being able to perform computations – the development by Tomonaga (1946), Schwinger (1948, 1949), Feynman (1949, 1950) and Dyson (1949) of the so-called 'renormalization' program was greeted by some physicists as the solution to most fundamental difficulties in QED. However, Dirac (1951) had stressed at the very beginning of this program, as well as ever since (Dirac, 1965, 1973, 1977, 1978), that he "finds the present quantum electrodynamics quite unsatisfactory" (Dirac, 1978b, p. 5) for both mathematical and physical reasons, which actually extend to all of LQFT.

Schwinger (1958, pp. xv–xvi) appeared to concur with Dirac's assessment when he stated in the fifties that

> . . . the observational basis of quantum electrodynamics is self-contradictory . . . . It seems that we have reached the limits of the quantum theory of measurement, which asserts the possibility of instantaneous observations, without reference to specific agencies. The localization of charge with indefinite precision requires for its realization a coupling with the electromagnetic field that can attain arbitrarily large magnitudes. The resulting appearance of divergences and contradictions, serves to deny the basic measurement hypothesis. We conclude that a convergent theory cannot be formulated consistently within the framework of present space-time concepts. To limit the magnitude of interactions while retaining this customary coordinate description is contradictory, since no mechanism is provided for precisely localized measurements.

In addition to the fundamental inconsistencies pointed out in the above quote of Schwinger, the renormalization procedures as applied in textbooks to QED and other LQFT models, including the recent nonabelian models (Rühl, 1980; Dita *et al*., 1982) are riddled with intrinsic inconsistencies, such as those pointed out by Landau and Pomeranchuk (1955), by Rohrlich (1980), and by many others. In fact, even a cursory glance at the renormalization program in its standard textbook form (Jauch and Rohrlich, 1955; Bogolubov and Shirkov, 1959; Schweber, 1961; Roman, 1969; Itzykson and Zuber, 1980) reveals that this program deals *only* with the question of assigning mathematical meaning to each term in the formal perturbation series of the S-matrix of a 'renormalizable' LQFT model. However, the convergence question for the entire series is left totally open. In fact, it has been conjectured (Dyson, 1952) that this series is only an asymptotic expansion, in which case successive partial sums would approach the (unknown) actual S-matrix only to within a certain (unknown) range, whereafter subsequent partial sums would gradually diverge more and more from the actual S-matrix. Consequently, taking into consideration (as is generally the practice) only the number of terms which provide the best agreement with experimental data introduces into the entire scheme an intrinsic element of partiality which is totally at odds with any ideal of scientific objectivity. Furthermore, the existence of an *actual* value of the S-matrix can be questioned in the absence of mathematically meaningful LQFT models for interacting fields, i.e. models in which meaning and existence are established for the quantum fields themselves as well as for the equations they obey, rather than just for the terms of *formal* perturbation expansions of a (possibly nonexistent) S-matrix.

Various axiomatic quantum field theoretic approaches (Streater and Wightman, 1964; Haag and Kastler, 1964; Jost, 1965; Bogolubov *et al.*, 1975) have attempted to deal with this last question by formulating mathematically meaningful and consistent general frameworks for LQFT. Subsequently, the constructive quantum field theory program (Glimm and Jaffe, 1981) attempted to implement some of these ideas in the context of actual models. Although some success has been encountered with simple models (ϕ^4 and Yukawa) in *two* spacetime dimensions (and, to a lesser extent, in three dimensions), almost two decades of intensive efforts have failed (Osterwalder, 1982) to produce *any* viable model of interacting 'local' relativistic quantum fields in *four* spacetime dimensions – i.e. for the world we actually live in.

However, even if such realistic and *mathematically* consistent LOFT models were to be eventually produced, an even more basic *physical* question in LQFT would remain unanswered: What is the meaning of a relativistic quantum field $\phi(x)$ at a point x (or at least in some bounded spacetime region R) in the absence of a corresponding meaning for the relativistic quantum particles at x, to which such a field supposedly gives rise? An attempt has been made by Wightman and Schweber (1955) to introduce quantum fields $\phi(\mathbf{y}, t)$ at values $\mathbf{y}$ (cf. (2.2.20)) in the spectra of the Newton–Wigner operators $\mathbf{X}_{NW}$ in (2.2.13) that purportedly, when appropriately smeared, could be related to localized relativistic particles. However, not only does such a procedure destroy the manifest relativistic covariance of a LQFT model, but as we have seen in Section 2.2, due to Hegerfeldt's theorem it cannot ultimately solve the localizability problem within the context of sharp particle localization.

Faced with this fundamental and unresolved problem, some of the adherents to LQFT subscribe to the opinion that although relativistic quantum particles cannot be sharply localized, nonetheless the quantum fields can be themselves localized in a sharp sense. This comforting belief appears to have been inspired by the well-known Bohr and Rosenfeld (1933, 1950) papers on the measurement of the quantum electromagnetic field. However, those papers were intended[1] to illustrate the limitations rather than the unconditional feasibility of localizing quantum fields. Moreover, the nature of the classical devices that are employed in those papers intrinsically limits the size of the domain over which measurements might be performed, even when the most outlandishly unrealistic assumptions are made about the physical implementability of such devices. In fact, after performing an exhaustive study of this subject, De Witt had to arrive at the conclusion that "10^{-32} cm constitutes an absolute limit on the domain of applicability of classical concepts, even as modified by the principle of complementarity" (De Witt, 1962, p. 371). Furthermore, in view of the nature of the operational procedures underlying the foundations of special and general relativity (discussed in Section 4.1 and 4.2), it is epistemologically unsound to discuss the question of arbitrarily accurate localizability for fields in isolation from that of test particles. Consequently, we conclude that since the problem of sharp localizability of relativistic quantum particles possesses no consistent solution (cf. Section 2.2), neither does that for sharply localized relativistic quantum fields.[2]

The various early nonlocal quantum field theories, proposing nonlocal interactions (Feynman, 1948b; Arnous *et al.*, 1960), or nonlocal fields (Yukawa, 1949, 1950; Pais and Uhlenbeck, 1950; Sen, 1958), were primarily aimed at removing divergences from LQFT models, rather than at solving any foundational questions beyond those of *formal* relativistic invariance, causality and unitarity. However, in later works on the subject, such as those by Markov (1959, 1960), Ingraham (1962, 1963, 1964), Katayama and Yukawa (1968), Katayama *et al.*, (1968), and many others,[3] the concern with the foundational question becomes more apparent, as the subjects of divisibility and stochasticity of spacetime at the micro-level are analyzed and used as guidelines for the introduction of nonlocal fields and nonlocal interactions.

The quantum spacetime framework of Chapter 4 provides a ready-made arena for the formulation of quantum field theory as field theory on quantum spacetime (FTQS), rather than as the *pro forma* submission of classical field theories to the procedure of 'second quantization'. On the other hand, as it will become evident in Section 5.1 at the free-field level, the resulting framework for FTQS does not represent, at the formal level, a radical departure from LQFT. Rather, in the case of Minkowski spacetime, superficially it can be viewed as a 'regularization' of 'local' quantum field singularities. However, at the deeper epistemological level, it is actually the fields in FTQS that can be justifiably viewed as the objects of direct physical significance, since it is only these latter fields that rely on a relativistically consistent solution of the localizability problem for the quantum particles to which they give rise in various quantum states. Furthermore, the mathematical consistency of LQFT for interacting fields in (4-dimensional) Minkowski spacetime is at best an unresolved problem (Osterwalder, 1982), in which the very notion of the interaction Lagrangian for 'local' fields, and the exact meaning of the equations of motion for those fields, is subject to somewhat arbitrary mathematical interpretation, and therefore is exposed to justifiable criticism (Segal, 1976). Naturally, this basic mathematical ambiguity in LQFT is a reflection of the loss of operational meaning of LQFT of the concept of (sharp) point event (Fierz, 1950; Landau, 1955; Ferretti, 1963).

On the other hand, no similar difficulties are encountered in FTQS (Prugovečki, 1981c, d, f; 1982b; 1983) where the concept of classical localizability has been altogether eliminated in favor of that of (stochastic) quantum localizability. Hence, FTQS does not recognize the concept of sharp-point event as appropriate in the quantum realm. By dispensing with this concept, which is both physically and mathematically ill-defined in quantum relativity, FTQS is capable of restoring mathematical sense to a subject known for its *ad hoc* manipulations of various kinds of 'infinities'. Although admittedly results of deeper consequence are still very few in number, we shall at least lay in this chapter the groundwork, and thus, it is hoped, stimulate the research-minded reader to develop methods of his own that might contribute to a better understanding of FTQS models, and therefore to quantum field theory in general.

5.1 Free Scalar Fields on Minkoswki Quantum Spacetime

We shall limit the considerations of this chapter to the case of Minkowski quantum spacetime, although many of the subsequent observations and conclusions remain valid in the general case. This restriction will facilitate, however, the drawing of comparisons with LQFT which, when formulated on curved spacetime (Parker, 1977), gives rise to additional fundamental difficulties (De Witt, 1975; Davies, 1980) not present in the Minkowski case. Furthermore, for the sake of simplicity we shall also limit ourselves to fields for single species of elementary particles – albeit the considerations of Section 4.5 and 4.6 suggest that realistic models should be formulated for fields representing families of particles that correspond to exciton states along entire Regge trajectories.

Let us, therefore, first consider coherent flows of only a single species of spin-zero A-particles without any internal degrees of freedom, but possessing distinct antiparticles. We shall denote by $\mathscr{S}^{\uparrow}$ the associated family of time-ordered reference hypersurfaces Σ_A described in Sections 2.3, 4.2 and 4.3. To each $\Sigma_A \in \mathscr{S}^{\uparrow}$ correspond the Hilbert spaces $L^2(\Sigma_A^+)$ and $L^2(\Sigma_A^-)$ for single particles and antiparticles, respectively. Out of these spaces we can construct the Fock space

$$\mathscr{F}(\Sigma_A) = \bigoplus_{n,n'=0}^{\infty} \mathscr{F}_{n,n'}(\Sigma_A), \qquad \mathscr{F}_{0,0}(\Sigma_A) = \{c\Psi_{0,0} \mid c \in \mathbb{C}^1\}, \tag{1.1}$$

in the usual manner (Bogolubov *et al.*, 1975), i.e. by inserting in the above direct sum of the appropriately symmetrized tensor products

$$\mathscr{F}_{n,n'}(\Sigma_A) = L^2(\Sigma_A^+)^{\circledS n} \otimes L^2(\Sigma_A^-)^{\circledS n'} \tag{1.2}$$

for $n + n' > 0$, and the one-dimensional space spanned by the Fock vacuum for $n = n' = 0$. The elements of $\mathscr{F}_{n,n'}$ can be deemed to be functions $\Psi_{n,n'}$ of $\zeta_1, \ldots, \zeta_n \in \Sigma_A^+$ and $\zeta'_1, \ldots, \zeta'_{n'} \in \Sigma_A^-$ that are symmetric under permutations of the variables $\zeta_1, \ldots, \zeta_n$ as well as of the variables $\zeta'_1, \ldots, \zeta'_{n'}$, and square integrable under the corresponding direct products

$$d\Sigma_A(\zeta_1, \ldots, \zeta_n; \zeta'_1, \ldots, \zeta'_{n'}) = \prod_{j=1}^{n} d\Sigma_A(\zeta_j) \prod_{j'=1}^{n'} d\Sigma_A(\zeta'_{j'}) \tag{1.3}$$

of the measures in (4.3.5).

In order to simultaneously consider the propagation of particles and antiparticles, let us introduce by analogy with the LQFT case (Schweber, 1961, p. 168) the function $\Delta_A^{(+)}$,

$$i\Delta_A^{(+)}(q, v; q', v') = \theta(v_0 v'_0) K_A(q, v; q', v'), \tag{1.4}$$

defined in terms of the propagator K_A in (4.3.10) at all $\zeta = (q, v)$ and $\zeta' = (q', v')$ in $\mathscr{M} = M(1, 3) \times \mathscr{V}$. In accordance with (4.3.5) and (4.3.7)

$$\mathrm{d}\Sigma(q, v) = 2\epsilon(v^0) v^\mu \mathrm{d}\sigma_\mu(q) \delta(v^2 - 1)\, \mathrm{d}^4 v. \tag{1.5}$$

On account of (4.3.13), for all $\zeta', \zeta'' \in \mathscr{M}$,

$$\Delta_A^{(+)}(\zeta'; \zeta'') = -\Delta_A^{(+)*}(\zeta''; \zeta') = -i \int_{\Sigma_A} \Delta_A^{(+)}(\zeta'; \zeta) \Delta_A^{(+)*}(\zeta''; \zeta)\, \mathrm{d}\Sigma(\zeta), \tag{1.6}$$

where the integration extends now over all of Σ_A. Furthermore, since, in accordance with (2.4.21), we have assigned in (4.3.10) the same proper wave function η_A to both A-particles and A-antiparticles, it follows from the reality of η_A that

$$K_A(\zeta; \zeta') = K_A^*(\zeta'; \zeta) = K_A(-\zeta'; -\zeta). \tag{1.7}$$

Consequently, by (1.4) we shall have that

$$\Delta_A^{(+)}(\zeta; \zeta') = -\Delta_A^{(+)*}(\zeta'; \zeta) = \Delta_A^{(+)}(-\zeta'; -\zeta), \tag{1.8}$$

and that $\Delta_A^{(+)}$ for antiparticles can be related also to K_A for particles:

$$i\Delta_A^{(+)}(q, v; q', v') = K_A(-q', -v'; -q, -v), \qquad v, v' \in \mathscr{V}^-. \tag{1.9}$$

Hence, later on we shall be able to limit the definitions of fields to the case of $v, v' \in \mathscr{V}^+$.

We can now define in $\mathscr{F}(\Sigma_A)$ A-particle annihilation operators $\phi_A^{(-)}(\zeta)$ at any $\zeta \in \mathscr{M}^+ = M(1, 3) \times \mathscr{M}^+$ by their action on all $\Psi_{n,n'} \in \mathscr{F}_{n,n'}(\Sigma_A)$:

$$[\phi_A^{(-)}(\zeta)\Psi_{n,n'}]_{n-1,n'}(\zeta_1, \ldots, \zeta_{n-1}; \zeta'_1, \ldots, \zeta'_{n'})$$

$$= i n^{1/2} \int_{\Sigma_A^{\pm}} \Delta_A^{(+)}(\zeta; \zeta_n) \Psi_{n,n'}(\zeta_1, \ldots, \zeta_n; \zeta'_1, \ldots, \zeta'_{n'})\, \mathrm{d}\Sigma_A(\zeta_n). \tag{1.10}$$

A totally analogous formula defines the antiparticle annihilation operator $\phi_A^{(-)}(\zeta')$ at any $\zeta' \in \mathscr{M}^-$. We note that in contradistinction to LQFT – where exactly the same defining procedure (Schweber, 1961, Sec. 7) for $\phi^{(-)}(x)$ is, strictly speaking, mathematically meaningless[4] and has to be replaced by one for operator-valued distributions (Bogolubov *et al.*, 1975, Sec. 12.1) $\phi^{(-)}(f)$, $f \in \mathscr{S}(\mathbb{R}^4)$ – the definition (1.10) is mathematically unambiguous due to the properties of the Hilbert spaces $L^2(\Sigma_A^{\pm})$ and of the propagators K_A, that have been established in Section 2.6. In fact, the definition (1.10) yields a (densely defined and closed) operator in $\mathscr{F}(\Sigma_A)$ which possesses an adjoint $\phi_A^{(-)*}(\zeta)$. On account of (1.6) we easily establish that the adjoint of $\phi_A^{(-)}(\zeta)$ coincides with the particle creation operator $\phi_A^{(+)}(\zeta)$ at $\zeta \in \mathscr{M}^+$,

$$[\phi_A^{(+)}(\zeta)\Psi_{n,n'}]_{n+1,n'}(\zeta_1, \ldots, \zeta_{n+1}; \zeta_1', \ldots, \zeta_{n'}')$$
$$= (n+1)^{-1/2} \sum_{j=1}^{n+1} i\Delta_A^{(+)}(\zeta_j; \zeta) \times$$
$$\times \Psi_{n,n'}(\zeta_1, \ldots, \zeta_{j-1}, \zeta_{j+1}, \ldots, \zeta_{n+1}; \zeta_1', \ldots, \zeta_{n'}'). \tag{1.11}$$

The same observation remains true for antiparticles, if we define $\phi_A^{(+)}(\zeta')$ at $\zeta' \in \mathscr{M}^-$ in a manner totally analogous to (1.11).

Due to the presence of the Bose-Einstein statistics, embedded in $\mathscr{F}(\Sigma_A)$ by means of the symmetric tensor products in (1.2), we have:

$$[\phi_A^{(\pm)}(\zeta), \phi_A^{(\pm)}(\zeta')] = 0, \quad \zeta, \zeta' \in \mathscr{M}. \tag{1.12}$$

On the other hand, by (1.6), (1.10) and (1.11)

$$[\phi_A^{(-)}(\zeta), \phi_A^{(+)}(\zeta')] = i\Delta_A^{(+)}(\zeta; \zeta'). \tag{1.13}$$

We can introduce now the free field $\phi_A(\zeta)$ at $\zeta = (q, \nu) \in \mathscr{M}$ by the formula

$$\phi_A(q, \nu) = \phi_A^{(+)}(q, \nu) + \phi_A^{(-)}(q, -\nu) \tag{1.14}$$

which is in keeping with the definition (Schweber, 1961, p. 197) of the LQFT free field $\phi(x)$. Due to the earlier made observations that

$$\phi_A^{(\pm)*}(\zeta) = \phi_A^{(\mp)}(\zeta), \ \zeta \in \mathscr{M}, \tag{1.15}$$

we shall have:

$$\phi_A^*(q, \nu) = \phi_A(q, -\nu), \quad (q, \nu) \in \mathscr{M}. \tag{1.16}$$

Consequently, the commutation relations

$$[\phi_A^*(\zeta), \phi_A(\zeta')] = i\Delta_A(\zeta; \zeta'), \quad \zeta, \zeta' \in \mathscr{M}, \tag{1.17}$$

$$\Delta_A(q, \nu; q', \nu') = \Delta_A^{(+)}(q, \nu; q', \nu') - \Delta_A^{(+)}(q', -\nu'; q, -\nu), \tag{1.18}$$

that follow from (1.12) and (1.13), incorporate as a special case the commutation relations

$$[\phi_A(\zeta), \phi_A(\zeta')] = [\phi_A^*(\zeta), \phi_A^*(\zeta')] = 0, \zeta, \zeta' \in \mathcal{M}^{\pm}, \tag{1.19}$$

in which both ζ and ζ' belong either to $\mathcal{M}^+$ or to $\mathcal{M}^-$. Upon introducing

$$\Delta_A^{(-)}(q, v; q', v') = -\Delta_A^{(+)}(-q, v; -q', v'), \tag{1.20}$$

we observe that on account of (1.4) and (1.7)

$$\Delta_A(\zeta; \zeta') = \Delta_A^{(+)}(\zeta; \zeta') + \Delta_A^{(-)}(\zeta; \zeta'). \tag{1.21}$$

In deriving the remaining FTQS counterparts of the LQFT formulae for charged scalar free fields (Schweber, 1961, Sec. 7f) we encounter remarkable formal analogies, as well as equally remarkable differences on a deeper mathematical level. Indeed, the derived FTQS expressions are meaningful as they stand, whereas their LQFT counterparts are often only formal expressions which require a careful mathematical reinterpretation if one adheres to the operator-valued distribution approach to the formulation to LQFT. This is already evident in the case of the particle and antiparticle number operators N_A^+ and N_A^-, respectively, where, by (1.6), (1.10) and (1.11),

$$N_A^{\pm} = \int_{\Sigma_A^{\pm}} \phi_A^{(+)}(\zeta)\phi_A^{(-)}(\zeta)\, \mathrm{d}\Sigma(\zeta), \tag{1.22}$$

$$N_A = N_A^+ + N_A^- = \int_{\Sigma_A} \phi_A^{(+)}(\zeta)\phi_A^{(-)}(\zeta)\, \mathrm{d}\Sigma(\zeta), \tag{1.23}$$

and the above Bochner integrals (cf. Appendix B) are well-defined on dense domains. The same will be true of the remaining Bochner integrals introduced in this and subsequent sections (cf. also Schroeck, 1983b).

For particles of charge e and antiparticles of charge $-e$, the charge operator is:

$$Q_A = e(N_A^+ - N_A^-) = e\int_{\Sigma_A} \epsilon(v^0)\phi_A^{(+)}(\zeta)\phi_A^{(-)}(\zeta)\, \mathrm{d}\Sigma(\zeta). \tag{1.24}$$

By (1.5), we can also write

$$Q_A = \int_{\sigma} \mathcal{J}_A^{\mu}(q)\, \mathrm{d}\sigma_{\mu}(q), \tag{1.25}$$

$$\mathcal{J}_A^{\mu}(q) = 2e\int_{\mathcal{V}} v^{\mu}\phi_A^{(+)}(q, v)\phi_A^{(-)}(q, v)\, \mathrm{d}\Omega(v), \tag{1.26}$$

where, as a consequence of (2.8.6), the operator-valued current $\mathscr{I}^{\mu}_A$ is conserved:

$$\frac{\partial}{\partial q^{\mu}} \mathscr{I}^{\mu}_A(q) = 0. \tag{1.27}$$

$\mathscr{I}^{\mu}_A$ is not the (formal) equivalent of the charge current in LQFT. To arrive at this equivalent, we use the counterparts of (2.6.7) and (2.6.12) in the dimensionless coordinates (4.3.6a) to deduce that

$$N^{\pm}_A = iZ'_A \int_{\Sigma^{\pm}} \phi^{(+)}_A(\zeta) \overleftrightarrow{\partial}_{\mu} \phi^{(-)}_A(\zeta)\, \mathrm{d}\sigma^{\mu}(q)\, \mathrm{d}\Omega(\nu), \tag{1.28}$$

$$Z'^{-1}_A = \frac{(2\pi)^3}{m_A{}^2} \int_{\mathscr{V}^+} |\eta_A(\nu^0)|^2\, \mathrm{d}\Omega(\nu). \tag{1.29}$$

Furthermore, we also note that

$$\int_{\Sigma_A} \phi^{(\pm)}(q, \nu) \overleftrightarrow{\partial}_{\mu} \phi^{(\pm)}(q, -\nu)\, \mathrm{d}\sigma^{\mu}(q)\, \mathrm{d}\Omega(\nu) = 0 \tag{1.30}$$

follows from the fact that for $\phi_1, \phi_2 \in L^2(\Sigma_A)$

$$\int_{\Sigma_A} \phi_1^*(q, \nu) \overleftrightarrow{\partial}_{\mu} \phi_2(q, -\nu)\, \mathrm{d}\sigma^{\mu}(q)\, \mathrm{d}\Omega(\nu) = 0. \tag{1.31}$$

The validity of this last observation is most easily established by working in a quantum Lorentz frame where σ is the q^0 = const hypersurface, then expressing $\phi_i(\zeta)$, i = 1, 2, in accordance with (2.5.5), and using, in the end, the Fourier-Plancherel transform theorem – i.e. in formal terms, performing first the integration in $\mathbf{dq}$ to obtain a $\delta(\mathbf{k}_1 - \mathbf{k}_2)$ function, which then integrated gives zero in conjunction with the $(k_1^0 + k_2^0)$ -factor resulting from the execution of $\overleftrightarrow{\partial}^0$ differentiation. Combining now (1.14), (1.29) and (1.30) with (1.25), we obtain:

$$Q_A = \int_{\sigma} \mathscr{I}^{\mu}_A(q)\, \mathrm{d}\sigma_{\mu}(q), \tag{1.32}$$

$$\mathscr{J}^{\mu}_A(q) = -ieZ'_A \int_{\mathscr{V}^+} :\phi^*_A(\zeta) \partial^{\mu} \phi_A(\zeta):\, \mathrm{d}\Omega(u). \tag{1.33}$$

The operator-valued 4-vector function $\mathscr{J}^{\mu}_A$ is not only in appearance the analogue of the charge current in LQFT, but due to (2.8.13) it is also conserved:

$$\frac{\partial}{\partial q^{\mu}} \mathscr{J}^{\mu}_A(q) = 0. \tag{1.34}$$

The relativistic covariance of $\mathscr{I}_A^\mu$ in (1.26) and $\mathscr{J}_A^\mu$ in (1.33) follows from the existence of the unitary representation

$$\mathscr{U}_A(a, \Lambda) = \bigoplus_{n,n'=0}^{\infty} U_A(a, \Lambda)^{\otimes(n+n')} \tag{1.35}$$

of the Poincaré group $\mathscr{P}$, defined in terms of the restriction U_A to $L^2(\Sigma_A)$ of the representation of $\mathscr{P}$ in (2.3.9), that is to be reexpressed in terms of its action on wave functions $\phi(\zeta)$ of the dimensionless (q, ν)-variables in (4.3.6). It is easily deduced from (1.10) and (1.12) that the infinitesimal generators P_A^μ of $\mathscr{U}_A(a, I)$,

$$\mathscr{U}_A(a, I) = \exp(ia_\mu P_A^\mu) \tag{1.36}$$

can be expressed as (densely defined in $\mathscr{F}_A(\Sigma_A)$) Bochner integrals:

$$P_A^\mu = i \int_{\Sigma_A} \phi_A^{(+)}(q, \nu) \frac{\partial}{\partial q_\mu} \phi_A^{(-)}(q, \nu)\, \mathrm{d}\Sigma(q, \nu)$$

$$= -i \int_{\Sigma_A} \left(\frac{\partial}{\partial q_\mu} \phi_A^{(+)}(q, \nu)\right) \phi_A^{(-)}(q, \nu)\, \mathrm{d}\Sigma(q, \nu). \tag{1.37}$$

To express P_A^μ in terms of the field $\phi_A(\zeta)$ in (1.14), we make the transition to a quantum Lorentz frame where Σ_A is the $q^0 = 0$ hypersurface, and then use the same technique as in deriving (1.28) to deduce that

$$P_A^\mu = Z_A' \int_{\substack{q^0=0 \\ \nu\in\mathscr{V}^+}} \left[\frac{\partial\phi_A^{(+)}(q, \nu)}{\partial q_\mu} \frac{\overleftrightarrow{\partial}}{\partial q^0} \phi^{(-)}(q, \nu) - \right.$$

$$\left. - \phi^{(+)}(q, -\nu) \frac{\overleftrightarrow{\partial}}{\partial q^0} \frac{\partial\phi_A^{(-)}(q, -\nu)}{\partial q_\mu}\right] \mathrm{d}\mathbf{q}\, \mathrm{d}\Omega(u). \tag{1.38}$$

Thus, by (1.31), we can write in an arbitrary frame:

$$P_A^\mu = -Z_A' \int_{\Sigma_A^+} : \phi_A^*(q, \nu) \frac{\overleftrightarrow{\partial}}{\partial q^\nu} \frac{\partial\phi_A(q, \nu)}{\partial q_\mu} : \mathrm{d}\sigma^\nu(q)\, \mathrm{d}\Omega(\nu). \tag{1.39}$$

Hence, for $\mu = 1, 2, 3$, we get in the frame where Σ_A is the $q^0 = 0$ hypersurface,

$$\mathbf{P}_A = Z'_A \int_{\substack{q^0=0 \\ v\in \mathscr{V}^+}} :\dot{\phi}_A^*(\zeta) \cdot \nabla_{\mathbf{q}}\phi_A(\zeta) + \nabla_{\mathbf{q}}\phi_A^*(\zeta) \cdot \dot{\phi}_A(\zeta): \, d\mathbf{q} \, d\Omega(v), \qquad (1.40)$$

where $\dot{\phi}_A = \partial\phi_A/\partial q^0$. In view of the fact that by (1.10), (1.11), (1.14) and (4.3.17)

$$(\Box_q + m_A^2)\phi_A(q, v) = 0, \qquad (1.41)$$

we obtain upon expressing P_A^0 in (1.39) in the same frame, and then integrating by parts:

$$P_A^0 = Z'_A \int_{\substack{q^0=0 \\ v\in \mathscr{V}^+}} :\dot{\phi}_A^*\dot{\phi}_A + \nabla_{\mathbf{q}}\phi_A^* \cdot \nabla_{\mathbf{q}}\phi_A + m_A^2\phi_A^*\phi_A: \, d\mathbf{q} \, d\Omega(v). \qquad (1.42)$$

To develop the theory of a neutral scalar field $\phi_B(\zeta)$ for which antiparticles are indistinguishable from particles, we carry out the particle–antiparticle identification by requiring that $\eta_B(\lambda)$, $1 \leqslant \lambda = u \cdot v$, $u, v \in \mathscr{V}^\pm$, should be also defined for $\lambda \leqslant -1$, i.e. for $u \in \mathscr{V}^\pm$ and $v \in \mathscr{V}^\mp$, so that

$$\eta_B(u \cdot v) = \eta_B(-u \cdot v), \quad u, v \in \mathscr{V}. \qquad (1.43)$$

The Fock space that replaces (1.1) is now

$$\mathscr{F}(\Sigma_B^+) = \bigoplus_{n=0}^{\infty} \mathscr{F}_n(\Sigma_B^+), \qquad \mathscr{F}_n(\Sigma_B^+) = L^2(\Sigma_B^+)^{\circledS n}, \qquad (1.44)$$

and the definition of $\phi_B^{(\pm)}(\zeta)$ proceeds as in (1.10) and (1.11). The neutral scalar field shall be

$$\phi_B(\zeta) = \phi_B^{(+)}(\zeta) + \phi_B^{(-)}(\zeta) = \phi_B^*(\zeta), \qquad (1.45)$$

and it will obey the commutation relation

$$[\phi_B(\zeta), \phi_B(\zeta')] = i\Delta_B(\zeta; \zeta'), \qquad (1.46)$$

with the definition of Δ_B in terms of $\Delta_B^{(\pm)}$ as in (1.20) and (1.21), since by (1.43)

$$\Delta_B^{(+)}(q, v; q', v') = \Delta_B^{(+)}(q, -v; q', -v'). \qquad (1.47)$$

Thus, in accordance with (4.3.10) we now have:

$$\Delta_B(\zeta; \zeta') = -i \int_{\mathscr{V}} \exp[im_A(q' - q) \cdot u] \; \eta_B(v \cdot u)\eta_B(v' \cdot u)\,\epsilon(u^0) \, d\Omega(u). \quad (1.48)$$

Clearly, due to the self-adjointness of $\phi_B(\zeta)$, $\mathcal{J}_B^\mu$ defined by analogy with (1.33) vanishes identically, whereas instead of (1.40) and (1.42) we now have, respectively,

$$\mathbf{P}_B = \frac{1}{2} Z'_B \int_{\substack{q^0=0 \\ v\in\mathcal{V}^+}} :\dot{\phi}_B \cdot \nabla_{\mathbf{q}}\phi_B + \nabla_{\mathbf{q}}\phi_B \cdot \dot{\phi}_B: \, \mathrm{d}\mathbf{q}\, \mathrm{d}\Omega(v), \tag{1.49}$$

$$P_B^0 = \frac{1}{2} Z'_B \int_{\substack{q^0=0 \\ v\in\mathcal{V}^+}} :\dot{\phi}_B^2 + \nabla_{\mathbf{q}}\phi_B \cdot \nabla_{\mathbf{q}}\phi_B + m_B^2\phi_B^2: \, \mathrm{d}\mathbf{q}\, \mathrm{d}\Omega(v). \tag{1.50}$$

The comparison of the derived FTQS formulae for charge, current and 4-momentum operators with their LQFT counterparts (Schweber, 1961, Sec. 7) reveals not only a striking formal similarity, but also the first indications as to the origin of the divergences that plague LQFT models. Indeed, in the sharp-point limit, $\phi_A(q, v)$ and $\phi_B(q, v)$ become asymptotically v-independent, so that the v-integrations in the above-derived Bochner integrals lead to divergences. Furthermore, the wave function normalization constant $\mathring{Z}_A$ in (4.5.25), as well as the renormalization constants Z'_A in (1.33), (1.40) and (1.41), also diverge for such choices of proper wave functions as those derived from the quantum metric operator model in Section 4.5. For example, for the proper wave function (4.5.25a), expressed in the dimensionless variables in (4.3.6),

$$\eta_A(u \cdot v) = Z_A^{-1/2} \exp(-l_A m_A u \cdot v), \tag{1.51}$$

we get by (4.3.10) and (1.4), when $v_0, v'_0 > 0$,

$$\Delta_A^{(+)}(q, v; q', v') = -iZ_A^{-1} \int_{\mathcal{V}^+} \exp\{u \cdot [im_A(q' - q) - \omega_A^{-1}(v + v')]\}\, \mathrm{d}\Omega(u), \tag{1.52a}$$

where the normalization constant Z_A is determined as in (4.5.24), so that:

$$Z_A = \frac{(2\pi)^4}{m_A^2} \frac{K_2(2l_A m_A)}{2l_A m_A}. \tag{1.52b}$$

By proceeding as we did in (2.9.14)–(2.9.19), we obtain for $\zeta, \zeta' \in \mathcal{M}^+$:

$$\Delta_A^+(\zeta; \zeta') = -2\pi i Z_A^{-1} \frac{K_1(f_A(\zeta; \zeta'))}{f_A(\zeta; \zeta')}, \tag{1.53a}$$

$$f_A^2(q, v; q', v') = -m_A^2 [q - q' - il_A(v + v')]^2. \tag{1.53b}$$

Consequently, by (4.3.10) and (1.29):

$$Z_A'^{-1} = \frac{(2\pi)^3}{m_A^2} \, i\Delta_A^{(+)}(\zeta;\zeta) = \frac{K_1(2l_A m_A)}{K_2(2l_A m_A)}. \tag{1.54}$$

Thus, we deduce from the behavior of the modified Bessel functions K_1 and K_2 at the origin (Butkov, 1968, p. 396) that $Z'_A \sim (l_A m_A)^{-1}$ as $\omega_A^{-1} = l_A m_A \to +0$.

5.2. LSZ Formalism and Causality on Quantum Spacetime

In this section we shall reproduce in the FTQS context the main results of the standard LSZ formalism presented in LQFT textbooks (Schweber, 1961; Roman, 1969; Itzykson and Zuber, 1980). For the sake of simplicity, we shall proceed in the manner of the aforementioned references and of the original papers by Lehmann *et al.* (1955, 1957), by considering a neutral scalar field $\phi(\zeta)$ along an initial-data reference hypersurface Σ_B^+ given by $q^0 = 0$ in the chosen quantum Lorentz frame. We then impose the LSZ asymptotic condition on $\phi(\zeta)$ in terms of arbitrary normalized and freely evolving $f = \mathcal{W}_\eta \tilde{f} \in L^2(\Sigma_B^+)$ (cf. e.g., (2.5.5)) by setting in accordance with (2.6.9) and (1.29)–(1.31)

$$\phi_{\text{ex}}^f = \underset{t\to\mp\infty}{\text{w-lim}}\, \phi^f(t)\Phi, \quad \text{ex} = \text{in, out}, \tag{2.1}$$

$$\phi_{\text{ex}}^f = \int_{\Sigma_B^+} f^*(q,\nu)\phi_{\text{ex}}^{(-)}(q,\nu)\, d\Sigma(q,\nu)$$

$$= iZ'_B \int_{q^0=ct} f^*(\zeta)\overleftrightarrow{\partial}_0\phi_{\text{ex}}(\zeta)\, d\mathbf{q}\, d\Omega(\nu), \tag{2.2}$$

$$\phi^f(t) = iZ'_B \int_{q^0=ct} f^*(\zeta)\overleftrightarrow{\partial}_0\phi(\zeta)\, d\mathbf{q}\, d\Omega(\nu). \tag{2.3}$$

Indeed, if $\phi_{\text{ex}}(\zeta)$ are to display the structure of the free fields $\phi_B(\zeta)$ in Section 5.1, then (2.2) holds, and ϕ_{ex}^f is t-independent. In fact, then we have

$$\phi_{\text{ex}}^{\eta_\zeta^B} = \int_{q^0=0} \eta_{\zeta'}^B(\zeta)\phi_{\text{ex}}^{(-)}(\zeta')\, d\Sigma(\zeta') = \phi_{\text{ex}}^{(-)}(\zeta), \tag{2.4}$$

where, in accordance with (2.6.23),

$$\eta_\zeta^B(\zeta') = K_B(\zeta';\zeta) = i\Delta_B^{(+)}(\zeta';\zeta). \tag{2.5}$$

Following the approach and notation of Lehmann *et al.* (1955), we introduce (in the Hilbert space $\mathscr{H}(\Sigma_B^+)$ in which $\phi(\zeta)$ acts) the orthonormal basis of states

$$\Phi_{\text{ex}}^{\alpha} = (n_1! \ldots n_r!)^{-\frac{1}{2}} \phi_{\text{ex}}^{\alpha_1} \ldots \phi_{\text{ex}}^{\alpha_r} \Phi_0, \quad \alpha = (\alpha_1, \ldots, \alpha_r), \tag{2.6a}$$

$$\phi_{\text{ex}}^{j} = \phi_{\text{ex}}^{f_j^*} = iZ_B' \int_{\Sigma_B^+} f_j(\zeta) \overleftrightarrow{\partial}^{\nu} \phi_{\text{ex}}(\zeta) \, \mathrm{d}\sigma_\nu(q) \, \mathrm{d}\Omega(\nu), \tag{2.6b}$$

that correspond to some choice $\{f_j\}_{j=1}^{\infty}$ of orthonormal basis in $L^2(\Sigma_B^+)$. Thus, in accordance with (2.6.27) and (2.5):

$$\sum_{j=1}^{\infty} f_j(\zeta) f_j^*(\zeta') = i\Delta_B^{(+)}(\zeta; \zeta'). \tag{2.7}$$

Hence, the $\mathscr{H}(\Sigma_B^+)$ matrix elements of the scattering operator S, namely

$$S_{\alpha\beta} = \langle \Phi_{\text{out}}^{\alpha} | \Phi_{\text{in}}^{\beta} \rangle = \langle \Phi_{\text{in}}^{\alpha} | S \Phi_{\text{in}}^{\beta} \rangle, \tag{2.8}$$

and the momentum-space representatives $\tilde{f}_j(k)$ of $f_j(\zeta)$, which in accordance with (2.4.2) and (2.4.22) are equal to

$$\tilde{f}_j(k) = m_B^{-1} \int_{\Sigma_B^+} \exp(iq \cdot k) \eta_B \left(\frac{\nu \cdot k}{m_B} \right) f_j(q, \nu) \, \mathrm{d}\Sigma(q, \nu), \tag{2.9}$$

can be used to compute the S-matrix in the momentum representation:

$$\langle k_1, \ldots, k_r | \, S \, | k_1', \ldots, k_s' \rangle = \sum_{\alpha_1, \ldots, \beta_s = 1}^{\infty} S_{\alpha\beta} \tilde{f}_{\alpha_1}(k_1) \ldots$$

$$\ldots \tilde{f}_{\alpha_r}(k_r) \tilde{f}_{\beta_1}^*(k_1') \ldots \tilde{f}_{\beta_s}^*(k_s'). \tag{2.10}$$

The key objects of the LSZ formalism are the τ-functions. In FTQS we set them equal to

$$\tau(\zeta_1, \ldots, \zeta_n) = \langle \Phi_0 \mid T[\phi(\zeta_1) \ldots \phi(\zeta_n)] \, \Phi_0 \rangle$$

$$= \Sigma \Theta(\zeta_{i_1}; \zeta_{i_2}) \ldots \Theta(\zeta_{i_{n-1}}; \zeta_{i_n}) \langle \Phi_0 | \phi(\zeta_{i_1}) \ldots \phi(\zeta_{i_n}) \Phi_0 \rangle, \tag{2.11}$$

where for $\zeta' \in \Sigma_B'^+$ and $\zeta'' \in \Sigma_B''^+$ the function $\Theta(\zeta'; \zeta'')$ equals 1 if $\Sigma_B'^+ \geqslant \Sigma_B''^+$, and zero if $\Sigma_B'^+ < \Sigma_B''^+$, in accordance with the time-ordering in the chosen reference family $\mathscr{S}^{\uparrow}$ to which $\Sigma_B'^{+}$ and $\Sigma_B''^{+}$ belong. We can then derive the fundamental formula

$$S_{\alpha\beta} = (iZ_B')^{r+s} \int d\mu(\zeta_1) \ldots d\mu(\zeta_{r+s}) f_{\alpha_1}^*(\zeta_1) \ldots f_{\alpha_r}^*(\zeta_r) \times$$

$$\times f_{\beta_1}(\zeta_{r+1}) \ldots f_{\beta_s}(\zeta_{r+s}) \mathscr{K}_{q_1}^B \ldots \mathscr{K}_{q_{r+s}}^B \tau(\zeta_1, \ldots, \zeta_{r+s}), \tag{2.12}$$

in which $d\mu$ is the relativistically invariant measure element

$$d\mu(q, v) = d^4 q\, \theta(v^0)\, d\Omega(v) = \theta(v^0) \delta(v^2 - 1)\, d^4 q\, d^4 v, \tag{2.13}$$

and $\mathscr{K}_q^B$ is the Klein–Gordon operator for B-particles:

$$\mathscr{K}_q^B = \frac{\partial^2}{\partial q_\mu\, \partial q^\mu} + m_B^2. \tag{2.14}$$

This can be achieved by first using mathematical induction to derive the reduction formulae

$$\langle \Phi_{\text{out}}^{\alpha} \mid T[\phi(\zeta_1) \ldots \phi(\zeta_n)] \, \Phi_{\text{in}}^{\beta} \rangle$$

$$= iZ_B' \int d\mu(\zeta) \mathscr{K}_q^B \langle \Phi_{\text{out}}^{\alpha} \mid T[\phi(\zeta_1) \ldots \phi(\zeta_n) \phi(\zeta)] \, \Phi_{\text{in}}^{\beta_1, \ldots, \beta_{s-1}} \rangle f_{\beta_s}(\zeta)$$

$$= iZ_B' \int d\mu(\zeta) \mathscr{K}_q^B \langle \Phi_{\text{out}}^{\alpha_1, \ldots, \alpha_{r-1}} \mid T[\phi(\zeta_1) \ldots \phi(\zeta_n) \phi(\zeta)] \, \Phi_{\text{in}}^{\beta} \rangle f_{\alpha_r}^*(\zeta), \tag{2.15}$$

as on pp. 212–214 of Lehmann *et al.* (1955), and then using (2.7) to derive the system

$$\tau(\zeta_1, \ldots, \zeta_n) = \Sigma \Theta(\zeta_1; \zeta_n) \ldots \Theta(\zeta_{n-1}; \zeta_n) \sum_{j=1}^{\infty} (iZ_B)^j \int d\mu(\zeta_1') \ldots d\mu(\zeta_j'') \times$$

$$\times \Delta_B^{(+)}(\zeta_1'; \zeta_1'') \ldots \Delta_B^{(+)}(\zeta_j'; \zeta_j'') \mathscr{K}_{q_1'}^B \ldots \mathscr{K}_{q_j'}^B \times$$

$$\times \tau(\zeta_1, \ldots, \zeta_{n-1}, \zeta_1', \ldots, \zeta_j') \mathscr{K}_{q_1''}^B \ldots \mathscr{K}_{q_j''}^B \tau(\zeta_n, \zeta_1'', \ldots, \zeta_j'') \tag{2.16}$$

of coupled integral equations for the τ-functions.

Let us introduce in accordance with (2.4) and (2.5)

$$\phi_{\text{ex}}^{(+)}(\zeta) = \phi_{\text{ex}}^{(-)*}(\zeta) = -i \int_{\Sigma_B^+} \Delta^{(+)}(\zeta'; \zeta) \phi_{\text{ex}}^{(+)}(\zeta')\, d\Sigma(\zeta'), \tag{2.17a}$$

$$\phi_{\text{ex}}^{(-)j} = \phi_{\text{ex}}^j, \qquad \phi_{\text{ex}}^{(+)j} = (\phi_{\text{ex}}^j)^*, \tag{2.17b}$$

so that by (2.7) we shall have

$$\phi_{\text{ex}}(\zeta) = \phi_{\text{ex}}^{(-)}(\zeta) + \phi_{\text{ex}}^{(+)}(\zeta) = \sum_{j=1}^{\infty} [f_j(\zeta)\phi_{\text{in}}^{(-)j} + f_j^*(\zeta)\phi_{\text{in}}^{(+)j}]. \tag{2.18}$$

The orthonormality of $\{f_j\}_{j=1}^{\infty}$ implies that

$$iZ_B' \int f_i^*(\zeta)\overleftrightarrow{\partial}^{\nu} f_j(\zeta)\, \mathrm{d}\sigma_\nu(q)\, \mathrm{d}\Omega(v) = \langle f_i | f_j \rangle_{\Sigma_B^+} = \delta_{ij}, \tag{2.19}$$

where the integration can be performed along any hypersurface in $\mathscr{S}^\uparrow$ with exactly the same outcome. On the other hand, the argument leading to (1.31) also yields

$$\int f_i(\zeta)\overleftrightarrow{\partial}^{\nu} f_j(\zeta)\, \mathrm{d}\sigma_\nu(q)\, \mathrm{d}\Omega(v) = 0. \tag{2.20}$$

Consequently, the commutation relations

$$[\phi_{\text{ex}}(\zeta), \phi_{\text{ex}}(\zeta')] = i\Delta_B(\zeta;\zeta'),\ [\dot{\phi}_{\text{ex}}(\zeta), \phi_{\text{ex}}(\zeta')] = i\partial_0\Delta_B(\zeta;\zeta'), \tag{2.21}$$

for the asymptotic fields are equivalent to

$$[\phi_{\text{ex}}^{(\pm)i}, \phi_{\text{ex}}^{(\pm)j}] = 0,\ [\phi_{\text{ex}}^{(-)i}, \phi_{\text{ex}}^{(+)j}] = \delta_{ij}. \tag{2.22}$$

By mathematical induction we then obtain from (2.6) and (2.8) that

$$S = \sum_{\alpha,\beta} S_{\alpha\beta}\phi_{\text{in}}^{(+)\alpha_1} \ldots \phi_{\text{in}}^{(+)\alpha_r}\, \phi_{\text{in}}^{(-)\beta_1} \ldots \phi_{\text{in}}^{(-)\beta_s}. \tag{2.23}$$

Upon inserting (2.12) into (2.23), we can perform the summations by using (2.18). Thus, after we take (2.4), (2.5) and (2.17) into account, we arrive at the FTQS counterpart of the Lehmann *et al.* (1955) expression for scattering operators (in the vacuum sector):

$$S = \sum_{n=0}^{\infty} \frac{(iZ_B')^n}{n!} \int \mathscr{K}_{q_1}^B \ldots \mathscr{K}_{q_n}^B \tau(\zeta_1, \ldots, \zeta_n) \times$$

$$\times :\phi_{\text{in}}(\zeta_1) \ldots \phi_{\text{in}}(\zeta_n): \mathrm{d}\mu(\zeta_1) \ldots \mathrm{d}\mu(\zeta_n). \tag{2.24}$$

If we define the functional differentiation of the S-operator in the conventional manner (Bogolubov *et al.*, 1975, Eq. (15.13)), then we immediately deduce from (2.16) and (2.24) the FTQS counterpart

$$\frac{\delta}{\delta\phi_{\text{in}}(\zeta')}\left[S^{-1}\frac{\delta S}{\delta\phi_{\text{in}}(\zeta'')}\right] = 0, \quad \zeta' \in \Sigma_B',\ \zeta'' \in \Sigma_B'',\ \Sigma_B' < \Sigma_B'', \tag{2.25}$$

of Bogolubov's 'causality' condition. Naturally, in LQFT this condition is supposed to be fulfilled at any x' and x'' for which $(x' - x'')^2 < 0$, since $\phi_{in}(x')$ and $\phi_{in}(x'')$ are supposed to be fields *at* the points $x', x'' \in M(1, 3)$ in classical Minkowski space. However, in Sections 2.1, 2.2, 4.1–4.4, and in the introduction to this chapter, we have reviewed the overwhelming evidence as to the inconsistency of sharp localizability for relativistic quantum particles and fields, which makes any such assumption physically unwarranted. On the other hand, in FTQS we deal only with nonsharp (stochastic) localizability, which provides probabilities for finding particles within Borel sets $\underset{\sim}{B} \subset \underset{\sim}{M}(1, 3)$, that are *never* equal to 1 as long as $\underset{\sim}{B}$ is bounded.[5] Thus, no categorical statements as to particle existence or field localization in a given state can be made for such stochastic regions, but rather only probabilistic statements reflecting a notion of *stochastic microcausality* (Greenwood and Prugovečki, 1984) can be valid. Hence, the *macrocausality* condition (2.25) reflects merely the macroscopic time-ordering imposed on the reference family $\mathscr{S}^\uparrow$ by the chosen coherent flow of test particles, i.e. in practical terms by the fact that the preparatory stages of any experimental procedure have to proceed its determinative stages in the natural ordering of the (mean-stochastic) time marked and measured with the chosen instrumentation.

The fact that *realistic* measurement procedures are intrinsically dependent on the adopted time-ordering, which in turn depends on the choice of coherent flow (i.e. in practical terms, on the kinematical status of the preparatory and detecting sections of the apparatus) can be illustrated by the time-of-flight measurement of 3-momentum. This measurement requires the determination of two *consecutive* (stochastic) position values $(\mathbf{q}', \chi_{\mathbf{q}'})$ and $(\mathbf{q}'', \chi_{\mathbf{q}''})$, and of the mean stochastic times t' and t'' of their registration – or at least of the mean time lapse $\Delta t = t'' - t'$. From this information a stochastic value $(\mathbf{p}, \chi_{\mathbf{p}})$ with $\mathbf{p} = m(\mathbf{q}'' - q')/\Delta t$ is obtained in the rest frames of the apparatus, whose confidence function $\overline{\chi}_{\mathbf{p}}$ can be easily deduced from the confidence functions $\chi_{\mathbf{q}'}$ and $\chi_{\mathbf{q}''}$. For example, a particle 'track' in a Wilson chamber or a bubble chamber can be viewed as a large sequence of such stochastic values, whose standard deviations are of macroscopic size, whereas in Heisenberg's (1930) *gedanken* experiments, by the fine tuning of the macroscopic parts of the detection apparatus to the chosen microdetectors (as discussed in Section 1.12) these standard deviations can be reduced to the order of magnitude of the proper radius of the employed microdetectors. In either case, since commonly encountered confidence functions $\chi_{\mathbf{q}'}$ and $\chi_{\mathbf{q}''}$, such as Gaussians, do not have bounded supports,[6] an observer in a frame boosted in relation to the apparatus cannot claim with *absolute* confidence that $\mathbf{q}'$ had been recorded either prior or posterior to $\mathbf{q}''$ in his own mean stochastic time. Thus, strictly speaking, experimental praxis is not conducive to *categorical* causality statements, but merely to statements of *practical* certainty as to causal ordering. In quantum spacetime, this practical certainty is achieved for regions whose size sufficiently exceeds by orders of magnitude the size of the microdetectors, so as to allow the resulting measures of confidence to approach the value 1 (i.e. statistical certainty) within stipulated limits (cf. Sec. 5 of Greenwood and Prugovečki, 1984).

As a consequence of the assumed existence of a unitary representation $\mathcal{U}(a, \Lambda)$ of the Poincaré group that leaves the vacuum Φ_0 invariant and is such that

$$\mathcal{U}(a, \Lambda)\phi(q, p)\mathcal{U}^{-1}(a, \Lambda) = \phi(a + \Lambda q, \Lambda p), \tag{2.26}$$

we immediately get that the S-matrix elements (2.10) are covariant due to (2.11)–(2.12) and to the covariance of (2.9). Naturally, this does not secure their invariance,[7] namely the exclusive dependence of (2.10) on the Minkowski inner products of the 4-vectors $k_1, \ldots, k'_s$. Thus, in general, we might be faced with an equivalence class of S-matrices associated with distinct coherent flows, whose elements are related by Lorentz transformations. Of course, since all flows are relativistically equivalent, so are all these S-matrix representatives, which transform covariantly into each other. Thus, we note that already the functions $\tilde{f}_j(k)$ in (2.9) are covariant without being invariant – as can be also operationally inferred from our earlier discussion of time-of-flight momentum measurements. On the other hand, the propagators given by (1.53) illustrate instances of FTQS functions which are invariants, so that the question of whether the S-matrix elements (2.10) are invariant, or merely covariant (i.e. scalars), is left open for the time being.

5.3. Lagrangian Field Theory on Quantum Spacetime

The comparison of the FTQS expressions for charge and 4-momentum operators derived in Section 5.1 with their LQFT counterparts suggests the consideration of action operators

$$W = \int_R L \, d\mu(\zeta), \quad d\mu(\zeta) = d^4 q \, d\Omega(v), \tag{3.1}$$

for Lagrangian densities $L(\phi)$, that are local in the sense of being constructed of field operators at one and the same stochastic point ζ in quantum spacetime.[8] For example, the expressions for the charged scalar field $\phi_A = \phi$ suggest the free Lagrangian

$$L_0(\phi) = Z_A [\phi_{,\nu}(\zeta)\phi^{,\nu}(\zeta) - m_A^2 \phi^2(\zeta)], \quad \phi_{,\nu} = \frac{\partial \phi}{\partial q^\nu}. \tag{3.2}$$

Varying W in (3.1), as is done in the Noether approach (Bogolubov and Shirkov, 1959) to classical fields, and keeping the spacetime component of the region $R \subset \mathcal{M}^+$ finite throughout,[9] we obtain the Euler–Lagrange equations

$$\frac{\partial L}{\partial \phi(\zeta)} - \frac{\partial}{\partial q^\nu} \frac{\partial \mathcal{L}}{\partial \phi_{,\nu}(\zeta)} = 0, \tag{3.3}$$

and the locally conserved stress-energy tensors

$$T^{\mu\nu}(\zeta) = \frac{\partial L}{\partial \phi_{,\mu}} \frac{\partial \phi}{\partial q_\nu} - g^{\mu\nu} L, \quad T^{\mu\nu}{}_{,\nu}(\zeta) = 0, \tag{3.4}$$

that are formally distinguished from their conventional counterparts only by the appearance of the mean stochastic 4-velocity variables v^μ in addition to the spacetime variables. Thus setting

$$P^\mu = \int_{\Sigma_A^+} :T^{\mu\nu}(\zeta): \, d\sigma_\nu(q) \, d\Omega(v), \tag{3.5}$$

we recover in case of the Lagrangian density in (3.2) the earlier derived expressions (1.39) and (1.42) for 4-momentum operators.

To illustrate on generally familiar examples some of the possibilities inherent in the FTQS approach to the nonlinear field problem, let us consider self-interacting neutral scalar field theories with Lagrangians

$$L = \frac{Z'_B}{2} [\phi_{,\mu}\phi^{,\mu} - m_B^2 \phi^2] + \sum_{n=1}^{\infty} a_{2n} \phi^{2n} \tag{3.6}$$

that incorporate as special cases the ϕ^4 and the sine-Gordon models (Rajaraman, 1975). The form of the free part L_0 of L in (3.6) is dictated by the requirement that when $L_I = 0$, i.e. when the coefficients a_{2n} are all zero, the Bochner integrals in (3.5) should coincide with the earlier derived expressions (1.49) and (1.50). However, this basic demand introduces the *finite* renormalization constant Z'_B into the field equations (3.3), which then assume the form:

$$(\Box_q + m_B^2)\phi = Z_B'^{-1} L'_I(\phi) = Z_B'^{-1} \sum_{n=1}^{\infty} 2n a_{2n} \phi^{2n-1}. \tag{3.7}$$

On the other hand, since in FTQS the fields are not singular, the nonlinear term in (3.7) is meaningful as it stands,[10] so that we are dealing with an already well-posed mathematical problem that can be approached by conventional mathematical methods.

To illustrate one such possibly useful method, let us employ the LSZ approach to FTQS presented in Section 5.2 to derive from (2.3) the relation

$$\phi^f(t) - \phi^f(t') = iZ'_B \int_{\mathscr{V}^+} d\Omega(v) \int_{\mathbb{R}^3} d\mathbf{q} \int_{ct'}^{ct} dq_0 \, [(\partial_0^2 f^*)\phi - f^* \partial_0^2 \phi]. \tag{3.8}$$

We proceed, as Lehmann *et al.* (1957), by substituting above

$$\partial_0^2 = \mathscr{K}_{\mathbf{q}}^B + \nabla_{\mathbf{q}}^2 - m_B^2. \tag{3.9}$$

We can then apply Green's theorem to transform the terms containing $\nabla_{\mathbf{q}}^2$ into surface integrals, which at spatial infinity yield zero. Hence, upon letting $t' \to -\infty$, we get by (2.1):

$$\phi^f(t) = \phi_{\text{in}}^f - iZ'_B \int d\Omega(v) \int_{-\infty}^{ct} d^4q f^*(q, v) (\mathcal{K}_q^B \phi)(q, v). \tag{3.10}$$

The interacting field ϕ cannot be split into creation and annihilation parts, but if we assume that the matrix elements of $\phi(\zeta)$ and $\partial_0\phi(\zeta)$ between states in $\mathcal{F}_{\text{in}}(\Sigma_B^+)$ as functions of ζ belong to $L^2(\Sigma_B^+)$, then by (1.30)

$$2Z'_B \int \phi(\zeta)\partial_0 f_j^*(\zeta)\, d\mathbf{q}\, d\Omega(v) = -\int f_j^*(\zeta)\phi(\zeta)\, d\Sigma(\zeta), \tag{3.11}$$

$$2Z'_B \int f_j^*(\zeta)\partial_0\phi(\zeta)\, d\mathbf{q}\, d\Omega(v) = \int f_j^*(\zeta)\, (m_B^2 - \nabla_{\mathbf{q}}^2)^{-1/2} \partial_0\phi(\zeta)\, d\Sigma(\zeta). \tag{3.12}$$

If we then use (2.7) and the basic relation

$$f(\zeta) = i \int \Delta_B^{(+)}(\zeta; \zeta') f(\zeta')\, d\Sigma(\zeta'), \quad f \in L^2(\Sigma_B^+), \tag{3.13}$$

which follows from (1.4) and (2.6.27), we obtain

$$\sum_{j=1}^{\infty} f_j(\zeta)\phi^{f_j} = \frac{1}{2}\, [\phi(\zeta) + (m_B^2 - \nabla_{\mathbf{q}}^2)^{-1/2} \partial_0\phi(\zeta)]\,. \tag{3.14}$$

Since $\{f_j^*\}_{j=1}^{\infty}$ constitutes an orthonormal basis in $L^2(\Sigma_B^+)$, we get by the same arguments that

$$\sum_{j=1}^{\infty} f_j^* \phi^{f_j^*} = \frac{1}{2}\, [\phi(\zeta) - (m_B^2 - \nabla_{\mathbf{q}}^2) \partial_0\phi(\zeta)]\,. \tag{3.15}$$

Consequently, by using (2.18) and (3.10), we obtain after adding together (3.14) and (3.15) a Yang–Feldman equation for $\phi(\zeta)$:

$$\phi(\zeta) = \phi_{\text{in}}(\zeta) + \int_{\mathcal{M}^+} \Delta_B^{(R)}(\zeta; \zeta') L'_I(\phi(\zeta'))\, d\mu(\zeta'), \tag{3.16}$$

$$\Delta_B^{(R)}(\zeta; \zeta') = -[\Delta_B^{(+)}(\zeta; \zeta') + \Delta_B^{(+)}(\zeta'; \zeta)]\, \Theta(\zeta; \zeta'). \tag{3.17}$$

In the case of classical fields, the existence of solutions for the nonlinear differential equation (3.7) can be established (Moravetz and Strauss, 1972) by studying the convergence properties of the sequence

$$\phi_n(\zeta) = \phi_{\text{in}}(\zeta) + N_B^{(R)}(\phi_{n-1})(\zeta), \quad n = 1, 2, \ldots, \tag{3.18}$$

of iterations of (3.16) that begin with the application to $\phi_0(\zeta) = \phi_{\text{in}}(\zeta)$ of the nonlinear operator

$$N_B^{(R)}(\phi)(\zeta) = \int \Delta_B^{(R)}(\zeta;\zeta') L_I'(\phi(\zeta'))\, d\mu(\zeta'). \tag{3.19}$$

However, this technique cannot be convincingly applied in the context of LQFT, since then all $L_I'(\phi_n)$ are ill-defined due to the nonexistence of $\phi_{\text{in}}(x)$ as *bona fide* operators (Bogolubov *et al.*, 1975, Sec. 10.4), so that their meaning requires a physically drastic reinterpretation (Rączka, 1975) in order for this whole procedure to make any mathematical sense. On the other hand, although the F-topologies used to establish the convergence of $\phi_1, \phi_2, \ldots$ in the classical case cannot be adapted to fields in quantum spacetime, each one of the terms in (3.18) is well defined in the context of FTQS, since neither $\phi_{\text{in}}(\zeta)$ nor $\Delta_B^{(R)}$ possesses any singularities. Hence, the field-theoretical problem based on Equation (3.7) is indeed mathematically well-posed, and the question of existence of a solution of the field equation (3.7) becomes one of technique, with the series (3.18) providing a practical method for arriving at such solutions in the vacuum sector.

The advances made in the sixties in the understanding and solutions of nonlinear field equations for classical fields (Whitham, 1974) have spurred extensive speculations (Jackiw, 1977; Faddeev and Korepin, 1978) as to the existence in LQFT of nonvacuum sectors related to kink and soliton solutions of classical problems. Most of these mathematically heuristic considerations can be adapted to FTQS, with the added possibility of eventually recasting them into a mathematically rigorous form that might provide new physical insights.[11] These possibility open the door on a multitude of totally uninvestigated problems, that require future considerations.

5.4. Dirac Bispinors and Free Fields on Quantum Spacetime

To extend the theory of stochastic phase-space representations $U(a, \Lambda)$ of $\mathcal{P}$, presented in Chapter 2 for the case of spin zero to spins $j = \frac{1}{2}, 1, \frac{3}{2}, 2, \ldots$, we shall introduce the Hilbert spaces

$$L_j^2(\Sigma) = L_j^2(\Sigma^+) \oplus L_j^2(\Sigma^-), \quad L_j^2(\Sigma^\pm) = \oplus L^2(\Sigma^\pm), \tag{4.1}$$

where the direct sum for $L_j^2(\Sigma^\pm)$ contains $2j + 1$ replicas of the Hilbert space $L^2(\Sigma^\pm)$ of functions $\psi_s(\zeta)$ of the dimensionless stochastic variables $\zeta = (q, v) \in \Sigma^\pm$. Upon introducing in $L_j^2(\Sigma^\pm)$ the (diagonal) direct sum of the operators

$$P^0 = \pm(\mathbf{P}^2 + m_A^2)^{1/2}, \qquad P^r = -i\frac{\partial}{\partial q^r} = i\frac{\partial}{\partial q_r}, \quad r = 1, 2, 3, \tag{4.2}$$

which we then denote by the same symbols, we can define time-evolution in accordance with (2.3.9b)–(2.3.11) by setting

$$\mathring{\psi}(q, v) = (\exp[-iP^\mu(q_\mu - q'_\mu)]\,\mathring{\psi})\,(q', v'), \quad (q', v') \in \Sigma^\pm, \tag{4.3}$$

where $\mathring{\psi}(\zeta)$ denotes the one-column matrix-valued function with upper $2j + 1$ components $\psi_s \in L^2(\Sigma^+)$, $s = -j, \ldots, +j$, and lower $2j + 1$ components $\psi_{\dot{s}} \in L^2(\Sigma^-)$, $\dot{s} = -j, \ldots, +j$, and the respective signs for P^0 in (4.2) are adopted in these two instances. The related representations of $\mathscr{P}$(interpreted as the semidirect product $\mathscr{T}_4 \wedge \mathrm{SL}(2, C)$ of the spacetime translations group and of $\mathrm{SL}(2, C)$) in $L_j^2(\Sigma^+)$, that correspond to the momentum-space representations first derived by Wigner (1939), are

$$(U_j(a, \mathring{A})\,\psi)_s(q, v) = \exp(im_A a \cdot v) \sum_{s'=-j}^{+j} D_{ss'}^{(j)}(A, P)\,\psi_{s'}(\Lambda^{-1}(q - a), \Lambda^{-1}v) \tag{4.4}$$

for the upper components, i.e. for one-particle states, with a similar formula involving $D_{\dot{s}\dot{s}'}^{(j)}$ holding for the lower components $\psi_{\dot{s}}$, i.e. for states of single antiparticles. In (4.4) we have taken advantage of the notation of Bogolubov *et al.* (1975, Sec. 6.3), according to which $D_{ss'}^{(j)}$ are functions of $\mathring{A} \in \mathrm{SL}(2, C)$ and $k \in \mathscr{V}_{m_A}^+$, corresponding to the matrix elements of the SU(2) representation $D^{(j)}(\mathring{A}_R)$ assigned to the SU(2) element $\mathring{A}_R$ equal to

$$D^{(1/2)}(\mathring{A}, k) = (k^\nu \sigma_\nu)^{-1/2}\,\mathring{A}\,(\mathring{A}^{-1} k^\nu \sigma_\nu \mathring{A}^{*-1})^{1/2}, \tag{4.5}$$

where σ_0 is the identity matrix acting on $l^2(2)$, and $\sigma_1, \sigma_2, \sigma_3$ are the Pauli matrices. For antiparticle states, $D_{\dot{s}\dot{s}'}^{(1/2)}$ corresponds to the complex conjugate representation of SU(2). In either case, $D_{ss'}^{(j)}(\mathring{A}, P)$ in (4.4) are bounded operator-valued functions of the self-adjoint operaors P^ν.

In the case of spin $j = \frac{1}{2}$ the irreducible subrepresentations of the representation (4.4) are unitarily equivalent to those obtained by Foldy and Wouthuysen (1950) for relativistic spin-$\frac{1}{2}$ particles and antiparticles.[12] This unitary equivalence can be established, as in Section 2.4, by means of unitary transforms defined as in (2.4.18). Furthermore, the infinitesimal generators P_{FW}^μ of spacetime translations for $U_{\mathrm{FW}} = U_{(1/2)}$ given by (4.4) can be written as

$$P_{\mathrm{FW}}^0 = \gamma^0 P^0, \qquad P_{\mathrm{FW}}^j = \gamma^0 P^j \gamma^0, \quad P^0 = (\mathbf{P}^2 + m_A^2)^{1/2}, \tag{4.6}$$

with P^i, $i = 1, 2, 3$, the same as in (4.2), and γ^0 corresponding to the γ-matrix representation

$$\gamma^0 = \begin{bmatrix} \sigma^0 & 0 \\ 0 & -\sigma^0 \end{bmatrix}, \qquad \gamma^i = \begin{bmatrix} 0 & \sigma^i \\ -\sigma^i & 0 \end{bmatrix}. \tag{4.7}$$

Let us denote by $L^2_{1/2}(\Sigma^\pm_A)$ the irreducible subspaces for U_{FW} that correspond to a proper wave function η_A, so that the orthogonal projector onto $L^2_{1/2}(\Sigma^\pm_A)$ equals $\mathbb{P}_A(\Sigma^\pm)$, where

$$(\mathbb{P}_A(\Delta)\psi_{\mathrm{FW}})(\zeta) = \int_\Delta K_A(\zeta;\zeta')\,\frac{1\pm\gamma^0}{2}\,\psi_{\mathrm{FW}}(\zeta')\,\mathrm{d}\Sigma(\zeta'), \quad \Delta\subset\Sigma^\pm. \tag{4.8}$$

The POV measures $\mathbb{P}_A(\Delta)$, taken in conjunction with the restriction to $L^2_{1/2}(\Sigma^\pm_A)$ of U_{FW}, give rise[13] to systems of covariance for $\mathscr{P}$,

$$U_{\mathrm{FW}}(a,\mathring{A})F_A(\Delta)U^{-1}_{\mathrm{FW}}(a,\mathring{A}) = F_A((a,\Lambda)\Delta), \quad \Delta\subset\mathscr{M}_A, \tag{4.9}$$

if F_A is constructed from $\mathbb{P}_A$ by analogy with (2.6.15)–(2.6.18). The propagator (4.3) can be expressed in $L^2_{1/2}(\Sigma_A)$ as

$$\psi_{\mathrm{FW}}(\zeta) = i\int \Delta^{(+)}_{\mathrm{FW}}(\zeta;\zeta')\psi_{\mathrm{FW}}(\zeta')\,\mathrm{d}\Sigma(\zeta'), \tag{4.10}$$

where the propagator $i\Delta^{(+)}_{\mathrm{FW}}$ is directly related by (1.4) and (1.9) to the one for corresponding spin-zero particles:

$$\Delta^{(+)}_{\mathrm{FW}}(\zeta;\zeta') = \frac{1+\gamma^0}{2}\,\Delta^{(+)}_A(\zeta;\zeta') + \frac{1-\gamma^0}{2}\,\Delta^{(+)}_A(-\zeta';-\zeta). \tag{4.11}$$

The transition to Dirac bispinors $\psi_{\mathrm{D}}(\zeta)$ is effected (Prugovečki, 1980) in two stages: first, we introduce in $L^2_{1/2}(\Sigma)$ the unitary operator

$$F_{\mathrm{W}} = [2P^0(P^0+m_A]^{-1/2}(\gamma_\nu P^\nu + m_A) \tag{4.12}$$

and use it to make the transition to a unitarily equivalent representation of U_{FW} in $L^2_{\frac{1}{2}}(\Sigma)$; afterwards, we make the transition to the new inner product:

$$(\Psi'\mid\Psi'') = \langle\Psi'\mid m_A P_0^{-1}\Psi''\rangle = \langle D\Psi'\mid D\Psi''\rangle, \qquad D = m_A^{1/2}P_0^{-1/2}. \tag{4.13}$$

We shall denote by $\hat{\mathscr{F}}_{1,1}(\Sigma)$ the Hilbert space with this inner product, and note that D can be extended to a unitary transformation $\mathbb{D}$ of $L^2_{1/2}(\Sigma)$ onto $\hat{\mathscr{F}}_{1,1}(\Sigma)$, so that

$$U_B(a,\mathring{A}) = \mathbb{D}F_{\mathrm{W}}\left[U_{\mathrm{FW}}(a,\mathring{A})\,\frac{1+\gamma^0}{2} + U_{\mathrm{FW}}(-a,\mathring{A})\,\frac{1-\gamma^0}{2}\right](\mathbb{D}F_{\mathrm{W}})^{-1} \tag{4.14}$$

provides a Dirac bispinor representation (with $\Lambda = \Lambda(\pm\mathring{A})$),

$$(U_B(a,\mathring{A})\Psi)(q,v) = S(\mathring{A})\Psi(\Lambda^{-1}(q-a),\Lambda^{-1}v), \tag{4.15}$$

in which $S(\mathring{A})$ is momentum-independent for all $A \in SL(2, C)$:

$$S(\mathring{A}) = R \begin{bmatrix} \mathring{A} & 0 \\ 0 & \mathring{A}^{*-1} \end{bmatrix} R, \qquad R = R^{-1} = 2^{-1/2} \begin{bmatrix} \sigma^0 & \sigma^0 \\ \sigma^0 & -\sigma^0 \end{bmatrix}. \tag{4.16}$$

This simplification is purchased at the price of having the positive and negative energy states for P_B^0,

$$P_B^\mu = \mathbb{D} F_{\mathrm{W}} P_{\mathrm{FW}}^\mu F_{\mathrm{W}}^{-1} \mathbb{D}^{-1}, \tag{4.17}$$

correspond no longer to vanishing upper or lower spinor components, but rather to eigenstates of the projectors

$$\mathbb{P}(\Sigma^\pm) = \mathbb{D} F_{\mathrm{W}} \frac{1 \pm \gamma^0}{2} F_{\mathrm{W}}^{-1} \mathbb{D}^{-1} = \frac{1}{2m} (m \pm \overline{\gamma_\mu P_B^\mu}). \tag{4.18}$$

Hence the splitting of $\hat{\mathscr{F}}_{1,1}(\Sigma)$ into particle and antiparticle components is now

$$\mathscr{F}_{1,1}(\Sigma) = \mathscr{F}_{1,0}(\Sigma) \oplus \mathscr{F}_{0,1}(\Sigma), \tag{4.19a}$$

$$\mathscr{F}_{1,0} = \mathbb{P}(\Sigma^+) \mathscr{F}_{1,1}, \qquad \mathscr{F}_{0,1} = \mathbb{P}(\Sigma^-) \mathscr{F}_{1,1}. \tag{4.19b}$$

On the other hand, for $\Psi', \Psi'' \in \mathscr{F}_{1,1}(\Sigma_A)$, where

$$\mathscr{F}_{1,1}(\Sigma_A) = \mathbb{P}_A(\Sigma) \mathscr{F}_{1,1}(\Sigma), \qquad \mathbb{P}_A(\Sigma) = \mathbb{D} F_{\mathrm{W}} \mathbb{P}_A(\Sigma) F_{\mathrm{W}}^{-1} \mathbb{D}^{-1}, \tag{4.20}$$

the inner product (4.13) can be written on account of (2.67) in the form

$$(\Psi' | \Psi'') = Z'_A \int \bar{\Psi}'(\zeta) \gamma^\nu \Psi''(\zeta) \, d\sigma_\nu \, d\Omega(\nu), \qquad \bar{\Psi} = \Psi^* \gamma^0. \tag{4.21}$$

This expression corresponds to the covariant current

$$J_A^\mu(q) = Z'_A \int \bar{\Psi}(q, \nu) \gamma^\mu \Psi(q, \nu) \, d\Omega(\nu), \tag{4.22}$$

which is conserved on account of the Dirac equation

$$\left(i\gamma^\nu \frac{\partial}{\partial q^\nu} - m_A \right) \Psi(q, \nu) = 0, \tag{4.23}$$

that follows from (4.3) or (4.10) after F_{W} is applied. In $\mathscr{F}_{1,1}(\Sigma_A)$ the propagation governed by (4.23) follows from (4.8):

$$\Psi(\zeta) = i \int S_A^{(+)}(\zeta; \zeta') \Psi(\zeta') \, d\Sigma(\zeta'), \tag{4.24}$$

$$S_A^{(\pm)}(q, \nu; q', \nu') = \left(i\gamma^\nu \frac{\partial}{\partial q^\nu} + m_A \right) \Delta_A^{(\pm)}(q, \nu; q', \nu'). \tag{4.25a}$$

For example, in case of the ground exciton proper wave function corresponding to (4.5.25a), we find:

$$S_A^{(+)}(\zeta;\zeta') = -2\pi i m_A Z_A \left[\frac{K_1(f_A)}{f_A} + i m_A(\zeta_\nu^* - \zeta_\nu')\gamma^\nu \frac{K_2(f_A)}{f_A^2}\right], \tag{4.25b}$$

$$f_A^2(\zeta;\zeta') = -m_A^2(\zeta_\nu^* - \zeta_\nu')(\zeta^{*\nu} - \zeta'^\nu),\ \zeta_\nu = q_\nu + i l_A v_\nu. \tag{4.25c}$$

Naturally, J_A^μ in (4.22) does not coincide with the probability currents

$$j_A^\mu(q) = \pm \int_{\mathscr{V}^\pm} v^\mu \overline{\Psi}'(q, v)\Psi'(q, v)\, \mathrm{d}\Omega(v), \quad \Psi' = F_{\mathrm{W}}^{-1}\mathbb{D}^{-1}\Psi, \tag{4.26}$$

that correspond to the system of covariance (4.9). Nevertheless, J_A^μ is related to the POV measures

$$\hat{F}_A^\pm(\Delta) = Z_A' \int_\Delta \mathrm{d}^4 q\Omega(v)\, |U_{\mathrm{D}}(q, \Lambda_v)\hat{\eta}_A^{(\pm)}\rangle\, n_\mu \gamma^\mu\, \langle U_{\mathrm{D}}(q, \Lambda_v)\hat{\eta}_A^{(\pm)}| \tag{4.27}$$

corresponding to the resolution generators

$$\hat{\eta}_A^{(\pm)} = \mathbb{D}F_{\mathrm{W}}\, \frac{1 \pm \gamma^0}{2}\, \eta_A, \tag{4.28}$$

which in turn can be deemed to correspond to the *charge* proper wave functions

$$\eta_A^{(\pm)}(u; s) = u_0^{-1}\,[2(u^0 + 1)]^{-1/2}(\gamma_\nu u^\nu + 1)\, \frac{1 \pm \gamma^0}{2}\, \eta_s^A(u^0) \tag{4.29}$$

for spin-$\frac{1}{2}$ particles and antiparticles, respectively. When taken in conjunction with $U_{\mathrm{D}}(a, \Lambda)$, the POV measures (4.27) give rise to Poincaré systems of covariance, that describe the stochastic localizability [14] for charge.

The Fock space that corresponds to (4.19) is

$$\hat{\mathscr{F}}(\Sigma_A) = \bigoplus_{n,n'=0}^{\infty} (\hat{\mathscr{F}}_{1,0}^{@n}) \otimes (\hat{\mathscr{F}}_{0,1}^{@n'}), \tag{4.30}$$

where @ denotes the antisymmetric tensor product of Hilbert spaces. The particle annihilation and creation operators can be defined by analogy with (1.10) and (1.11):

$$[\psi_A^{(-)}(\zeta, r)\Psi_{n,n'}]_{n-1,n'}(\zeta_1, r_1, \ldots, \zeta_{n-1}, r_{n-1}; \zeta_1', r_1', \ldots, \zeta_{n'}', r_{n'}')$$

$$= -i n^{1/2} \sum_{r_n} \int S_A^{(+)}(\zeta, r; \zeta_n, r_n)\Psi_{n,n'}(\zeta_1, r_1, \ldots, \zeta_n, r_n, \zeta_1', r_1', \ldots, \zeta_{n'}', r_{n'}') \times$$

$$\times\, \mathrm{d}\Sigma(\zeta_n), \tag{4.31}$$

$$[\psi_A^{(+)}(s, r)\Psi_{n,n'}]_{n+1,n'}(\zeta_1, r_1, \ldots, \zeta_{n+1}, r_{n+1}; \zeta_1', r_1', \ldots, \zeta_{n'}', r_{n'}')$$

$$= i(n+1)^{-1/2} \sum_{j=1}^{n+1} (-1)^{j-1} \overline{S}_A^{(+)}(\zeta, r; \zeta_j, r_j) \times$$

$$\times \Psi_{n,n'}(\zeta_1, r_1, \ldots, \zeta_{j-1}, r_{j-1}, \zeta_{j+1}, r_{j+1}, \ldots, \zeta_{n+1}, r_{n+1}; \zeta_1', r_1', \ldots, \zeta_{n'}', r_{n'}'). \tag{4.32}$$

The antiparticle annihilation and creation operators can be defined accordingly in terms of $S_A^{(-)}$, so that (Prugovečki, 1981d):

$$\{\psi_A^{(\pm)}(\zeta, r), \psi_A^{(\pm)}(\zeta', r')\} = 0, \tag{4.33a}$$

$$\{\psi_A^{(-)}(\pm\zeta, r), \overline{\psi}_A^{(+)}(\pm\zeta', r')\} = iS_A^{(\pm)}(\zeta, r; \zeta', r'). \tag{4.33b}$$

The consistency of the above definitions and the fact that

$$\overline{\psi}^{(\pm)}(\zeta) = \psi^{(\mp)\dagger}(\zeta) = \psi^{(\pm)T}(\zeta)\gamma^0 \tag{4.33c}$$

follow from the propagator properties

$$\int S_A^{(+)}(\zeta; \zeta'')\overline{S}_A^{(+)}(\zeta'; \zeta'')\, d\Sigma(\zeta'') = iS_A^{(+)}(\zeta; \zeta')\gamma^0 \tag{4.34}$$

which $S_A^{(+)}$ inherits from $\Delta_A^{(+)}$ due to (1.6) and (4.25).

The Dirac free field at $\zeta \in \mathcal{M}$

$$\psi_A(q, v) = \psi_A^{(-)}(q, v) + \overline{\psi}_A^{(+)}(q, -v) \tag{4.35}$$

satisfies the Dirac equation

$$\left(i\gamma^\nu \frac{\partial}{\partial q^\nu} - m_A\right) \psi_A(q, v) = 0 \tag{4.36}$$

on account of (4.31), (4.32) and the fact that by (4.25)

$$\left(\pm i\gamma^\nu \frac{\partial}{\partial q^\nu} - m_A\right) S_A^{(\pm)}(\zeta; \zeta') = -\frac{1}{2m_A}\left(\frac{\partial^2}{\partial q_\nu\, \partial q^\nu} + m_A^2\right) \Delta_A^{(\pm)}(\zeta; \zeta') = 0. \tag{4.37}$$

Furthermore, the anticommutation relations (4.33) for creation and annihilation operators yield the following anticommutation relations free-field operators:

$$\{\psi_A(\zeta, r), \overline{\psi}_A(\zeta'; r')\} = iS_A(\zeta, r; \zeta', r'), \tag{4.38}$$

$$S_A(\zeta; \zeta') = S_A^{(+)}(\zeta; \zeta') + S_A^{(-)}(\zeta; \zeta'). \tag{4.39}$$

The particle and antiparticle number operators $N_A^{(\pm)}$ can be expressed as:

$$N_A^{(\pm)} = \int \psi_A^{(+)T}(\zeta)\gamma^0\psi_A^{(-)}(\zeta)\,\mathrm{d}\Sigma(\zeta). \tag{4.40}$$

From these expressions we can derive, by proceeding as in (1.24)–(1.33), that the total charge operator Q equals

$$Q_A = -e(N_A^{(+)} - N_A^{(-)}) = \int \hat{\mathscr{J}}_A^{\nu}(q)\,\mathrm{d}\sigma_\nu(q), \tag{4.41}$$

where the charge operator current is:

$$\hat{\mathscr{J}}_A^{\nu}(q) = -eZ'_A \int :\bar{\psi}_A(q,\nu)\gamma^0\psi_A(q,\nu):\,\mathrm{d}\Omega(\nu). \tag{4.42}$$

Similarly, the infinitesimal generators $\hat{P}_A^{\mu}$ of spacetime translations for the representation of $\mathscr{T}_4 \wedge \mathrm{SL}(2, C)$ in $\hat{\mathscr{F}}(\Sigma_A)$ that results from (4.15) equal

$$\hat{P}_A^{\mu} = \frac{i}{2}\int \psi_A^{(+)T}(q,\nu)\gamma^0\overleftrightarrow{\partial}^{\mu}\psi_A^{(-)}(q,\nu)\,\mathrm{d}\Sigma(q,\nu), \tag{4.43}$$

and a procedure that parallels the one leading from (1.37) to (1.39) yields:

$$\hat{P}_A^{\mu} = \frac{i}{2}Z'_A \int :\bar{\psi}_A(q,\nu)\gamma^{\nu}\frac{\overleftrightarrow{\partial}}{\partial q_\mu}\psi_A(q,\nu):\,\mathrm{d}\sigma_\nu(q)\,\mathrm{d}\Omega(\nu). \tag{4.44}$$

This suggests the symmetrized Lagrangian density

$$L_0^A(\bar{\psi},\psi) = Z'_A\left[\frac{i}{2}\bar{\psi}_A\gamma^{\mu}\frac{\overleftrightarrow{\partial}}{\partial q^{\mu}}\psi_A - m_A\bar{\psi}_A\psi_A\right] \tag{4.45}$$

in which the (finite) renormalization constant Z'_A makes its appearance.

Thus, all the FTQS counterparts of all the LQFT expressions (Schweber, 1961, Sec. 8) for free Dirac fields are recovered. We see that, just as in the scalar case, the formal similarities are very striking, and yet the physical and mathematical substance is very disparate: in the FTQS we are working with *bona fide* operators $\psi_A(\zeta)$ and $\bar{\psi}_A(\zeta)$ that represent fields at stochastic points marked by stochastically extended test particles; on the other hand, their local counterparts $\psi(x)$ and $\bar{\psi}(x)$ are mathematically ill defined and require 'smearing' with test functions f, whereby operators are obtained; but nevertheless their physical meaning cannot be consistently maintained to be related to sharply localized particles.

5.5. Photon Localizability and the Electromagnetic Field on Quantum Spacetime

The problem of photon localizability has historically encountered a series of fundamental difficulties, in which all the problems faced by the localizability question for massive relativistic quantum particles, that were surveyed in Sections 2.1 and 2.2, were compounded by new ones peculiar to mass-zero particles and to the absence of longitudinal polarization modes. The *formal* solutions proposed in the past varied from discarding some of the conditions imposed on the space-coordinate observables, such as their commutativity (Pryce, 1948) – possibly coupled with physically new notions of localization, such as front localizability[15] (Acharya and Sudarshan, 1960) – to the dismissal of the requirement of σ-additivity of PV measures (Jauch and Piron, 1967; Amrein, 1968) that had been discussed in Section 1.2, and was found to be generally unsatisfactory. In fact Hegerfeldt's theorem (cf. Section 2.2) precludes any notion of *sharp* localizability for relativistic quantum particles of *any* mass, so that even if one introduces POV measures for sharply localized photons (Kraus, 1971) inconsistencies will still result.

To formulate a physically and mathematically consistent concept of stochastic localizability for photons, we first observe that the concept of proper wave function has no operational meaning associated with position and momentum uncertainties in reference frames in which mass-zero particles are stochastically at rest, since there is no (classical or quantum) rest frame for such particles. Thus, a photon or any other mass-zero particle can be localized only in relation to massive test particles, but cannot itself serve as a constituent of a quantum Lorentz frame, or as part of a coherent flow of test particles marking spacetime events.[16] Given the conventional description of photon states by wave functions $\tilde{\Phi}^{\mu}(k)$ that satisfy the (covariant) Lorentz gauge condition

$$k_{\mu}\tilde{\Phi}^{\mu}(k) = 0, \quad k \in \mathscr{V}_0^+, \tag{5.1}$$

we might localize them by means of massive spinless particles of proper wave function η_A by extrapolating from (2.5.5) and (4.3.23) to

$$\hat{\Phi}^{\mu}(q, v) = \tilde{Z}_A^{(0)-1/2} \int_{\mathscr{V}_0^+} \exp(-iq \cdot k)\eta_A(v \cdot k)\tilde{\Phi}^{\mu}(k)\, d\Omega_0(k), \tag{5.2}$$

where k is measured in units of $\hat{m}_0\hat{c} = \hat{\hbar}\hat{l}_0^{-1}$. Upon adopting the (noncovariant) Coulomb gauge

$$\mathbf{k} \cdot \tilde{\mathbf{\Phi}}(k) = 0, \qquad \tilde{\Phi}^0(k) = 0, \tag{5.3}$$

we see immediately that

$$P_{(0)}(B) = -\int_B \hat{\Phi}_{\mu}^*(\zeta)\hat{\Phi}^{\mu}(\zeta)\, d\Sigma(\zeta) \tag{5.4}$$

is positive definite on the Borel sets B in any reference hypersurface Σ_A^+. If (5.4) were to represent a probability measure, then the mapping $\tilde{\Phi}^\mu(k) \mapsto \hat{\Phi}^\mu(\zeta)$ should be an isometric transformation from the Hilbert space of wave functions $\tilde{\Phi}(k)$ that satisfy (5.1) and have the squared norm

$$\|\tilde{\Phi}\|^2_{\mathscr{V}_0^+} = - \int_{\mathscr{V}_0^+} \tilde{\Phi}^*_\mu(k)\tilde{\Phi}^\mu(k)\, \mathrm{d}\Omega_0(k), \tag{5.5}$$

into the Hilbert space $\oplus L^2_\mu(\Sigma_A^+)$ of wave function $\hat{\Phi}(\zeta)$ with components $\hat{\Phi}^\mu(\zeta)$, that have the squared norm

$$\|\hat{\Phi}\|^2_{\Sigma_A^+} = - \int_{\Sigma_A^+} \hat{\Phi}^*_\mu(\zeta)\hat{\Phi}^\mu(\zeta)\, \mathrm{d}\Sigma(\zeta), \tag{5.6}$$

and satisfy the gauge condition

$$\frac{\partial}{\partial q^\mu}\, \hat{\Phi}^\mu(q, v) \equiv 0. \tag{5.7}$$

By duplicating steps (2.4.4) to (2.4.10), we conclude that will be the case if and only if

$$\int \mathrm{d}\mathbf{v}\, |\eta_A(v \cdot k)|^2 = 2k^0(2\pi)^{-3}\tilde{Z}_A^{(0)} \tag{5.8}$$

for all $k \in \mathscr{V}_0^+$, and with $v = (v^0, \mathbf{v}) \in \mathscr{V}^+$.

Formally, the integral in (5.8) is rotationally invariant, and therefore only a function of $k^0 = |\mathbf{k}|$. However, since k assumes values on the forward light cone $\mathscr{V}_0^+$, not only are formulae (2.4.11)–(2.4.17) not reproducible in the present case, but that integral is generally divergent. That means that the expression (5.4) is not a probability measure over Σ^+, but rather gives rise to relative probabilities of the same nature as those in (1.3.7) and (1.3.8b), or those corresponding to the expectation values in (1.6.23b) or (2.6.14). On the other hand, in contradistinction to those previously encountered instances of relative probabilities, the present ones can be renormalized by means of a wave function renormalization involving the removal of the infinite constant $\tilde{Z}_A^{(0)}$ in (5.2) by a formally covariant limiting process. We can carry out this limiting process by formally imparting to photons a small rest mass $\epsilon > 0$. Then, instead of equating (5.6) with (5.5), we write

$$\int_{\Sigma_A^+} \Phi^*_\mu(\zeta; \epsilon)\Phi^\mu(\zeta; \epsilon)\, \mathrm{d}\Sigma(\zeta) = \int_{\mathscr{V}_\epsilon^+} \tilde{\Phi}^*_\mu(k)\tilde{\Phi}^\mu(k)\, \mathrm{d}\Omega_\epsilon(k), \tag{5.9}$$

where $k = (k_\epsilon^0, \mathbf{k})$, with $k_\epsilon^0 = (\mathbf{k}^2 + \epsilon^2)^{1/2}$. As in (2.4.4)–(2.4.10), we find that the above equality is obeyed iff

$$\tilde{Z}_A^{(\epsilon)} = \frac{(2\pi)^3}{2\epsilon} \int_{\mathscr{V}^+} |\eta_A(\epsilon v^0)|^2 \, \mathrm{d}\mathbf{v} \sim \frac{(2\pi)^3}{2\epsilon^4} \int_{\mathscr{V}_0^+} |\eta_A(k^0)|^2 \, \mathrm{d}\mathbf{k}. \tag{5.10}$$

Hence, we arrive at the renormalized probability amplitudes

$$\Phi^\mu(\zeta) = \lim_{\epsilon \to +0} \tilde{Z}_A^{(\epsilon)} \Phi^\mu(\zeta; \epsilon) = \int_{\mathscr{V}_0^+} \exp(-iq \cdot k) \eta_A(v \cdot k) \tilde{\Phi}^\mu(k) \, \mathrm{d}\Omega_0(k), \tag{5.11}$$

which constitute, as $\tilde{\Phi}^\mu$, $\mu = 0, \ldots, 3$, vary over $L^2(\mathscr{V}_0^+)$, a vector space $\mathring{\mathscr{F}}_1(\Sigma^+)$ with the indefinite scalar product

$$\langle \Phi_{(1)} | \Phi_{(2)} \rangle_{\Sigma_A^+} = - \int_{\mathscr{V}_0^+} \tilde{\Phi}^*_{(1)\mu}(k) \tilde{\Phi}^\mu_{(2)}(k) \, \mathrm{d}\Omega_0(k). \tag{5.12}$$

The Hilbert space $\mathscr{F}_1^{\text{e.m.}}(\Sigma_A^+)$ of photon probability amplitudes is the subspace of those $\Phi^\mu(q, v)$ which satisfy the Lorentz gauge condition (5.7). The associated renormalized probabilities can be expressed in terms of a probability current $j^\mu(q)$,

$$P_{\text{ren}}(B) = \int_B j^\mu(q) \, \mathrm{d}\sigma_\mu(q), \quad j^\mu(q) = -2 \int v^\mu \Phi_\nu^*(\zeta) \Phi^\nu(\zeta) \, \mathrm{d}\Omega(v) \tag{5.13}$$

which is covariant and conserved, but involves the stochastic 4-velocity v^μ of the test particles rather than an intrinsic *stochastic* 4-momentum p^μ of photons – which is undefined in the absence of proper wave functions for mass-zero particles.

The renormalization procedure (5.11) secures the two basic propagator features – namely the counterparts of (4.2.2) – for the 4 × 4 matrix-valued function $\mathscr{K}^A$ with elements $g_{\mu\nu}\mathscr{K}^A$, where

$$\mathscr{K}(q, v; q', v') = \int_{\mathscr{V}_0^+} \exp[ik \cdot (q' - q)] \, \eta_A(v \cdot k) \eta_A(v' \cdot k) \, \mathrm{d}\Omega_0(k). \tag{5.14}$$

The photon propagator in the A-standard is then

$$\mathscr{K}^A_{\text{e.m.}} = \mathbb{P}_{\text{e.m.}} \mathscr{K}^A \mathbb{P}_{\text{e.m.}}, \tag{5.15}$$

there $\mathbb{P}_{\text{e.m.}}$ is the projection operator from $\mathring{\mathscr{F}}_1$ onto $\mathscr{F}_1^{\text{e.m.}}$. Thus, if $e_{(r)}(k)$, $r = 1, 2, 3$, are 4-vector functions for which

$$e^*_{(r)}(k) \cdot e_{(r)}(k) = -\delta_{rr'}, \; k \cdot e_{(r)}(k) \equiv 0, \tag{5.16}$$

then we can write the matrix components of $\mathscr{K}^A_{\text{e.m.}}$ as:

$$\mathscr{K}^{\mu\nu}_{\text{e.m.}}(\zeta;\zeta') = \sum_{r=1}^{3} \int \exp[ik \cdot (q' - q)]\, \eta_A(v \cdot k)\eta_A(v' \cdot k) e^{\mu}_{(r)}(k) \times$$

$$\times\, e^{\nu *}_{(r)}(k)\, d\Omega_0(k). \tag{5.17}$$

The theory of the free electromagnetic field on quantum spacetime can be arrived at by analogy with the conventional case (Pandit, 1959; Schweber, 1961, Sec. 9) by constructing the Gupta–Bleuler spaces

$$\mathscr{F}^{\text{e.m.}}(\Sigma_A^+) = \bigoplus_{n=0}^{\infty} \mathscr{F}_n^{\text{e.m.}}(\Sigma_A^+) \subset \mathring{\mathscr{F}}(\Sigma_A^+) = \bigoplus_{n=0}^{\infty} \mathring{\mathscr{F}}_n(\Sigma_A^+) \tag{5.18}$$

from the spaces $\mathring{\mathscr{F}}_n = \mathring{\mathscr{F}}_1^{\circledS n}$ with indefinite scalar product. Annihilation and creation operators can be defined as follows,

$$[A_\mu^{(-)}(\zeta)\Phi_n]_{n-1}^{\nu_1,\ldots,\nu_{n-1}}(\zeta_1,\ldots,\zeta_{n-1})$$

$$= n^{1/2} \int \mathscr{K}_{\mu\nu_n}(\zeta;\zeta_n)\Phi_n^{\nu_1,\ldots,\nu_n}(\zeta_1,\ldots,\zeta_n)\, d\Sigma(\zeta_n), \tag{5.19}$$

$$[A_\mu^{(+)}(\zeta)\Phi_n]_{n+1}^{\nu_1,\ldots,\nu_{n+1}}(\zeta_1,\ldots,\zeta_{n+1}) = (n+1)^{-1/2} \sum_{j=1}^{n+1} \mathscr{K}_{\nu_j\mu}(\zeta_j;\zeta) \times$$

$$\times\, \Phi_n^{\nu_1,\ldots,\nu_{j-1},\nu_{j+1},\ldots,\nu_{n+1}}(\zeta_1,\ldots,\zeta_{j-1},\zeta_{j+1},\ldots,\zeta_{n+1}), \tag{5.20}$$

with the renormalization procedure implicit in the integral in (5.19). The resulting field operators are

$$A_\mu(\zeta) = A_\mu^{(+)}(\zeta) + A_\mu^{(-)}(\zeta), \tag{5.21}$$

and a Gupta–Bleuler (1950) type of gauge condition has to be imposed:

$$\frac{\partial}{\partial q_\nu} A_\nu^{(-)}(\zeta)\Phi = 0,\ \Phi \in \mathscr{F}^{\text{e.m.}}(\Sigma_A^+). \tag{5.22}$$

From (5.19) and (5.20) we find

$$[A_\mu(\zeta), A_\nu(\zeta')] = ig_{\mu\nu}D(\zeta;\zeta'), \tag{5.23}$$

$$iD(q,v;q',v') = \mathscr{K}(q,v;q',v') - \mathscr{K}(-q,v;-q',v'). \tag{5.24}$$

For example, if the test particles are the ground excitons with proper wave function in (1.51), then by proceeding as in (2.9.14) to (2.9.18), but with $k \in \mathscr{V}_0^+$, we obtain:

$$\mathscr{K}(q, v; q', v') = -2\pi \left\{ Z_A \left[q - q' - \frac{i}{\omega_A} (v + v') \right]^2 \right\}^{-1}. \tag{5.25}$$

The generators of spacetime translations in $\mathring{\mathscr{F}}(\Sigma_A^+)$ can be written in the form

$$P^\nu_{\text{e.m.}} = i \int A^{(+)}_\mu(\zeta) \partial^\nu A^{(-)\mu}(\zeta) \, d\Sigma_{\text{ren}}(\zeta) \tag{5.26}$$

if the renormalization procedure (5.11) is made implicit in integrals of the above type. The transition to the form

$$P^0_{\text{e.m.}} = -\frac{1}{2} Z''_A \sum_{\nu=0}^{3} \int : \partial_\nu A_\mu(\zeta) \partial_\nu A^\mu(\zeta) : \, dq \, d\Omega(v) \tag{5.27}$$

for the time translation infinitesimal generator can be achieved by reverting to the amplitudes in (5.9), and then proceeding as in (2.6.7)–(2.6.12) to derive

$$\langle \Phi^{(1)} | \Phi^{(2)} \rangle_{\Sigma_A^+} = -\int \tilde{\Phi}^{(1)*}_\mu(k) \tilde{\Phi}^{(2)\mu}(k) \, d\Omega_0(k)$$

$$= -i Z_A^{(\epsilon)} \int \Phi^{(1)*}_\nu(\zeta; \epsilon) \overleftrightarrow{\partial}_0 \Phi^{(2)\nu}(\zeta; \epsilon) \, dq \, d\Omega(v), \tag{5.28}$$

$$Z_A^{(\epsilon)-1} = (2\pi)^3 \int_{\mathscr{V}^+} |\eta_A(v \cdot k)|^2 \, d\Omega(v) = \frac{(2\pi)^3}{\epsilon^2} \int_{\mathscr{V}_\epsilon^+} |\eta_A(p^0)|^2 \, d\Omega_\epsilon(p). \tag{5.29}$$

Upon renormalizing by $(2\pi)^{-3}\epsilon^2$ the element of measure on the hyperboloids $\mathscr{V}_\epsilon$ and then letting $\epsilon \to +0$, we get, according to (5.19) and (5.20),

$$P^0_{\text{e.m.}} = Z''_A \int [A^{(+)}_\mu (\partial^0 \partial_0 A^{(-)\mu}) - (\partial^0 A^{(+)}_\mu)(\partial_0 A^{(-)\mu})] \, dq \, d\Omega(v), \tag{5.30}$$

$$Z''^{-1}_A = \int_{\mathscr{V}_0^+} |\eta_A(k^0)|^2 \, d\Omega_0(k), \tag{5.31}$$

so that we obtain normalized probability distributions

$$\chi_{(v)}(k) = Z''_A \, |\eta_A(v \cdot k)|^2 \tag{5.32}$$

on the forward light cone $\mathscr{V}_0^+$. Since $\partial_\nu \partial^\nu A_\mu^{(-)} \equiv 0$, integration by parts yields

$$P^0_{\text{e.m.}} = Z_A'' \sum_{\nu=0}^{3} \int \partial_\nu A_\mu^{(+)} \partial_\nu A^{(-)\mu} \, \mathrm{dq} \, \mathrm{d}\Omega(\nu). \tag{5.33a}$$

Finally we arrive at (5.27) by taking into account that

$$\int \partial_\nu A_\mu^{(\pm)} \partial_\nu A^{(\pm)\mu} \, \mathrm{dq} \, \mathrm{d}\Omega(\nu) = 0, \; \nu = 0, \ldots, 3, \tag{5.33b}$$

due to the type of argument used in deriving (1.30).

We note that due to the presence of the normalization constant Z_A'' in (5.32), $\chi_{(\nu)}$ can be interpreted as the confidence function of a stochastic value $(0, \chi_{(\nu)})$, heuristically representing the uncertainty that an A-particle of stochastic 4-velocity $\underset{\sim}{\nu}$ has emitted or absorbed an infinitely soft photon. Alternatively, $\chi_{(\nu)}$ could be viewed to describe the fluctuation in momentum transfer due to photon exchange during electromagnetic interactions between an A-particle and some other particle.

From the point of view of the Lagrangian field theory of Section 5.3, the expression (5.27) suggest the Lagrangian density in the Feynman 'gauge' (Itzykson and Zuber, 1980, p. 128):

$$L^A_{\text{e.m.}} = -\frac{1}{2} Z_A'' \left[\frac{1}{2} (\partial_\mu A_\nu - \partial_\nu A_\mu)(\partial^\mu A^\nu - \partial^\nu A^\mu) - (\partial^\mu A_\mu)^2 \right]. \tag{5.34}$$

Upon imposing the subsidiary condition (5.23) we arrive at the equations of motion

$$\partial^\mu F_{\mu\nu} = 0, \; F_{\mu\nu} = \partial_\nu A_\mu - \partial_\mu A_\nu, \tag{5.35}$$

in the usual manner (Schweber, 1961, Sec. 9a).

We note that $L^A_{\text{e.m.}}$ contains the *finite* renormalization constant Z_A'' in (5.31), which is determined by the proper wave function η_A of test particles only by extrapolation. Indeed, $\eta_A(\lambda)$ in Section 5.2 or 5.4 is defined only for $\lambda = u \cdot \nu$ with $u, \nu \in \mathscr{V}^\pm$, i.e. only for $|\lambda| \geqslant 1$. Consequently, we need some fundamental physical principles to extrapolate $\eta_A(\lambda)$ to all $\lambda \in \mathbb{R}^1$. Born's reciprocity principle and the related quantum metric operator studied in Section 4.5 supplies the required extrapolation in unambiguous terms, since it provides us with proper wave functions analytic in λ.

5.6. Reciprocally Invariant Quantum Electrodynamics for Extended Fermions

The standard formulation of quantum electrodynamics (to which we shall refer as LQED) is depicted in all textbooks on LQFT as the crowning achievement of that discipline. And indeed, the agreement between some of the predictions of LQED and experiment appears, at least on the surface, very impressive. Yet, at the same time LQED epitomizes all the intrinsic inconsistencies that plague conventional relativistic quantum theories, and which were discussed in Sections 1.1–1.2 as well as in the introductory paragraphs of this chapter: the lack of a consistent notion of particle or field localizability, *ad hoc* infinite 'renormalization' rules for S-matrix elements (which make the terms of its *formal* perturbation series finite, but say absolutely nothing about its convergence properties),[17] and a host of other mathematical as well as physical difficulties usually not mentioned in LQFT textbooks (cf., e.g., Strocchi and Wightman, 1974, p. 2199; Ferrari *et al.*, 1974, p. 28). In particular, after noting that "our present formulation [of LQED] is in important respects *physically* incorrect", Rohrlich (1980) lists the following problems: the deficient treatment of the Coulomb self-interaction; the need for a correct treatment of the 'photon cloud' of the electron, for which "one needs coherent states or some other representation of the commutation relations" (Rohrlich, 1980, p. 159); the nonexistence of a 'clean proof' that LQED merges into classical electrodynamics in a suitably formulated classical limit; the nonexistence in LQED (or in the Weinberg–Salam theory of electro-weak interactions) of a theoretically predicted value for the renormalized coupling constant of the ϕ^4 term required to ensure the renormalizability of LQED for spin-zero particles. These and other mathematical and physical unresolved problems demonstrate that the current state of the art is correctly summarized by the following statement of Dirac: "Quantum electrodynamics ... was built up from physical ideas that were not correctly incorporated into the theory and has no sound mathematical foundation" (Dirac, 1978b, p. 5).

Thus, if one adopts an objective view of the history of the subject, keeping in mind also the facts presented in the introductory paragraphs of this chapter while assessing the strengths and the weaknesses of LQED and its nonabelian offspring, one is very much reminded of some such once popular theories as the Ptolemaic theory of planetary motions, the phlogiston theory of combustion, or the ether theory of light propagation. Indeed, from such a historical perspective, there is a striking community of features between all these theories: they all originally sprang from very common-sensical applications of prevalent conceptual frameworks to new realms of phenomena, whereupon they were found to give rise to long chains of structural and/or observational inconsistencies. These inconsistencies were then handled by *ad hoc* new postulates[18] that hid the basic difficulties under a morass of technical complexities, rather than by a systematic reevaluation of the fundamental epistemic premises on which the respective frameworks rested. These historical parallels are especially evident if one compares LQED to the theory of ether, which by the later half of the nineteenth century came to embrace all electromagnetic phenomena known at that time, hence treating the same general group of physical phenomena as LQED. Thus, instead of a postulated all-pervasive 'ether' selectively exhibiting features (Whittaker, 1951, Ch. V) not encountered in ordinary matter, in LQED we encounter a 'physical vacuum' envisaged as an all-pervasive sea of 'virtual' particles, whose presence also manifests itself very selectively – e.g. by vacuum polarization effects, but not by gravitational effects (Feynman and Hibbs, 1965, p. 246). If one includes in this comparative historical analysis the nonabelian extrapolations of LQED, additional qualitative parallels emerge: the ultimate replacement of a plurality of ethers by a single ether *vis-à-vis* 'grand unification' attempts, replacing a plurality of schemes for weak, strong and electromagnetic interactions by a single scheme – juxtaposed with the failure to develop an ether theory of gravitation *vis-à-vis* the failure to derive 'renormalizable' theories of quantum gravity. And, most interestingly, the idea of a 'conspiracy of natural laws' as the cause of a convenient contraction of solid objects moving through ether – *vis-à-vis* the idea that Nature obligingly conforms only to those laws which suitably fit conventional renormalization schemes,[19] thus implicitly conspiring to match the computational limitations inherent in the conventional (so-called 'local') formulation of QFT.

In drawing such comparisons it has to be also recognized that, as a whole, the theory of ether contained many basically valid features which survived as part of classical electromagnetism even after the removal of its metaphysical superstructure. By analogy, it can be hoped that the agreement of LQED predictions with observation is not due entirely to coincidences resulting from computational expediency, but that it does signify the presence of basically sound premises that lie hidden in the thicket of formal manipulations used to extract numerical predictions from renormalizable LQFT models.

Clearly, the striking formal similarities between LQFT and the FTQS expressions for free fields derived in the preceding five sections present one possible explanation as to how renormalizable LQFT models might be the limits of physically and mathematically consistent FTQS models: the latter would contain one or more parameters representing fundamental lengths related to the proper radii of fundamental particles (i.e. of 'standards') in those models, whose S-matrix elements would then approach those of their LQFT counterparts in a *formal* sharp-point limit. As discussed in the introductory paragraphs of Chapter 4, the idea of fundamental lengths is almost as old as LQFT itself, but in FTQS its introduction reflects well-established physical principles that afforded the solution of the localizability problem for relativistic quantum particles in the form of the Poincaré system of covariance derived in Chapter 2. In combination with Born's reciprocity principle, this approach can produce a coherent program for developing FTQS models which might eventually be treated by legitimate mathematical methods (such as those mentioned in Section 5.2) that are beyond the scope of their LQFT counterparts.

To illustrate the relation between an LQFT model of interacting fields and its FTQS counterpart, let us consider the most natural FTQS counterpart of LQED suggested by the free-field models considered in Sections 5.4 and 5.5. The total Lagrangian density at $\zeta \in \mathscr{M}$ for this model is

$$L^A = L^A_{\text{mat}} + L^A_{\text{rad}} + L^A_{\text{int}}, \tag{6.1}$$

where L^A_{mat} and L^A_{rad} are given by (4.45) and (5.34), respectively, but with free fields replaced by interacting fields, and with

$$L^A_{\text{int}} = j_\mu(\zeta)A^\mu(\zeta),\ j_\mu = -\tfrac{1}{2}e\,\{\overline{\psi}\gamma_\mu, \psi\}. \tag{6.2}$$

The resulting Heisenberg-picture field equations at any $\zeta = (q, v)$ are therefore

$$(i\gamma^\mu \partial_\mu - m_A)\psi(\zeta) = -eZ_A'^{-1}\gamma^\mu A_\mu(\zeta)\psi(\zeta), \tag{6.3}$$

$$\partial_\mu \partial^\mu A_\nu(\zeta) = -Z_A''^{-1}\frac{1}{2}e\{\overline{\psi}(\zeta)\gamma_\nu, \psi(\zeta)\}, \tag{6.4}$$

subject to the Gupta–Bleuler subsidiary condition (5.23) on asymptotic fields. A truly rigorous investigation of this model requires the investigation of existence of solutions of these coupled system of nonlinear partial differential equation for operator-valued functions by some mathematically meaningful method – such as the interative procedure described in Section 5.2. Establishing the convergence of such an interative procedure is, however, a mathematically difficult task even though the quantum field operators, as well as the nonlinear terms in (6.3) and (6.4), are in the zeroth approximation mathematically well defined, and the non-linear integral operators exhibit no singularities, so that to each finite order the approximations are well defined.

We shall therefore limit ourselves to comparing the above model of electro-dynamics on quantum spacetime to LQED at the interaction-picture formal per-turbation series level. We shall also restrict ourselves to the case of the ground exciton proper wave function η_A in (1.51), which leads to a manifestly *reciprocally invariant quantum electrodynamics* (RQED) since the free fermion and photon propagators from (q, v) to (q', v') then display a manifestly reciprocally invariant (and relativistically covariant) form (4.25) and (5.25), respectively.

The usual formal manipulations (Schweber, 1961, Secs. 11f and 13a) in LQFT adapted to FTQS yield, in general, the formal perturbation series

$$S = \sum_{n=0}^{\infty} \frac{(-i)^n}{n!}\int d\mu(\zeta_1)\ldots\int d\mu(\zeta_n)\, T[L_{\text{int}}(\zeta_1)\ldots L_{\text{int}}(\zeta_n)], \tag{6.5}$$

in which the integrations with respect to the measure (2.13) extend over $\mathscr{M}^+$. Clearly, the time-ordering in (6.5) is supplied, as in (2.11), by the chosen reference family $\mathscr{S}^\uparrow$, so that we can set

$$T[a(\zeta)b(\zeta')] = \Theta(\zeta;\zeta')a(\zeta)b(\zeta') \pm \Theta(\zeta';\zeta)b(\zeta')a(\zeta), \tag{6.6}$$

with the plus or minus sign adopted in case of boson or fermion field operators, respectively.

In RQED the starting Lagrangian (6.1) already contains[20] the (finite) renormal-ization constants Z_A' and Z_A'', by virtue of the fact that these constants had to be introduced in order to have equality between the generators for time translations P_0 for the asymptotically free matter and radiation fields, and the respective integrals of the densities $T_{00}(\zeta)$, derived from the respective free Lagrangian densities in accordance with (3.4). Thus, mimicking the jargon of conventional renormalization theory (Itzykson and Zuber, 1980, p. 389), we can refer to L^A

in (6.1) as the *renormalized* Langrangian for RQED. In order to obtain a 'bare' Lagrangian, which contains no renormalization constants and is of the form

$$L^{(l)} = (i\bar{\psi}^{(l)} \cdot \overleftrightarrow{\partial}\psi^{(l)} - m_A \bar{\psi}^{(l)}\psi^{(l)} - \frac{1}{2}[F^{(l)^2} - (\partial A^{(l)})^2] + L^{(l)}_{\mathrm{int}}, \tag{6.7}$$

where $F^{(l)}_{\mu\nu} = \partial_\nu A^{(l)}_\mu - \partial_\mu A^{(l)}_\nu$, and $L^{(l)}_{\mathrm{int}}$ is of the form

$$L^{(l)}_{\mathrm{int}} = -e^{(l)}\bar{\psi}^{(l)}\gamma \cdot A^{(l)}\psi^{(l)}, \quad l = l_A, \tag{6.8}$$

we have to perform an 'inverse'[21] renormalization by introducing the 'bare' field operators

$$\psi^{(l)}(\zeta) = Z_2^{(l)^{1/2}}\psi(\zeta), \qquad A^{(l)}(\zeta) = Z_3^{(l)1/2}A(\zeta), \tag{6.9a}$$

as well as the 'bare' charge

$$e^{(l)} = Z_2^{(l)-1}Z_3^{(l)-1/2}e, \qquad Z_2^{(l)} = Z'_A, \quad Z_3^{(l)} = Z''_A. \tag{6.9b}$$

The *finite* renormalization constants $Z_3^{(l)} = Z'_A$ and $Z_3^{(l)} = Z''_A$ are (formally) the RQED counterparts of the infinite renormalization constants Z_2 and Z_3 in LQED (Schweber, 1961, p. 634; Itzykson and Zuber, 1980, p. 413). In fact, since in the present RQED model η_A is given by (1.51), $Z_2^{(l)}$ assumes the form (1.54), whereas in accordance with (1.52), (5.25) and (5.31)

$$Z_3^{(l)} = [\mathcal{K}(\zeta;\zeta)]^{-1} = 2(2\pi)^3 \frac{l_A}{m_A} K_2(2l_A m_A), \tag{6.10}$$

so that $Z_2^{(l)} \to \infty$ and $Z_3^{(l)} \to \infty$ in the sharp-point limit $l = l_A \to 0$. Furthermore, $Z_1^{(l)} = Z_A^{3/2}$ is the RQED counterpart of the infinite LQED renormaliztion constant Z_3 that originates from the vertex part of each Feynman diagram. Indeed, it will turn out in the sequel that $Z_1^{(l)}$ enters as a multiplicative factor in the cutoff vertex factor (6.28), but that it can be eliminated from there if we multiply the right-hand side of (6.9b) by $Z_1^{(l)} = Z_A^{3/2}$. We note that for η_A in (1.51) Z_A equals (1.52b), so that $Z_1^{(l)} \to \infty$ in the sharp-point limit $l \to +0$. Thus, it can be argued that in LQED the formal introduction of Z_1, Z_2 and Z_3 serves to compensate for $Z_1^{(l)}$, $Z_2^{(l)}$ and $Z_3^{(l)}$ in the sharp-point limit of RQED, rather than being due to any kind of self-interactions that resemble those in classical electrodynamics. Indeed, the pursuit of the latter type of analogy has not led to a better understanding of the problem of LQED divergences (Schweber, 1961, p. 515). The fact that, contrary to the formal LQED equality $Z_1 = Z_2$, in RQED we have $Z_1^{(l)} \neq Z_2^{(l)}$, can be easily understood in the light of the fact (Bogolubov and Shirkov, 1959, Ch. VIII) that in LQED we can always add finite parts to Z_1, Z_2 and Z_3 without altering observable effects, since according to conventional wisdom all 'bare' quantities can be suitably redefined without changing the physically observable predictions of LQFT models.

On the other hand, the (finite) values of Z_A, Z'_A and Z''_A depend in general on the choice of η_A, and therefore the manner in which they diverge in the sharp-point limit might also depend on the chosen class of proper wave functions.

The renormalization (6.9a) of the field operators brings about a corresponding renormalization of all n-point functions. For example, the causal two-point function for fermions is

$$S^{(l)}_{\alpha\beta}(\zeta;\zeta') = \langle 0|\, T\bar{\psi}^{(l)}_{\alpha}(\zeta)\psi^{(l)}_{\beta}(\zeta')\,|0\rangle = (i\gamma^{\mu}\partial_{\mu} - m_A)\,\Delta^{(l)}(\zeta;\zeta'), \tag{6.11}$$

where $\Delta^{(l)}$ acquires the factor $Z^{(l)}_2$ in relation to $\Delta^{(\pm)}_A$ given by (1.9) and (1.20),

$$\Delta^{(l)}(\zeta;\zeta') = Z^{(l)}_2\,[\Theta(\zeta;\zeta')\,\Delta^{(+)}_A(\zeta;\zeta') - \Theta(\zeta';\zeta)\,\Delta^{(-)}_A(\zeta;\zeta')]\,. \tag{6.12}$$

Similarly, in case of photons we have:

$$\begin{aligned} D^{(l)}_{\mu\nu}(\zeta;\zeta') &= \langle 0|\, TA^{(l)}_{\mu}(\zeta)A^{(l)}_{\nu}(\zeta')\,|0\rangle \\ &= Z^{(l)}_3 g_{\mu\nu}[\Theta(\zeta;\zeta')\,D^{(+)}(\zeta;\zeta') - \Theta(\zeta';\zeta)\,D^{(-)}(\zeta;\zeta')]\,, \end{aligned} \tag{6.13}$$

$$iD^{(+)}(q,v;q',v') = \mathcal{K}(q,v;q',v') = -iD^{(-)}(-q,v;-q',v'). \tag{6.14}$$

We note that when we are dealing with the ground exciton state η_{A} for $l_{\mathrm{A}} = \hat{l}_{\mathrm{A}}/\hat{l}_0 = 1$ and $m_{\mathrm{A}} = \hat{m}_{\mathrm{A}}/\hat{m}_0 = 1$, the exclusive dependence of the resulting two-point functions on the relativistic and reciprocal invariant

$$f_{\mathrm{A}}(\zeta,\zeta') = -(\zeta^*_{\nu} - \zeta'_{\nu})\,(\zeta^{*\nu} - \zeta'^{\nu}),\; \zeta_{\nu} = q_{\nu} + iv_{\nu}, \tag{6.15}$$

secures the relativistic covariance as well as reciprocal invariance of the RQED theory for fundamental fermions with such proper wave functions.

In general, the Feynman rules for stochastic phase-space RQED diagrams with n vertices are only slight variations of the configuration space LQED Feynman rules (Schweber, 1961, p. 471). The only basic differences are: in RQED each vertex is marked by $\zeta \in \mathcal{M}^+$, rather than by $x \in M(1,3)$; the final integrations are with respect to the invariant measure $d\mu$ in (2.13), rather than with respect to the Lebesgue measure d^4x. Since the two-point functions (6.11) and (6.13) are devoid of singularities, it can be expected that no ultraviolet divergences will occur. The fact that this is indeed the case can be most easily established in the momentum representation, to which we turn next.

Concentrating again on ground exciton wave functions (1.51), we obtain by adapting the well-known method (Schweber, 1961, Sec. 13d; Roman, 1969, Sec. 1.2) of representing Green's functions to the present case:

$$\begin{aligned} \Delta^{(\pm)}_A(\zeta;\zeta') = -\frac{Z^{(l)}_1}{2\pi}\int d\mathbf{p}\int_{C^{(\pm)}_0}\frac{dp^0}{p^2 - m^2_A}\; &\exp\{il_A p\cdot[q' - q + i\epsilon(\mathrm{Re}\,p^0)\times \\ &\times(v + v')]\},\; q^0 \gtrless q'^0,\; \epsilon(\mathrm{Re}\,p^0) = \theta(\mathrm{Re}\,p^0) - \theta(-\mathrm{Re}\,p^0). \end{aligned} \tag{6.16}$$

In (6.16) the integration in p^0 is to be performed in the complex plane along contours $\pm C_\delta^{(\pm)}$, $\delta \to +0$, that consist of the following straight lines: from $(+\infty + i)\delta$ to $(1 + i)\delta$, from $(1 + i)\delta$ to $(1 \mp i\infty)\delta$, from $(-1 \mp i\infty)\delta$ to $(-1 - i)\delta$, and from $(-1 - i)\delta$ to $(-1 - i\infty)\delta$. Indeed, the integrand in (6.16) is obviously analytic in the complex p^0-plane everywhere except along the imaginary axis, due to the presence of $\epsilon(\mathrm{Re}\, p^0)$. Furthermore, for $q_0 \gtrless q_0'$ in $\Delta_A^{(\pm)}$, respectively, this integrand rapidly approaches zero when confined to semicircles of radius $R \to +\infty$ in the respective lower and upper p^0 half-planes. Hence, by a standard application (Schweber, 1961, Sec. 13d) of the calculus of residues, we find that $\Delta^{(l)}$ in (6.12) can be displayed as a sum of two distinct parts: one invariant *primary part* $\Delta'_{(l)}$, in which the p^0-integration is performed along the familiar contour C_F and can be, therefore, alternatively written as

$$\Delta'_{(l)}(\zeta;\zeta') = -\frac{iZ_A Z_2^{(l)}}{2\pi}\int \frac{\mathrm{d}^4 p}{p^2 - m_A^2 + i0}\exp\{il_A p \cdot [q' - q + i\epsilon(p^0)(v' + v)]\},\tag{6.17}$$

and a covariant *secondary part* $\Delta''_{(l)}$ originating from the discontinuity of the integrand along the imaginary p^0-axis, which can be written in the form

$$\Delta''_{(l)}(\zeta;\zeta') = -\frac{iZ_A Z_2^{(l)}}{2\pi}\int_{\mathrm{Re}\, p^0 = 0} \mathrm{d}^4 p\, \frac{\epsilon(\mathrm{Im}\, p^0)}{p^2 - m_A^2} \times$$

$$\times \exp\{il_A p \cdot [\epsilon(q;q')\epsilon(\mathrm{Im}\, p^0)(q - q') - i(v + v')]\},\tag{6.18}$$

where $\epsilon(q;q')$, $q \in \sigma$, $q' \in \sigma'$, equals either +1 or −1, depending on whether $\Sigma^+ \gtrsim \Sigma'^+$ or $\Sigma^+ < \Sigma'^+$, respectively, for $\Sigma^+ = \sigma \times \mathscr{V}^+$ and $\Sigma'^+ = \sigma' \times \mathscr{V}^+$. Hence, by (6.11) we have a corresponding decomposition of $S^{(l)}$ for which its primary part equals

$$S'_{(l)}(\zeta;\zeta') = -\frac{iZ_1^{(l)}}{2\pi}\int S_{(l)}(p)\exp\{il_A p \cdot [q' - q + i\epsilon(p^0)(v' + v)]\}\,\mathrm{d}^4 p,\tag{6.19}$$

where $S_{(l)}(p)$ is a renormalized version of the 'bare' LQED fermion propagator in the momentum representation

$$S_{(l)}(p) = \frac{Z_2^{(l)}}{2\pi}\frac{1}{\gamma \cdot p - m + i0}.\tag{6.20}$$

A similar splitting can be effected upon $D^{(l)}$ in (6.13), with the result that the primary part shall be

$$D'_{(l)}(\zeta;\zeta') = -i\int D_{(l)}(k)\exp\{il_A k \cdot [q' - q + i\epsilon(k^0)(v' + v)]\}\,\mathrm{d}^4 k,\tag{6.21}$$

where $D_{(l)}$ is the renormalized version of the bare LQED photon propagator in the chosen Feynman 'gauge':

$$D_{(l)}(k) = \frac{Z_3^{(l)}}{2\pi} \frac{1}{k^2 + i0}. \tag{6.22}$$

It is the primary terms that supply the formal part of RQED S-matrix expansion that can be compared, diagram by diagram, with the corresponding LQED S-matrix perturbation series, and leads to the conclusion that in RQED ultraviolet divergences are absent. Indeed, in the momentum representation, for each incoming RQED fermion line at ζ corresponding to 4-momentum p and spin component s, the annihilation operator $\psi^{(-)}(\zeta)$ gives rise to

$$Z_A^{1/2} w^s(p) \exp(-il_A p \cdot \zeta^*) \tag{6.23}$$

that contributes to the integral with respect to the measure μ. Similar factors result from other particle or antiparticle creation and annihilation operators, whereas for a photon of 4-momentum k and given polarization, the corresponding factor equals

$$Z_A^{1/2} e_{(r)}^{\mu}(k) \exp(-il_A k \cdot \zeta^*) \tag{6.24}$$

for choices of $e_{(r)}^{\mu}(k)$ as given by (5.16). Consequently, if at a vertex labelled by $\zeta = q + iv$ in an RQED diagram we perform the μ-integration first in the variable v with respect to

$$\mathrm{d}\Omega(v) = \theta(v^0)\delta(v^2 - 1)\,\mathrm{d}^4 v, \tag{6.25}$$

we shall obtain from primary terms the contribution

$$V_A(p, p', k) = Z_A^{3/2} \int_{\mathscr{V}^+} \exp\{-l_A v \cdot [\epsilon(p_0)p + \epsilon(p_0')p' + \epsilon(k_0)k]\}\,\mathrm{d}\Omega(v) \tag{6.26}$$

if at that vertex fermion lines carrying 4-momenta p and p' meet a photon line carrying 4-momentum $\pm k$. The above integration in $v \in \mathscr{V}^+$ can be performed by means of (2.9.12) since $\mathrm{d}\Omega(v)$ is an invariant and

$$y = [\epsilon(p_0)p + \epsilon(p_0')p' + \epsilon(k_0)k]^2 > 0 \tag{6.27}$$

for all $p, p' \in \mathscr{V}_A$, $k \in \mathscr{V}_0$, so that we can always execute a Lorentz transformation which will set the space components of the time-like 4-vector in (6.27) equal to zero. The outcome of this integration is the invariant *vertex factor*

$$V_A(p, p', k) = \frac{2\pi Z_1^{(l)}}{l_A} \frac{K_1(l_A y^{1/2})}{y^{1/2}}, \quad Z_1^{(l)} = Z_A^{3/2}, \tag{6.28}$$

whose exponential decrease as $y \to \infty$ secures the kind of cutoff that prevents the occurrence of ultraviolent divergences in RQED diagrams. On the other hand, in the sharp-point limit $l = l_A \to +0$, this vertex factor asymptotically behaves as $l_A^{-2} y^{-1}$, so that ultraviolet divergences re-emerge.

From the (formal) LQED point of view, contributions from secondary parts, such as (6.18), represent noncovariant contributions since they depend on the choice of reference hypersurfaces – i.e., of coherent flow of test particles – and can be totally neglected in relation to the primary ones only in the limit $l_A \to +0$. In fact, 'noncovariant' quantities of this type already occurs in LQED when the so-called 'naive time-ordering' (Itzykson and Zuber, 1980, p. 222)

$$Tj_\mu(x)j_\nu(x') = \theta(x_0 - x_0')j_\mu(x)j_\nu(x') + \theta(x_0' - x_0)j_\nu(x')j_\mu(x) \tag{6.29}$$

of current operators is adopted. This 'naive' time-ordering is actually the *natural* time-ordering that corresponds to (6.6), but once it is found (Itzykson and Zuber, 1980, p. 530) that

$$\langle 0|\, Tj_\mu(x)j_\nu(x')\,|0\rangle = \langle 0|\, T'j_\mu(x)j_\nu(x')\,|0\rangle - i(g_{\mu\nu} - g_{\mu 0}g_{\nu 0})\,\delta^4(x - x'), \tag{6.30}$$

where the term containing T' is manifestly covariant, whereas the last in (6.30) is evidently not, the time-ordering is *conveniently redefined* as the one given by T' by the *ad hoc* introduction of Schwinger counterterms. Naturally, in LQED the presence of terms containing $g_{\mu 0}g_{\nu 0}$ is an acute embarrassment, since the (supposedly) pointlike LQED fermions propagating in classical spacetime should abide by the criteria of classical causality, which dictates that such propagation is insensitive to the adopted choice of reference family $\mathscr{S}^\uparrow$. On the other hand, in RQED the depence of *observable* propagation on the mode of observation, namely on the choice of coherent flow of test particles, and therefore of the (time-ordered) family $\mathscr{S}^\uparrow$ of reference hypersurfaces, is mandatory in view of the stochastically extended nature of the particles whose electromagnetic interactions RQED models are designed to describe.

It is hoped that future research dealing with LQED as the sharp-point limit of RQED will lead to a better understanding of the *true* physical and mathematical nature of the host of arbitrary manipulations (such as (6.30), or the 'substraction of infinities'), in which the conventional renormalization program abounds. However, it is RQED, and not LQED, that rests on a physical and mathematical foundation of *consistently interrelated* ideas. Thus, in the long run, the search should proceed in the direction of pinpointing, either by theoretical arguments or by experimental means, the values of the fundamental parameters characterizing extended fermions (leptons and baryons, or possibly quarks, treated as exciton states) which interact electromagnetically in accordance with RQED. This might first require the formulation of FTQS models for electroweak interactions, and a total integration of the results of Sections 4.5 and 4.6 with the FTQS framework.

In any event, the present chapter has barely laid down a skeletal groundwork for FTQS, and much remains to be done before even all the basic stages of FTQS are completed. It is therefore hoped that those readers who find some merit in the present approach to quantum field theory will join the search after FTQS models that would not only produce results in agreement with experiments, but would also be based on mathematically sound techniques, as well as on an epistemologically consistent systems of physical ideas.

In fact, it is the conviction of the present author that physical theories should be considered at best tentative until they reach the stage where they strictly conform to demands of scientific objectivity that incorporate established standards of mathematical meaning and correctness, as well as traditional criteria of truth in science. As illustrated by the quotes heading the chapters of these monograph, such standards constitute an essential part of the legacy left to us by the founding fathers of relativity and quantum theory, which in turn is part of an old tradition without which science loses its moral force and raison d'être. It is a tradition rooted in the belief that the criteria for accepting scientific theories should not be based exclusively on momentary popularity and "agreement with observation" (which, as emphasized by Einstein (1949, pp. 21–22), can be always conveniently achieved), but also on such essential ingredients as internal consistency, scope, structural simplicity, unifying potential, mathematical correctness of theoretical conclusions (derived without ad hoc deductive steps), epistemic soundness – in short, that they should reflect a true spirit of scientific objectivity and astuteness. This is a spirit described by Bertrand Russell in compelling words, which we adopt as the underlying motto of the present work:

> "What is number? What are space and time? What is mind, and what is matter? I do not say that we can here and now give definitive answers to all these ancient questions, but I do say that a method has been discovered by which . . . we can make successive approximations to the truth, in which each new stage results from an improvement, not a rejection, of what has gone before.
>
> In the welter of conflicting fanaticisms, one of the few unifying forces is scientific truthfulness, by which I mean the habit of basing our beliefs upon observations and inferences as impersonal, and as much divested of local and temperamental bias, as is possible for human beings."
>
> RUSSELL (1945, pp. 835–836)

Notes

1 The main conclusion of Bohr and Rosenfeld (1950) was that only averages of quantum fields over spacetime regions can be measured, and that the boundaries of those regions cannot be assumed to be sharply delineated, but rather have to be fuzzy in order to avoid infinite fluctuations in field and charge current distributions (Corinaldesi, 1951, 1953). Thus, the Bohr–Rosenfeld studies strongly hinted at the inconsistency of the concept of sharp localizability for relativistic quantum fields. However, the later mathematically precise embodiment of that conclusion (Streater and Wightman, 1964) was in the form of quantum fields as operator-valued distributions over Schwartz spaces that included test functions of arbitrarily small compact supports, and therefore allowed for *precise* localizability within arbitrarily small regions of spacetime.

[2] The reader lacking first-hand exposure to the controversies surrounding this issue might well wonder at the rather widespread passive acceptance of conventional LQED since the inception of the renormalization program, despite obvious inconsistencies and lack of foundational justification which this program had inherited from the underlying LQFT concepts. Commenting on this state of affairs, Dirac has observed: "People are, I believe, too complacent in accepting a theory which contains basic imperfections" (Dirac, 1978a, p. 20). Thus, the reasons for this apparent acceptance belongs to the realm of sociology rather than philosophy of science. Perhaps the sociological causes of such 'complacency' are to be found in a professional climate conditioned (Mitroff, 1974, pp. 65–77; Yaes, 1974) by the high-pressure contemporary manner of promoting fashionable theories, which is more conducive to the production of results that can win rapid establishment approval, rather than to analytic thinking and critical evaluation. As a consequence, the kind of open and meaningful debates over fundamental issues that were so essential to the early development of quantum theory (cf., e.g., the famous Bohr–Einstein debate, as later recounted by Bohr, 1961) do not tend to take place any longer, even when incisive and well-founded criticism, that strikes at the very foundations of fashionable physical ideas and techniques, is raised in print by such luminaries of twentieth-century physics as Dirac (1951, 1965, 1973, 1977, 1978), Born (1949, 1955) or Heisenberg (1976).

[3] See the conference proceedings edited by Blokhintsev *et al.* (1968), and the review article by Kirzhnitz (1969). The monographs of Ingraham (1967) and Efimov (1977) contain many additional references related to their own respective approaches to nonlocal QFT. In addition, the possibility of momentum cutoffs and other nonlocal techniques had been previously considered by Chretien and Peierls (1953, Pauli (1953, 1956), Stueckelberg and Vanders (1954), Landau (1955), Deser (1957), and many others.

[4] In LQFT this ambiguity stems from the basic mathematical fact that the elements of such Hilbert spaces as the spaces $L^2(\mathscr{V}_m^{\pm})$ described in Section 2.1 are not functions, but rather equivalence classes of almost everywhere equal functions (cf. Appendix A), and that we can change the value of any of these functions at any point without leaving its equivalence class, i.e. without changing the quantum state represented by that function. Of course, this mathematical inconsistency is of a superficial nature, and therefore, totally rectifiable by mathematical 'tricks', but it reflects some of the deep physical inconsistencies (cf. Sections 2.1, 2.2, 4.1–4.4) that are inherent in conventional relativistic quantum mechanics, and therefore also in LQFT.

[5] And by the same token, never exactly equal to zero outside B. In theories which routinely exploit analyticity, as is the case in LQFT, the difference between 'arbitrarily small' and 'exactly equal to zero' can be crucial in view of the elementary fact that there are no non-zero complex analytic functions that vanish on open sets R, whereas the opposite is true when only *approximate* equality to zero is required.

[6] If the supports are bounded, our conclusions still remain true for those frames in which the t = const hypersurfaces intersect both supports – as will always be the case when $(ct'', \mathbf{q}'')$ is sufficiently close to the light cone with apex at $(ct', \mathbf{q}')$.

[7] In physical literature, the terms 'invariance' and 'covariance' are often used interchangeably. However, a scalar field is in general not an invariant (i.e. form-independent under Lorentz transformation) except if it is a function of Minkowski products of 4-vectors. Once the distinction is made, it is only covariance, rather than invariance, that is required by the relativity principle.

[8] Indeed, relativistic covariance alone would allow more general expressions, such as the FTQS counterparts of nonlocal theories with invariant cutoffs considered by Ingraham (1957), Efimov (1977) and others.

[9] The restriction to finite spacetime regions is required even in the classical case in order to avoid divergences (Whitham, 1974, Sec. 11.4).

[10] In LQFT, mathematical meaning has to be imparted by *ad hoc* procedures, such as cutoff methods which can be questioned physically as well as mathematically (Segal, 1976). The

cutoff method can be also combined with either time-dependent (Kato and Mugibayashi, 1963) or time-independent (De Witt, 1955) methods [as applied by various authors (Hoegh-Krohn, 1970; Kato and Mugibayashi, 1971; Schroeck, 1971, 1973, 1975; Dimock, 1972; Prugovečki and Manoukian, 1972; Albeverio, 1974) to LQFT] in the context of FTQS (Prugovečki, 1981f), but it remains liable to the same basic objections as in LQFT.

11 For example, the 'collective coordinates' (Jackiw, 1977) that emerge in the process of relating classical kink or soliton solutions to field form factors in LQFT (Goldstone and Jackiw, 1975) can be reinterpreted as the mean stochastic coordinates of quantum spacetime (Prugovečki, 1981c, p. 377).

12 The historical development has been precisely the opposite to the one adopted in Section 5.4, with the Dirac (1928) equation and the Dirac representation discovered first, and with the Foldy–Wouthuysen (1950) transformation and representation discovered much later – although the second is in many ways more natural than the first, as it emerges from Wigner's (1939) classification of all irreducible representations of the Poincaré group $\mathscr{P}$ (interpreted as $\mathscr{T}_4 \wedge SL(2, C)$) for all spins and masses; in turn, in the modern context the latter follows naturally from Mackey's imprimitivity theorem and induced representations theory for $\mathscr{P}$ (Barut and Rączka, 1977, pp. 513–525). The reason for the historical anomaly lies in Dirac's original motivation (discussed in Section 2.1) of solving the localizability problems for spin-½ particles, and the apparent success achieved by deriving the 'probability' current (2.1.33) – which later turned out not to be (cf. Section 2.1) after all a probability current for sharply localized point particles.

13 The proof of this statement, as well as of all the subsequent statements about the relativistic stochastic quantum mechanics of spin-½ particles and antiparticles, can be found in Prugovečki (1980). The present procedure can be generalized to arbitrary spin by using the spinor approach of Weidlich and Mitra (1963). For the nonrelativistic case, a formulation of Galilei-like spinors has been given by Brooke (1978, 1980).

14 This stochastic localizability cannot be made sharp by taking the sharp-point limit, or by any other similar means, as a consequence of the fact that the projectors onto the positive or negative energy states do not commute with the spectral measure of X^j_D, $j = 1, 2, 3$, in (2.1.34).

15 Stochastic 'front localizability' of the photon can be achieved by means of POV measures for phase space Galilei (Prugovečki, 1976c) as well as Poincaré (Prugovečki, 1978e) systems of covariance. However, both these approaches have to be dismissed as physically unrealizable, since there are no massive test particles that are of finite extension in one spatial direction, but not the rest.

16 On first sight, this might appear to preclude the existence of photon–photon scattering in quantum spacetime. It has to be recalled, however, that in order to establish that such scattering has taken place, the photon momenta before and after scattering have to be measured, and for that one requires massive test particles. These arguments do not preclude, however, the possibility of photon proper wave functions determined on theoretical grounds – such as by mass formula arguments of the kind employed by Born (1949b, p. 468).

17 It has already been suggested (Johnson *et al.*, 1967; Adler, 1972) "that a reordering of the [LQED] perturbation series might eliminate ultraviolet divergences for some appropriate value of the bare coupling" (Itzykson and Zuber, 1980, p. 423). However, by elementary calculus, a series that converges conditionally and not absolutely can produce by a suitable rearrangement of terms *any* desired value as its sum. The conjectured (Dyson, 1952) asymptotic convergence of the LQED perturbation series would make this question of arbitrariness even more poignant, since (Zinn-Justin, 1982) partial sums of asymptotic expansions for *actually existing* functions $f(\lambda)$, even when summed up in the given order, do not approach $f(\lambda)$ arbitrarily closely, but only optimally closely for a *conveniently* chosen number of terms. This observation can raise the question of *objective prediction* versus *convenient manipulation* in the minds of astute assessors of LQED successes, and give credence to the possibility, pointed out by Dirac (1978b, p. 5), that "the agreement [of LQED predictions] with observation is presumably a coincidence".

[18] Such as the *ad hoc* addition of epicycles to the Ptolemaic system, the eventually postulated negative weight of phlogiston, the *ad hoc* suppression of transversal waves and the later postulation of vortices in ether, the convenient addition of 'infinite counterterms' (equivalent to an *ad hoc* substraction of corresponding infinite terms) in 'renormalizable' LQFT models, etc.

[19] According to Weinberg (1981, p. 3), "renormalizability might be the key criterion, which also in a more general context would impose a precise kind of simplicity on our theories and help us to pick out the one true physical theory out of the infinite variety of conceivable quantum field theories". It should be noted, however, that Weinberg's criterion of 'simplicity' is actually one of *computational feasibility within the confines imposed by conventional approaches*, and has nothing to do with criteria of simplicity previously used as guidelines in physics – such as the criterion of *structural simplicity* used by Einstein in developing special and general relativity (Einstein, 1949, p. 69). Hence, a somewhat more convincing argument in favor of 'renormalizability' is the one which appeals to the so-called "zero mass decoupling theorem" (Nanopoulos, 1981, p. 15).

[20] In the first paper on RQED (Prugovečki, 1982b) these two renormalization constants were not explicitly displayed, since that approach still involved an *ad hoc* and inconsistent infinite renormalization of the photon propagator, which was later (Prugovečki, 1983) removed by the renormalization of the invariant measure on $\mathscr{V}_\epsilon^+$ presented in the preceding section. In this context, note that a nonzero mass for the photon remains a possibility (Schwinger, 1962) both physically (Itzykson and Zuber, 1980, p. 138) and mathematically.

[21] From the point of view of RQED, ψ, A and e are the actual fields and the true charge, respectively, whereas $\psi^{(l)}$, $A^{(l)}$ and $e^{(l)}$ are their renormalized counterparts. The present terminology is chosen exclusively for the sake of facile comparison with LQED.

Appendix A

Elements of Measure Theory

In this appendix we shall present and explain the most basic concepts and results on measure theory that are essential to the theory of stochastic spaces. Details and proofs can be found in books on measure theory (Halmos, 1950; Ash, 1972), as well as in some of the texts on probability theory (Billingsley, 1979) or on quantum theory (Prugovečki, 1981a).

A *measurable space* $\{\mathscr{X}, \mathscr{A}\}$ consists of a set $\mathscr{X}$, and a Boolean σ-algebra $\mathscr{A}$ of subsets of $\mathscr{X}$. By definition, a family $\mathscr{A}$ of subsets of $\mathscr{X}$ constitutes a Boolean σ-algebra iff: (1) $\mathscr{X} \in \mathscr{A}$; (2) whenever $\Delta \in \mathscr{A}$, then its complement $\Delta' = \mathscr{X} \backslash \Delta$ also belongs to $\mathscr{A}$ (so that, in particular, $\emptyset = \mathscr{X} \backslash \mathscr{X} \in \mathscr{A}$); (3) whenever $\{\Delta_i\}_{i=1}^{\infty}$ is a countable family of sets $\Delta_i \in \mathscr{A}$, then the union $\cup \Delta_i$ of all these sets also belongs to $\mathscr{A}$ (and, therefore, by point (2), the intersection $\cap \Delta_i$ also belongs to $\mathscr{A}$). In case that we are dealing with measurable spaces used in probability theory, $\mathscr{X}$ is called the *sample space*, and the element of $\mathscr{A}$ are called *events*.

A *measure* μ on the measurable space $\{\mathscr{X}, \mathscr{A}\}$ is a function that assigns to each $\Delta \in \mathscr{A}$ either a non-negative number $\mu(\Delta)$ or $+\infty$, in such a manner that $\mu(\emptyset) = 0$, and that

$$\mu(\Delta_1 \cup \Delta_2 \cup \ldots) = \mu(\Delta_1) + \mu(\Delta_2) + \ldots \tag{A.1}$$

for any choice of disjoint sets $\Delta_1, \Delta_2, \ldots \in \mathscr{A}$. The triple $\{\mathscr{X}, \mathscr{A}, \mu\}$ is called a *measure space*. The consistency of the definition of measure is based on the fact that one can set, by definition, $a + \infty = \infty + a = \infty$ if $a \geqslant 0$ or $a = +\infty$, without running into inconsistencies or encountering ambiguities. However, in the case of a *complex measure* μ, (A.1) is still retained, but the values $\mu(\Delta)$ have to be restricted to complex numbers, since the addition of such infinities would lead to ambiguities – e.g. if $a = -\infty$, so that we are faced with the consistently undefineable value of $\infty - \infty$.

A measure μ is called *finite* if $\mu(\mathscr{X}) < \infty$; it is called *$\sigma$-finite* if $\mathscr{X}$ is equal to countable union $\cup_{k=1}^{\infty} S_k$ of sets of finite measure $\mu(S_k)$. By definition, a *probability measure* is a finite measure for which $\mu(\mathscr{X}) = 1$, and for which $\mu(\Delta)$ represents the probability of the event $\Delta \in \mathscr{A}$, i.e. the probability of observing a sample value $x \in \mathscr{X}$ that belongs to Δ. The simplest type of probability measure on a probabilistic measurable space $\{\mathscr{X}, \mathscr{A}\}$ is the δ-measure μ_{δ_x} at each $x \in \mathscr{X}$, defined as follows for all $\Delta \in \mathscr{A}$:

$$\mu_{\delta_x}(\Delta) = \begin{cases} 0 & \text{if } x \notin \Delta, \\ 1 & \text{if } x \in \Delta. \end{cases} \tag{A.2}$$

A *topological space* $\{\mathscr{X}, \mathscr{T}\}$ consists of a set $\mathscr{X}$, and a family $\mathscr{T}$ of subsets S of $\mathscr{X}$ such that: (1) the empty set $\emptyset$ as well as $\mathscr{X}$ belong to $\mathscr{T}$; (2) if $\{S_i \mid i \in I\}$ is *any* family of sets $S_i \in \mathscr{T}$, then their union $\cup S_i$ also belongs to $\mathscr{T}$; (3) if $\{S_i\}_{i=1}^n$ is any *finite* family of sets in $\mathscr{T}$, then their intersection $\cap S_i$ also belongs to $\mathscr{T}$. The elements of $\mathscr{T}$ are called *open sets*; by definition, a subset T of $\mathscr{X}$ is a *closed set* in the topology provided by $\mathscr{T}$ iff it equals the complement $S' = \mathscr{X} \backslash S$ of an open set S. An *open neighborhood* of a point $x \in \mathscr{X}$ is any set $\mathscr{N}_x \in \mathscr{T}$ that contain x, and a *neighborhood* of x is any subset of $\mathscr{X}$ that contains an open neighborhood of x. A subset C of $\mathscr{X}$ is called *compact* iff every open covering $\{S_i \mid i \in I\}$ of C (i.e. $S_i \in \mathscr{T}$ for all $i \in I$, and $\cup S_i \supset C$) has a finite subcovering $\{S_{i_1}, \ldots, S_{i_n}\}$ (i.e. $S_{i_1} \cup \ldots \cup S_{i_n}$ for some choice $i_1, \ldots, i_n \in I$ contains C).

A very general amalgamation (Ash, 1972) of topological and measure-theoretical structures on one and the same set $\mathscr{X}$ is achieved by starting from a family $\mathscr{T}$ of open sets, and then considering the Boolean σ-algebra $\mathscr{B} = \mathscr{A}(\mathscr{T})$ *generated* by $\mathscr{T}$, i.e. the smallest Boolean σ-algebra $\mathscr{A}(\mathscr{T})$ that contains the family $\mathscr{T}$. The elements of $\mathscr{B}$ are called *Borel sets* (in the given topology $\mathscr{T}$).

For us, the most important case is that when $\mathscr{X}$ equals the n-dimensional real Euclidean space $\mathbb{R}^n$, with the topology supplied by the Euclidean metric in $\mathbb{R}^n$. Then $S \in \mathscr{A}(\mathbb{R}^n)$ – i.e. S is an open set in $\mathbb{R}^n$ – iff for any $x \in S$ there is a sphere centered at x whose interior lies entirely within S. It turns out that in this Euclidean topology a subset C of $\mathbb{R}^n$ is compact iff it is bounded and closed. The family $\mathscr{B}^n$ of all Borel sets in $\mathbb{R}^n$, generated by this Euclidean topology, is extremely rich: it contains all the open sets, all the closed sets, all conceivable intervals in $\mathbb{R}^n$, all finite sets, as well as other more exotic kinds of subsets of $\mathbb{R}^n$ – so much so that it is a rather nontrivial task to construct examples (Ash, 1972) of subsets of $\mathbb{R}^n$ which are *not* Borel sets. In this monograph we shall denote by $\mathscr{B}^n$ the family of all subsets of $\mathbb{R}^n$ that are Borel sets in the Euclidean topology on $\mathbb{R}^n$, and whenever we shall say that B is a Borel set in $\mathbb{R}^n$, we shall mean that B belongs to $\mathscr{B}^n$.

Whereas the concept of topology on a set $\mathscr{X}$ is required in order to give a precise and yet most general meaning to the concept of continuity of functions $f(x)$ on $\mathscr{X}$, that of measure space is needed to impart a general meaning to the concept of integral $\int f \, d\mu$ of f over *measurable sets* Δ in $\mathscr{X}$ (i.e. subsets Δ such that $\Delta \in \mathscr{A}$). The procedure for such a general construction of integrals starts by setting

$$\int_\Delta s(x) \, d\mu(x) \overset{\text{def}}{=} \sum_{i=1}^{n} a_i \mu(\Delta_i \cap \Delta) \tag{A.3}$$

for so-called *simple functions*. These are functions of the form

$$s(x) = \sum_{i=1}^{n} a_i \chi_{\Delta_i}(x), \quad \Delta_i \cap \Delta_j = \emptyset, \; i \neq j, \tag{A.4}$$

where $\Delta_i \in \mathscr{A}$, $i = 1, \ldots, n$, and χ_S denotes the *characteristic function* of any subset S of $\mathscr{X}$, i.e. $\chi_S(x) = 1$ if $x \in S$ and $\chi_S(x) = 0$ if $x \notin S$. The definition then progresses in several stages (Prugovečki, 1981a, pp. 80–94) to the entire family of *measurable functions*, i.e. real functions for which $f^{-1}(I) \in \mathscr{A}$ for any interval $I \subset \mathbb{R}^1$, or complex functions whose real and imaginary parts are real measurable functions. This family is extremely rich, and in case $\mathscr{A}$ consists of Borel sets, it incorporates all piecewise continuous functions. The details of this construction of integrals are not that important from the point of view of the applications considered in this monograph, beyond the observations that a measurable function $f(x)$ is μ-integrable on Δ iff $|f(x)|$ is μ-integrable on Δ, so that both sides of the generally valid inequality

$$\left| \int_\Delta f(x) \, d\mu(x) \right| \leqslant \int_\Delta |f(x)| \, d\mu(x), \tag{A.5}$$

are defined whenever f is μ-integrable on Δ, and that in that case there are sequences $s_1(x), s_2(x), \ldots$ of simple functions such that

$$\int_\Delta f(x) \, d\mu(x) = \lim_{i\to\infty} \int_\Delta s_i(x) \, d\mu(x). \tag{A.6}$$

The key theorem resulting from this kind of definition of integrals is presented below (cf., e.g., Prugovečki, 1981a, p. 287, for its proof), and in it the important concept of 'almost everywhere on a measure space $\{\mathscr{X}, \mathscr{A}, \mu\}$' makes its appearance: a relationship depending on $x \in \mathscr{X}$ is true *almost everywhere* in $\Delta \in \mathscr{A}$ iff it is true for all x from a subset Δ_0 of Δ, which is such that $\mu(\Delta \backslash \Delta_0) = 0$.

LEBESGUE'S DOMINANTED CONVERGENCE THEOREM. *Suppose $f_1(x), f_2(x), \ldots$ is a sequence of functions that are integrable on some $\Delta \in \mathscr{A}$, and that there is a function $g(x)$ that is integrable on Δ, and such that $|f_n(x)| \leqslant g(x)$, $x \in \Delta$, for $n = 1, 2, \ldots$. If*

$$f(x) = \lim_{n\to\infty} f_n(x) \tag{A.7}$$

almost everywhere in Δ, then $|f(x)|$ is integrable on Δ and

$$\lim_{n\to\infty} \int_\Delta f_n(x) \, d\mu(x) = \int_\Delta f(x) \, d\mu(x). \tag{A.8}$$

The above theorem turns out to be a powerful tool in many applications. For its validity the existence of a function $g(x)$ with the required properties is very essential. For example, if $\{f_n\}_{n=1}^{\infty}$ is a δ-sequence on $\mathbb{R}^1$, so that $f_n(x) \to 0$ for $x \neq 0$, but at the same time

$$\int_{\mathbb{R}^1} f_n(x)\, dx = 1, \quad n = 1, 2, \ldots, \tag{A.9}$$

then we have that $\lim f_n(x) = 0$ almost everywhere with respect to the Lebesgue measure on $\mathbb{R}^1$, and yet by (A.9) the integrals $\int f_n\, dx$ do not converge to zero.

The Lebesgue measure on $\mathbb{R}^n$ is the measure that extends the theory of Riemann integrals on intervals on $\mathbb{R}^n$ to Lebesgue integrals

$$\int_{\Delta} f(x)\, dx, \quad \Delta \in \mathscr{B}^n, \tag{A.10}$$

over arbitrary Borel sets Δ in $\mathbb{R}^n$ for the class of *Lebesgue integrable* functions $f(x)$ – which incorporates Riemann (absolutely) integrable functions as a proper subset, and for which the Riemann integral coincides with the Lebesgue integral when the former exists in an absolute sense. The definition of the *Lebesgue measure* on $\mathbb{R}^1$ proceeds by assigning to each (open, closed or semiclosed) interval from $a \in \mathbb{R}^1$ to $b > a$ its length $b - a$, and then extending the definition in a natural manner (Prugovečki, 1981a, p. 78) to all Borel sets $\Delta \in \mathscr{B}^1$. The definition of the Lebesgue measure in $\mathbb{R}^n$ can be carried out in the same manner, by starting with the Lebesgue measure of n-dimensional intervals:

$$\mu_{\mathscr{L}}([a_1, b_1] \times \ldots \times [a_n, b_n]) \stackrel{\text{def}}{=} \prod_{i=1}^{n} (b_i - a_i). \tag{A.11}$$

Alternatively, one can first define the concept of Cartesian product of measure spaces, and then define the Lebesgue measure in n-dimensions as the Cartesian product of n one-dimensional Lebesgue measures.

The *Cartesian product* of two measure spaces $\{\mathscr{X}_i, \mathscr{A}_i, \mu_i\}$, $i = 1, 2$, is defined as the measure space $\{\mathscr{X}, \mathscr{A}, \mu\}$ in which $\mathscr{X} = \mathscr{X}_1 \times \mathscr{X}_2$ (i.e. $\mathscr{X}$ is the Cartesian production of the sets $\mathscr{X}_1$ and $\mathscr{X}_2$, and as such it consists of all ordered pairs $(x_1, x_2) = x$, with $x_1 \in \mathscr{X}_1$ and $x_2 \in \mathscr{X}_2$), $\mathscr{A}$ is the Boolean σ-algebra generated by the family $\mathscr{A}_1 \times \mathscr{A}_2$ of subsets of $\mathscr{X}$, and μ is the measure on $\mathscr{A}$ uniquely determined the requirement that

$$\mu(\Delta_1 \times \Delta_2) \stackrel{\text{def}}{=} \mu_1(\Delta_1)\mu_2(\Delta_2), \quad \forall \Delta_i \in \mathscr{A}_i,\ i = 1, 2, \tag{A.12}$$

where the symbol $\forall$ signifies 'for all'. The definition of the Cartesian product of n measure spaces proceeds in a totally analogous manner. The key theorem in the theory of integrals with respect to Cartesian products of measures is given below (cf. e.g., Prugovečki, 1981a, p. 96, for proof), which finds frequent applications in this monograph.

FUBINI'S THEOREM. *Suppose* $\{\mathscr{X}, \mathscr{A}, \mu\}$ *is the Cartesian product of two measure space* $\{\mathscr{X}_i, \mathscr{A}_i, \mu_i\}$, $i = 1, 2$, *and that* μ *is* σ*-finite. If* $f(x_1, x_2)$ *is integrable on a set* $\Delta \in \mathscr{A}$, *then we have*

$$\int_\Delta f(x_1, x_2)\, d\mu(x_1, x_2) = \int_{\Delta_1} d\mu_1(x_1) \int_{\Delta_{x_1}} f(x_1, x_2)\, d\mu_2(x_2)$$

$$= \int_{\Delta_2} d\mu_2(x_2) \int_{\Delta_{x_2}} f(x_1, x_2)\, d\mu_1(x_1) \tag{A.13}$$

where the above integrals over Δ_{x_i}, $i = 1, 2$,

$$\Delta_{x_1} = \{x_2 \mid (x_1, x_2) \in \Delta\},\ \Delta_1 = \{x_1 \mid (\{x_1\} \times \mathscr{X}_2) \cap \Delta \neq \emptyset\}, \tag{A.14}$$

$$\Delta_{x_2} = \{x_1 \mid (x_1, x_2) \in \Delta\},\ \Delta_2 = \{x_2 \mid (\mathscr{X}_1 \times \{x_2\}) \cap \Delta \neq \emptyset\}, \tag{A.15}$$

exist, and define functions that are integrable over Δ_i *with respect to* μ_i, $i = 1, 2$.

Sometimes, when one tries to interchange orders of integration, one is faced with the second or third form of integral in (A.13), rather than the first form. In that case one has to apply the following theorem (Prugovečki, 1981a, p. 98) before one can use Fubini's theorem.

TONELLI'S THEOREM. *If* Δ_1 *and* Δ_{x_1} *are as in* (A.14) *and*

$$\int_{\Delta_1} d\mu_1(x_1) \int_{\Delta_{x_1}} |f(x_1, x_2)|\, d\mu_2(x_2) \tag{A.16}$$

exists (so that, implicitly, the integral over Δ_{x_1} *exists for* μ_1*-almost all* $x \in \Delta_1$), *then* $f(x_1, x_2)$ *is* $\mu_1 \times \mu_2$*-integrable over* Δ.

In applying the above theorem, one has to recall that in measure theory integrability implies and is implied by absolute integrability. Thus, in general, if we consider the set $L^1_\mu(\mathscr{X})$ (sometimes denoted by $L^1(\mathscr{X}, \mu)$) of all functions $f(x)$ that are integrable on some measure space $\{\mathscr{X}, \mathscr{A}, \mu\}$, this set can be also characterized as the set of all functions f for which

$$\|f\|_1 \overset{\text{def}}{=} \int_{\mathscr{X}} |f(x)|\, d\mu(x) \tag{A.17}$$

exists. The nonnegative numbers $\|f\|_1$ play the role of norms, and the set $L^1_\mu(\mathscr{X})$ (in which any two almost everywhere equal functions are identified) is a complete normed space, i.e. a Banach space. It is not, however, a Hilbert space.

To extract a Hilbert space from all the functions that are measurable on the measure space $\{\mathscr{X}, \mathscr{A}, \mu\}$, we have to consider those functions which are square-integrable on $\mathscr{X}$, i.e. for which

$$\|f\|^2 = \int_{\mathscr{X}} |f(x)|^2\, d\mu(x) \tag{A.18}$$

exists. The above numbers $\|f\|$ have the properties of norms that correspond to the inner products

$$\langle f_1 | f_2 \rangle \stackrel{\text{def}}{=} \int_{\mathscr{X}} f_1^*(x) f_2(x) \, d\mu(x), \tag{A.19}$$

provided that we identify any two functions that are equal almost everywhere on $\mathscr{X}$, i.e. work with the set $L^2_\mu(\mathscr{X})$ of equivalence classes of almost everywhere equal square-integrable functions (so that $\|f\| = 0$ unambiguously characterizes the equivalence class of those functions which are almost everywhere equal to zero).

In the present monograph the cases of paramount importance correspond to $\mathscr{X}$ equaling a Borel subset of $\mathbb{R}^n$ (such as $\mathscr{X} = \mathbb{R}^3$, $\mathscr{X} = \mathbb{R}^6 = \Gamma$, $\mathscr{X} = \mathscr{V}^+_m$, etc.), to $\mathscr{A}$ equaling the family of all Borel sets in $\mathbb{R}^n$ that are subsets of $\mathscr{X}$, and to μ finite or σ-finite on $\mathscr{X}$. Then $L^2_\mu(\mathscr{X})$ is a separable Hilbert space (Prugovečki, 1981a, pp. 101–114) – i.e. it is a vector space of (equivalence classes of) functions with inner product (A.19) – which is complete in the sense that if $\|f_m - f_n\| \to 0$, $m, n \to 0$, then $f_1, f_2, \ldots$ converges to a function $f \in L^2_\mu(\mathscr{X})$, and it is separable in the sense that it possesses dense (cf. Appendix B) and countable subsets. We shall also encounter the situation that $\mathscr{X} = \mathscr{X}_1 \times \mathscr{X}_2$ and $\mu = \mu_1 \times \mu_2$, in which case we shall carry out the identification

$$L^2_{\mu_1 \times \mu_2}(\mathscr{X}_1 \times \mathscr{X}_2) \cong L^2_{\mu_1}(\mathscr{X}_1) \otimes L^2_{\mu_2}(\mathscr{X}_2), \tag{A.20}$$

by identifying $f_1 \otimes f_2$, $f_i \in L^2_{\mu_i}(\mathscr{X}_i)$, with $f_1 \cdot f_2 \in L^2_\mu(\mathscr{X})$ represented by the product $f_1(x_1) f_2(x_2)$.

A special but very important case of $L^2_\mu(\mathscr{X})$ occurs when $\mathscr{X} = \mathbb{R}^n$ and μ is the Lebesgue measure on $\mathbb{R}^n$. In that case the resulting Hilbert space is customarily denoted by $L^2(\mathbb{R}^n)$, so that (A.20) assumes the form

$$L^2(\mathbb{R}^{m+n}) \cong L^2(\mathbb{R}^m) \otimes L^2(\mathbb{R}^n). \tag{A.21}$$

Another important case occurs when μ is the δ-measure μ_{δ_x} at x, that is

$$\mu_{\delta_x}(\{x\}) = 1, \quad \mu(\mathbb{R}^n \backslash \{x\}) = 0, \tag{A.22}$$

or is the superposition of δ-measures at $x_1, x_2, \ldots \in \mathbb{R}^n$, so that

$$\mu(\{x_i\}) = 1, \quad \mu(\mathbb{R}^n \backslash \{x_1, x_2, \ldots\}) = 0. \tag{A.22}$$

In the later, general case, (A.19) assumes the form

$$\langle f_1 | f_2 \rangle = \sum_i f_1^*(x_i) f_2(x_i), \tag{A.24}$$

so that $L^2_\mu(\mathbb{R}^n)$ can be identified with an l^2-space of one-column matrices.

It should be noted that although a δ-measure has much in common with a δ-function in the sense of distributions or generalized functions (Gel′fand and Shilov, 1964), the two concepts are mathematically very distinct. In fact, any measure can be regarded as a functional

$$\mu(f) = \int_{\mathcal{X}} f(x)\, \mathrm{d}\mu(x) \tag{A.25}$$

on the space $L^1_\mu(\mathcal{X})$, whereas a generalized function is also a functional, but on spaces of functions (so-called test functions) which do not constitute Banach spaces, but rather belong to the more general class of countably normed spaces (Gel′fand and Vilenkin, 1968). The formal and heuristic manipulations of δ-'functions' in typical quantum-mechanical textbooks and physics papers can be mathematically justified by sometimes interpreting the δ-symbol as a generalized function (on an appropriate space of test functions, such as the Schwartz $\mathscr{S}$-space), and at other times as a δ-measure. For example, it was found out that in the mathematically rigorous development of the foundation of nonrelativistic quantum scattering theory (Prugovečki, 1981a, Ch. V), the δ-symbol sometimes requires the interpretation of δ-measure, and at other times it can be altogether avoided by simple limiting procedures, which do not use the topologies of the theory of generalized functions. The same remains true to a large extent of the use of the δ-symbol in the present monograph.

A measure μ is *absolutely continuous* with respect to another measure ν over the same measurable space $\{\mathcal{X}, \mathscr{A}\}$ if $\mu(\Delta) = 0$ for any $\Delta \in \mathscr{A}$ such that $\nu(\Delta) = 0$. It turns out (cf., e.g., Ash, 1972, p. 63) that μ is absolutely continuous with respect to a σ-finite measure ν if and only if there is a nonnegative measurable function $\rho(x), x \in \mathcal{X}$, such that

$$\mu(\Delta) = \int_{\Delta} \rho(x)\, \mathrm{d}\nu(x), \qquad \Delta \in \mathscr{A}, \tag{A.26}$$

for any measurable set Δ of finite μ measure. The function $\rho(x)$ with these properties is ν-almost everywhere unique, and it is called the *Radon–Nikodym derivative* $\mathrm{d}\mu/\mathrm{d}\nu$ of μ with respect to ν. For example, the confidence functions in (3.2.9) are Radon–Nikodym derivatives of the measures μ_q' and μ_q'' with respect to the Lebesgue measure on $\mathbb{R}^N$. On the other hand, no δ-measure in $\mathbb{R}^N$ is absolutely continuous with respect to the Lebesgue measure on $\mathbb{R}^N$.

Appendix B

Elements of Operator Theory on Hilbert Spaces

The elementary introductions to Hilbert space theory presented in standard textbooks on (nonrelativistic and/or relativistic) quantum mechanics suffer from one basic deficiency: they totally ignore the fact that since almost all nontrivial quantum mechanical models are formulated on infinite-dimensional (separable) Hilbert spaces, the naive matrix-algebra type of formal manipulations which they tend to advocate are often demonstrably mathematically inconsistent, since the theory of these spaces displays many features that are not present in the finite-dimensional case. For example, the impression[1] is left that such basic quantum mechanical operators as Hamiltonians (e.g. (1.2.16)), position and momentum operators (e.g. (1.2.2) and (1.2.12)), etc., are definable on the entire Hilbert space $\mathscr{H}$ (e.g. $L^2(\mathbb{R}^3)$) of state-vectors for a given quantum-mechanical model, whereas standard functional analysis, as presented in textbooks on that subject as well as in several more mathematically oriented textbooks on quantum theory (von Neumann, 1955; Glimm and Jaffe, 1981, Prugovečki, 1981a; Schechter, 1981), has to cope with the incontestable mathematical fact that precisely the opposite is *true*. Indeed, the impossibility of *consistently* extending such symmetric operators to the entire Hilbert space is due to the *Hellinger–Toeplitz theorem*, which states:[2] *Any symmetric operator A defined on the entire Hilbert space $\mathscr{H}$ is bounded.*

The *bound* of a linear operator A with domain of definition $\mathscr{D}_A$ is, by definition, equal to the supremum (i.e. least upper bound) $\|A\|$ of

$$\|Af\| \stackrel{\text{def}}{=} (\langle Af | Af \rangle)^{1/2} \tag{B.1}$$

for normalized vectors $f \in \mathscr{D}_A$. If $\|A\|$ is finite, A is said to be *bounded*; if $\|A\| = \infty$, then A is called *unbounded*. All the aforementioned operators representing quantum observables are demonstrably unbounded (Prugovečki, 1981a, pp. 180, 181, 224, 342, 358, 368); they are also symmetric (i.e. loosely speaking, 'Hermitian'), in the sense that a linear operator A is said to be *symmetric* iff $\langle Af|g\rangle = \langle f|Ag\rangle$ for all $f, g \in \mathscr{D}_A$. Consequently, the Hellinger–Toeplitz theorem applies to all the aforementioned quantum-mechanical operators, usually defined as differential operators which can be extended (in a unique manner, if their domains are judiciously chosen) to self-adjoint operators, defined on maximal domains of symmetry.

These domains are dense in $\mathscr{H}$, but never equal to $\mathscr{H}$. In this context, it should be recalled that a set $S \subset \mathscr{H}$ is *dense* in $\mathscr{H}$ iff, given any vector $f \in \mathscr{H}$, there is a sequence $f_1, f_2, \ldots \in S$ such that $\|f - f_n\| \to 0$ as $n \to \infty$.

In order to concentrate on essential concepts and results, no attempt has been made in this monograph to deal with domain questions, or with similar technical aspects. It is hoped, however, that this appendix will clarify the basic issues, and that by providing the interested reader with specific instructions as to where details can be found, it will enable him to deal on his own with those questions of technique not discussed in this monograph.

One such question is whether all symmetric (i.e. 'Hermitian') operators A are self-adjoint, i.e. whether the equality $A = A^*$ is automatically true for symmetric operators, and therefore, implicitly, that $\mathscr{D}_A = \mathscr{D}_{A^*}$. In fact, not only is the answer to this question negative, but the standard theory of Cayley transforms of symmetric operators (cf., e.g., Prugovečki, 1981a, pp. 219–225) shows that most symmetric operators cannot even be extended to self-adjoint operators – in the sense that B is called on *extension* of A (or A a *restriction* of B) iff $\mathscr{D}_B \supset \mathscr{D}_A$ and $Bf = Af$ for all $f \in \mathscr{D}_A$. In this context, it has to be realized that the *adjoint* A^* of any densely defined linear operator A is uniquely determined (Prugovečki, 1981a, p. 187) by the equations

$$\langle g^\dagger | f\rangle = \langle g | Af\rangle, \; f \in \mathscr{D}_A, \tag{B.2}$$

as equal to the mapping $M: g \mapsto g^\dagger$ with domain $\mathscr{D}_M$ consisting of all $g \in \mathscr{H}$ for which (B.2) has a solution $g^\dagger$ valid for all $f \in \mathscr{D}_A$.

The simplest type of operators are the projectors – of which the zero operator $\mathbb{O}$ and the identity operator $\mathbb{1}$ are special cases. A linear operator E is a *projector* iff it has domain $\mathscr{D}_E = \mathscr{H}$ and satisfies the equalities:

$$E = E^* = E^2. \tag{B.3}$$

It turns out (Prugovečki, 1981a, p. 200) that (B.3) is satisfied iff E is the orthogonal projection operator onto some closed subspace $\mathbf{M}$ of $\mathscr{H}$, i.e. iff $Ef \in \mathbf{M}$ and $(f - Ef) \perp \mathbf{M}$ for all $f \in \mathbf{M}$. It should be recalled that a *closed subspace* of $\mathscr{H}$ is any linear subset $\mathbf{M}$ of $\mathscr{H}$ such that if $f_1, f_2, \ldots$ is any sequence of vectors belonging to $\mathbf{M}$ and $\|f_m - f_n\| \to 0$ as $m, n \to \infty$, then its limit f in the norm,

$$f = \operatorname*{s-lim}_{n\to\infty} f_n \Leftrightarrow \|f - f_n\| \to 0, \quad n \to \infty, \tag{B.4}$$

belongs to $\mathbf{M}$. It should be noted that the vector f, also called the *strong limit* of $\{f_n\}_{n=1}^{\infty}$, exists (due to the completeness of $\mathscr{H}$) and is unique. If f_λ depends on a continuous family of variables, such as is the case when $\lambda = \zeta \in \Gamma$, then f_λ is called *norm-continuous* iff s-lim $f_\lambda = f_{\lambda'}$ as $\lambda \to \lambda'$ at all λ' at which f_λ is defined.

In addition to strong limits, we also require in this monograph *weak limits*, defined by

$$f = \operatorname*{w-lim}_{n\to\infty} f_n \Leftrightarrow \langle g \,|\, f - f_n\rangle \to 0, \quad \forall g \in \mathscr{H}, \tag{B.5}$$

i.e. by $\langle g|f\rangle = \lim \langle g|f_n\rangle$ for all $g \in \mathscr{H}$. If $f_1, f_2, \ldots$ converges strongly to f_n, it also converges weakly to f, but the converse is by no means true, as witnessed by the case of an orthonormal system $\{e_n\}_{n=1}$, for which $\langle g|e_n\rangle \to 0$ (Prugovečki, 1981a, p. 38) for any $g \in \mathscr{H}$, but for which $\|e_n\| = 1$. On the other hand, we have (Prugovečki, 1981a, p. 334):

$$f = \text{w-lim}\, f_n \;\&\; \|f_n\| \leqslant \|f\| \Rightarrow f = \text{s-lim}\, f_n. \tag{B.6}$$

For sequences $A_1, A_2, \ldots$ of bounded operators with $\mathscr{D}_{A_n} = \mathscr{H}$ we similarly define:

$$A = \text{w-lim}\, A_n \Leftrightarrow Af = \text{w-lim}\, A_n f, \qquad \forall f \in \mathscr{H} \tag{B.7}$$

$$A = \text{s-lim}\, A_n \Leftrightarrow Af = \text{s-lim}\, A_n f, \qquad \forall f \in \mathscr{H} \tag{B.8}$$

$$A = \text{u-lim}\, A_n \Leftrightarrow \|A - A_n\| \to 0. \tag{B.9}$$

Clearly, (B.8) implies (B.7), but the opposite is not true. Similarly, (B.9) implies (B.8), since by the definition of operator bounds

$$\|(A - A_n)f\| \leqslant \|A - A_n\|\, \|f\|, \tag{B.10}$$

but the opposite is also not true, as witnessed by the definition

$$E = \sum_{k=1}^{\infty} E_k \stackrel{\text{def}}{=} \underset{n\to\infty}{\text{s-lim}} \sum_{k=1}^{n} E_k \tag{B.11}$$

of an infinite sum of mutually orthogonal nonzero projectors $E_1, E_2, \ldots$ (i.e. $E_iE_j = 0$ for $i \neq j$, but $E_i \neq 0$), in which the replacement of s-lim by u-lim would lead to a mathematical inconsistency (Prugovečki, 1981a, p. 230)

In addition to projectors, isometries constitute another physically and mathematically important class of bounded operators on Hilbert spaces. An operator $U: \mathscr{H}_1 \to \mathscr{H}_2$, with domain $\mathscr{D}_U = \mathscr{H}_1$, is an *isometry* iff $\|Uf\|_2 = \|f\|_1$ (where $\|\cdot\|_i$ denotes the norm in the Hilbert space $\mathscr{H}_i$) or, equivalently (Prugovečki, 1981a, p. 212), iff it preserves inner products; U is *unitary* if in addition its range

$$\mathscr{R}_U = \{Uf \mid f \in \mathscr{D}_U\} \tag{B.12}$$

equals $\mathscr{H}_2$. However, if U is an isometry from $\mathscr{H}_1$ into $\mathscr{H}_2$, without being unitary, its range $\mathscr{R}_U$ is still a closed subspace of $\mathscr{H}_2$, and therefore $\mathscr{R}_U$ is a Hilbert space $\mathscr{H}_3$ in its own right. Hence the mapping $U' : \mathscr{H}_1 \to \mathscr{H}_3 = \mathscr{R}_U$ is a unitary operator from $\mathscr{H}_1$ onto $\mathscr{H}_3$, which therefore has an inverse U'^{-1} defined on all of $\mathscr{H}_3$.

We note that if U is an isometry on $\mathscr{H}$ (i.e. if $\mathscr{H}_1 = \mathscr{H}_2 = \mathscr{H}$), we have $\mathscr{D}_U = \mathscr{H}$, but if U is unitary we also have $\mathscr{R}_U = \mathscr{H}$. Hence, for isometries on $\mathscr{H}$ we have only $U^*U = \mathbb{1}$, whereas for unitary operators on $\mathscr{H}$ we have the further-reaching relations

$$U^*U = UU^* = \mathbb{1}, \tag{B.13}$$

so that U is unitary iff $U^* = U^{-1}$ (Prugovečki, 1981a, p. 214). Textbooks on quantum mechanics leave the impression that all isometries are unitary operators, but that is the case *only* for the case of finite-dimensional Hilbert spaces, which play only a very minor role in quantum theory. In case that dim $\mathscr{H} = \infty$ there are (Prugovečki, 1981a, p. 213) in a sense more isometric operators that are *not* unitary, than there are unitary ones. The earlier mentioned theory of Cayley transforms

$$C_A = (A - i)(A + i)^{-1} \tag{B.14}$$

of symmetric operators A establishes that C_A is an isometry from $\mathscr{R}_{A+i}$ into $\mathscr{R}_{A-i}$, that A has a self-adjoint extension iff

$$\dim \mathscr{R}^{\perp}_{A-i} = \dim \mathscr{R}^{\perp}_{A+i}, \tag{B.15}$$

and that, in particular, A is self-adjoint iff C_A is unitary.

The *spectral theorem* for self-adjoint operators (Prugovečki, 1981a, p. 250) assigns to each $A = A^*$ a unique PV-measure (cf. (1.2.20), and the related definition) $E^A(B)$ on the Borel sets $B \in \mathscr{B}^1$ in $\mathbb{R}^1$, so that

$$A = \int_{\mathbb{R}^1} \lambda \, dE^A_\lambda . \tag{B.16}$$

$E^A(B)$ is called the *spectral measure* of A, and the symbolic relation (B.16) signifies in detail that

$$\langle f \mid Ag \rangle = \int_{\mathbb{R}^1} \lambda \, d\langle f \mid E^A_\lambda g \rangle, \quad \forall f \in \mathscr{H}, \ \forall g \in \mathscr{D}_A, \tag{B.17}$$

where the integration is carried out with respect to the complex measure

$$\mu_{f,g}(B) = \langle f \mid E^A(B) g \rangle. \tag{B.18}$$

If a set $\{A_1, \ldots, A_n\}$ of commuting[3] self-adjoint operators is given, then a unique PV-measure $E^{A_1, \ldots, A_n}(B)$ on the Borel sets $B \in \mathscr{B}^n$ in $\mathbb{R}^n$ can be assigned to them by requesting that

$$E^{A_1, \ldots, A_n}(B_1 \times \ldots \times B_n) = E^{A_1}(B_1) \ldots E^{A_n}(B_n) \tag{B.19}$$

for all $B_1, \ldots, B_n \in \mathscr{B}^1$. $E^{A_1, \ldots, A_n}$ is called the *joint spectral measure* of $A_1, \ldots, A_n$. In principle, all the physical information contained in conventional nonrelativistic quantum mechanical models can be extracted from the probability measures

$$P_\psi^{A_1, \ldots, A_n}(B) = \langle \psi | E^{A_1, \ldots, A_n}(B) \psi \rangle, \qquad \| \psi \| = 1, \tag{B.20}$$

corresponding to all of the above expectation values of all joint spectral measures of all sets of compatible observables. This observation provided the first step in the step-by-step extrapolation from conventional to stochastic quantum mechanics.

The second step was provided by the fact that in practice this information is related to complete sets $\{A_1, \ldots, A_n\}$ of observables, which are represented by infinitesimal generators of groups of physical operations. The following is a general and yet mathematically rigorous definition of such complete sets – although several equivalent and equally useful criteria exist (Prugovečki, 1981a, pp. 315–318): $\{A_1, \ldots, A_n\}$ is a *complete set* of self-adjoint operators in $\mathscr{H}$ iff there is a measure μ on $\mathscr{B}^n$, and a unitary operator $U: \mathscr{H} \to L^2_\mu(\mathbb{R}^N)$, such that

$$(UA_k \psi)(y) = y_k (U\psi)(y), \quad k = 1, \ldots, n, \tag{B.21}$$

for μ-almost all $y = (y_1, \ldots, y_n) \in \mathbb{R}^n$. The Hilbert space $L^2_\mu(\mathbb{R}^n)$ is called a *spectral representation space* for $\{A_1, \ldots, A_n\}$.

The above definition of a complete set of observables avoids the inconsistencies of the typical quantum mechanics textbook definition in terms of 'ket-vectors' $|y\rangle$ which 'satisfy'

$$A_k |y\rangle = y_k |y\rangle, \quad k = 1, \ldots, n, \tag{B.22}$$

albeit they are not elements of the Hilbert space $\mathscr{H}$ (in which A_k is defined) whenever y_k belongs to the continuous part of the spectrum of A_k. Of course, by using such devices as equipped Hilbert space, one can extend A_k to larger Hilbert space $\mathscr{H}^\dagger$ into which $|y\rangle$ in (B.22) can be incorporated (Prugovečki, 1973b). A mathematically more straightforward and physically more relevent approach (Prugovečki, 1981a, pp. 491–501) is arrived at, however, when $\mathscr{H}$ is a spectral representation space $L^2_\nu(\mathbb{R}^s)$ for some $\{B_1, \ldots, B_s\}$, so that $\mathscr{H}$ consists of functions $\psi(x)$, and $|y\rangle$ is regarded as a representative Φ_y of an eigenfunction expansion in terms of which the action of U in (B.21) can be written as

$$(U\psi)(y) = \text{l.i.m.} \int \Phi_y^*(x) \psi(x) \, d\nu(x), \tag{B.23}$$

where l.i.m. stands for 'limit-in-the-mean'.

To elucidate these ideas with an example related to the considerations in Section 1.2, let us start from the configuration representation (1.2.2), i.e. let us work in the spectral representation space $\mathscr{H}_{\text{conf}} = L^2(\mathbb{R}^3)$ of the complete set $\{X^1, X^2, X^3\}$ of the position operators of a spinless quantum particle. Upon introducing the 'plane waves'

$$\Phi_{\mathbf{k}}(\mathbf{x}) = (2\pi\hbar)^{-3/2} \exp\left(\frac{i}{\hbar}\, \mathbf{x}\cdot\mathbf{k}\right), \tag{B.24}$$

we can make the transition to the momentum representation (i.e. to the spectral representation space $\mathscr{H}_{\text{mom}} = L^2(\mathbb{R}^3) = \mathscr{H}_{\text{conf}}$) as follows: if $\psi \in L^2(\mathbb{R}^3)$, then there is in general a sequence $\psi_1, \psi_2, \ldots \in L^2(\mathbb{R}^3)$ such that

$$\psi = \text{s-lim}\ \psi_n, \quad \psi_n \in L^1(\mathbb{R}^3). \tag{B.25}$$

Since $\psi_n(x)$, $n = 1, 2, \ldots$, are integrable, their Fourier transforms

$$\tilde{\psi}_n(\mathbf{k}) = \int_{\mathbb{R}^3} \Phi^*_{\mathbf{k}}(\mathbf{x})\,\psi_n(\mathbf{x})\,d\mathbf{x} \stackrel{\text{def}}{=} \langle \Phi_{\mathbf{k}} | \psi_n \rangle \tag{B.26}$$

exist. Due to (B.25), and to the linearity and isometry properties of Fourier transforms,

$$\|\tilde{\psi}_m - \tilde{\psi}_n\| = \|\psi_m - \psi_n\| \to 0, \quad m, n \to \infty, \tag{B.27}$$

so that the sequence $\tilde{\psi}_1, \tilde{\psi}_2, \ldots$ possesses a strong limit $\tilde{\psi}$:

$$\tilde{\psi} = \text{s-lim}\ \tilde{\psi}_n \stackrel{\text{def}}{=} U_{\text{F}}\psi. \tag{B.28}$$

Since in an L^2-space, such as $L^2(\mathbb{R}^3)$, the strong limit coincides with the limit-in-the-mean, one writes symbolically:

$$(U_{\text{F}}\psi)(\mathbf{k}) = \text{l.i.m.} \int \Phi^*_k(\mathbf{x})\,\psi(\mathbf{x})\,d\mathbf{x}. \tag{B.29}$$

Naturally, if $\psi \in L^2(\mathbb{R}^3)$ but $\psi \notin L^1(\mathbb{R}^3)$, then the integral in (B.28) does not exist, since according to Appendix A, if it existed, then

$$|\Phi^*_k(\mathbf{x})\,\psi(\mathbf{x})| = (2\pi\hbar)^{-3/2}\,|\psi(\mathbf{x})| \tag{B.30}$$

would have to be integrable over $\mathbb{R}^3$, i.e. $\psi \in L^1(\mathbb{R}^3)$. Hence the meaning of the right-hand side of (B.29) is imparted exclusively by the procedure (B.25)–(B.28), which is then symbolized by l.i.m. in (B.29). This procedure is easily seen to be independent of the sequence $\psi_1, \psi_2, \ldots$ chosen in (B.25), so that it unambiguously defines a (unitary) operator on $L^2(\mathbb{R}^3)$, which is called in mathematical literature the *Fourier–Plancherel transform* (to be distinguish from the Fourier transform, whose natural domain of definition is $L^1(\mathbb{R}^3)$, and not $L^2(\mathbb{R}^3)$).

We see from (1.6.29) and (1.6.30) that

$$\Phi_{\mathbf{k}}(\mathbf{x}) = \lim_{l \to +\infty} \; (2\pi l^2)^{3/4} \, \xi^{(l)}_{\mathbf{0},\mathbf{k}}(\mathbf{x}), \tag{B.31}$$

$$\Phi^*_{\mathbf{q}}(\mathbf{k}) = \lim_{l \to +0} \left(\frac{\pi\hbar^2}{2l^2}\right)^{3/4} \tilde{\xi}^{(l)}_{\mathbf{q},\mathbf{p}}(\mathbf{k}), \tag{B.32}$$

so that the eigenfunction expansions $\Phi_{\mathbf{k}}$ of free particles are the renormalized limits of stochastic phase-space continuous resolutions of the identity. It is this type of relation that underlies the third, and last, step in the extrapolation from conventional to quantum mechanics. In fact, a study of the manner in which most computations that extract experimentally verifiable information from quantum mechanical models are performed in practice reveals that the pivotal objects in these computations are not so much the (Hilbert space) operators that represent observables, but rather the eigenfunction expansions related to those operators – which are themselves limits of corresponding continuous resolutions of the identity in stochastic quantum mechanics. It is this fact that also provides an explanation as to why most practical computations are not riddled with errors and inconsistencies, despite the extremely lax standards with which the theory of operators in Hilbert space is treated in physics textbooks and papers.[4]

Let $\{\mathscr{X}, \mathscr{A}, \mu\}$ be a measure space defined in Appendix A, and let $A(x)$, $x \in \Delta$, be an *operator-valued function* on a measurable set Δ in this measure space. Thus, to each $x \in \Delta$ is assigned a bounded or unbounded operator $A(x)$ on a Hilbert space $\mathscr{H}$. If $\|A(x)\| < \infty$, we assume that $\mathscr{D}_{A(x)} = \mathscr{H}$, whereas in case of unbounded operator we generally assume that

$$\mathscr{D}(\Delta) = \bigcap_{x \in \Delta} \mathscr{D}_{A(x)} \tag{B.33}$$

is dense in $\mathscr{H}$. Then we define

$$A = \int_\Delta A(x) \, d\mu(x) \tag{B.34}$$

as the operator with domain $\mathscr{D}_A$ consisting of all those $f \in \mathscr{D}(\Delta)$ for which

$$\int_\Delta \|A(x) f\| \, d\mu(x) < \infty, \tag{B.35}$$

and which for all such f, as well as for all $g \in \mathscr{H}$, satisfies the equation

$$\langle g | Af \rangle = \int_\Delta \langle g | A(x) f \rangle \, d\mu(x). \tag{B.36}$$

It can be easily shown (Prugovečki, 1981a, p. 480) that A is a well-defined linear operator, known in mathematical literature as the *Bochner integral* of $A(x)$ over Δ with respect to the measure μ.

Notes

[1] Since the foundations of the modern theory of Hilbert spaces were laid down at about the same time as (but independently of) the foundations of modern quantum mechanics. the mathematically loose approach of early textbooks on quantum mechanics to their subject-matter is quite understandable. However, it is much harder to understand in case of texts of recent vintage, especially since such carelessness with the mathematics can give rise to *physically erroneous* conclusions even at an elementary level (cf., e.g., Exer. 3.6 on pp. 301 and 644 of Prugovečki, 1981a).

[2] For proof, see Prugovečki (1981a, p. 195), or standard textbooks on functional analysis. This theorem is actually a special case of one of the most central theorems in the theory of topological vector spaces, namely the *closed-graph theorem*, which for the special Hilbert space case can be found on p. 210 of Prugovečki (1981a).

[3] Commuting in the sense that their spectral measures commute. This kind of commutativity is equivalent to $[A_i, A_j] = 0$ if $A_1, \ldots, A_n$ are bounded and everywhere defined (Prugovečki, 1981a, p. 268), but in the unbounded case the former implies the latter, with the converse not being generally true – as shown by a counterexample of Nelson (1959).

[4] By the same token, such inconsistencies can be found in some physics papers on scattering theory for long-range potentials, since in that case Hilbert space methods are more essential in the handling of the corresponding (modified) eigenfunction expansions. A short survey, as well as references for the mathematically rigorous approach to that theory, can be found on pp. 513–516 of Prugovečki (1981a).

References

Abraham, R., and J. E. Marsden: 1978, *Foundations of Mechanics* (2nd edn), Benjamin/Cummings, Reading, Mass.

Acharya, R., and E. C. G. Sudarshan: 1960, *J. Math. Phys.* **1**, 532.

Adler, R., M. Bazin and M. Schiffer: 1975, *Introduction to General Relativity*, McGraw-Hill, New York.

Adler, S. L.: 1972, *Phys. Rev.* **D5**, 3021.

Agarwal, G. S., and E. Wolf: 1970, *Phys. Rev.* **D2**, 2187.

Aharanov, Y., and D. Bohm: 1961, *Phys. Rev.* **122**, 1649.

Aharanov, Y., D. Z. Albert and C. K. Au: 1981, *Phys. Rev. Lett.* **47**, 1029.

Aichelburg, P. C., and Sexl, R. U. (eds): 1979, *Albert Einstein: His Influence in Physics, Philosophy and Politics*, Friedr. Vieweg, Braunschweig/Wiesbaden.

Albeverio, S.: 1974, in *Scattering Theory in Mathematical Physics*, J. A. LaVita and J.-P. Marchand (eds), Reidel, Dordrecht.

Albeverio, S., and R. J. Hoegh-Krohn: 1976, *Mathematical Theory of Feynman Path Integrals*, Springer, Berlin.

Albeverio, S., *et al.* (eds): 1979, *Feynman Path Integrals*, Lecture Notes in Physics, Vol. 106, Springer, Berlin.

Alfvén, H.: 1977, in *Cosmology, History and Theology,* W. Yourgrau and A. D. Breck (eds), Plenum, New York.

Ali, S. T.: 1979, *J. Math. Phys.* **20**, 1385.

Ali, S. T.: 1980, *J. Math. Phys.* **21**, 818.

Ali, S. T.: 1981, in *Differential Geometric Methods in Mathematical Physics*, H. D. Doebner (ed.), Lecture Notes in Physics, vol. 139, Springer, Berlin.

Ali, S. T.: 1982a, *Hadronic J.* **5**, 1001.

Ali, S. T.: 1982b, in *Differential Geometric Methods in Mathematical Physics*, H. D. Doebner, S. J. Anderson, and H. R. Petry (eds), Lecture Notes in Mathematics, Springer, Berlin.

Ali, S. T.: 1982c, 'Commutative Systems of Covariance and a Generalization of Mackey's Imprimitivity Theorem' (Concordia University preprint).

Ali, S. T.: 1982d, *Czech, J. Phys.* **B32**, 609.

Ali, S. T., and H. D. Doebner: 1976, *J. Math. Phys.* **17**, 1105.

Ali, S. T., and G. G. Emch: 1974, *J. Math. Phys.* **15**, 176.

Ali, S. T., R. Gagnon and E. Prugovečki: 1981, *Can. J. Phys.* **59**, 807.

Ali, S. T., and N. Giovannini: 1983, 'On Some *K*-Representations of the Poincaré and Einstein Groups', *Helv. Phys. Acta* (to appear).

Ali, S. T., and E. Prugovečki: 1977a, *J. Math. Phys.* **18**, 219.

Ali, S. T., and E. Prugovečki: 1977b, *Physica* **89A**, 501.

Ali, S. T., and E. Prugovečki: 1977c, *Int. J. Theor. Phys.* **16**, 689.

Ali, S. T., and E. Prugovečki: 1980, in *Mathematical Methods and Applications of Scattering*

Theory, J. A. De Santo, A. W. Saenz, and W. W. Zachary (eds), Lecture Notes in Physics, vol. 130, Springer, Berlin.
Ali, S. T., and E. Prugovečki: 1981, *Nuovo Cimento* **63A**, 171.
Ali, S. T., and E. Prugovečki: 1983a, 'Extended Harmonic Analysis of Phase Space Representations of the Galilei Group' (University of Toronto preprint).
Ali, S. T., and E. Prugovečki: 1983b, 'Harmonic Analysis and Systems of Covariance for Phase Space Representations of the Poincaré Group' (University of Toronto preprint).
Altarelli, G.: 1982, *Phys. Reports* **81**, 1.
Amrein, W. O.: 1969, *Helv. Phys. Acta* **42**, 149.
Araki, H., and M. Yanase: 1961, *Phys. Rev.* **120**, 666.
Arnous, E., W. Heitler, and Y. Takahashi: 1960, *Nuovo Cimento* **16**, 671.
Aronsajn, N.: 1950, *Trans. Amer. Math. Soc.* **68**, 337.
Arzeliès, H.: 1966, *Relativistic Kinematics*, Pergamon Press, Oxford.
Ash, R. B.: 1972, *Measure, Integration, and Functional Analysis*, Academic Press, New York.
Atiyah, M. F.: 1979, *Geometry of Yang-Mills Fields*, Accademia Nazionale dei Lincei, Pisa.
Avis, S. J., C. J. Isham and D. Storey: 1978, *Phys. Rev.* **D18**, 3565.
Balazs, N. I.: 1980, *Physica* **102A**, 236.
Balescu, R.: 1975, *Equilibrium and Non-equilibrium Statistical Mechanics*, Wiley, New York.
Barber, D. P. *et al.*: 1979, *Phys. Rev. Lett.* **43**, 1915.
Bargmann, V.: 1947, *Ann. Math.* **48**, 568.
Bargmann, V.: 1954, *Ann. Math.* **59**, 1.
Bargmann, V.: 1961, *Comm. Pure Appl. Math.* **24**, 187.
Bargmann, V.: 1967, *Comm. Pure Appl. Math.* **20**, 1.
Barut, A. O.: 1964, *Electrodynamics and Classical Theory of Fields and Particles*, Macmillan, New York.
Barut, A. O.: 1968, in *Lectures in Theoretical Physics*, A. O. Barut and W. E. Britten (eds), Vol. X–B, Gordon and Breach, New York.
Barut, A. O.: 1980, in *Foundations of Radiation Theory and Quantum Electrodynamics*, A. O. Barut (ed.), Plenum, New York.
Barut, A. O.: 1981, in *Group Theory and Its Applications in Physics*, T. H. Seligman (ed.), Amer. Inst. of Physics, New York.
Barut, A. O., and A. J. Bracken: 1981, *Phys. Rev.* **D23**, 2454.
Barut, A. O., and A. Böhm: 1965, *Phys. Rev.* **139B**, 1107.
Barut, A. O., and R. Rączka: 1976, *Nuovo Cimento* **31B**, 19.
Barut, A. O., and R. Rączka: 1977, *Theory of Group Representations and Applications*, PWN-Polish Scientific Publisher, Warsaw.
Bates, D. R. (ed.): 1961, *Quantum Theory*, Vol. 1, Academic Press, New York.
Berestetskiĭ, V. B., E. M. Lifshitz and L. P. Pitaevskiĭ: 1979, *Relativistic Quantum Theory*, Part I, J. B. Sykes and J. S. Bell (trans.), Pergamon, Oxford.
Billingsley, P.: 1979, *Probability and Measure*, Wiley, New York.
Bjorken, J. D., and S. D. Drell: 1964, *Relativistic Quantum Mechanics*, McGraw-Hill, New York.
Bleuler, K.: 1950, *Helv. Phys. Acta* **23**, 567.
Blokhintsev, D. I.: 1966, *Sov. Phys. Dokl.* **11**, 23, 503.
Blokhintsev, D. I.: 1968, *The Philosophy of Quantum Mechanics*, Reidel, Dordrecht.
Blokhintsev, D. I.: 1972, *Teor. Mat. Fiz* **11**, 3.
Blokhintsev, D. I.: 1973, *Space and Time in the Microworld*, Reidel, Dordrecht.
Blokhintsev, D. I.: 1974, *Sov. Phys. Usp.* **16**, 485.
Blokhintsev, D. I.: 1975, *Sov. J. Particles Nucl.* **5**, 243.
Blokhintsev, D. I., *et al.* (eds): 1968, *Proceedings of the International Conference on Nonlocal Quantum Field Theory July 4–7, 1967*, Dubna, JINR.
Bogolubov, N. N., A. A. Logunov and I. T. Todorov: 1975, *Introduction to Axiomatic Quantum Field Theory*, S. A. Fulling and L. G. Popova (trans.), Benjamin, Reading, Mass.

Bogolubov, N. N., and D. V. Shirkov: 1959, *Introduction to the Theory of Quantized Fields*, Interscience Publishers, New York.
Böhm, A.: 1966, *Phys. Rev.* **145**, 1212.
Böhm, A.: 1968a, in *Lectures in Theoretical Physics*, in A. O. Barut and W. E. Britten (eds), vol. X–B, Gordon and Breach, New York.
Böhm, A.: 1968b, *Phys. Rev.* **175**, 1767.
Böhm, A., R. B. Teese, A. Garcia and J. S. Nilsson: 1977, *Phys. Rev.* **D15**, 684.
Böhm, D.: 1952, *Phys. Rev.* **85**, 166.
Böhm, D., and J. Bub: 1966, *Rev. Mod. Phys.* **38**, 453, 470.
Böhm, D. and B. J. Hiley: 1981, *Found. Phys.* **11**, 179.
Bohr, N.: 1928, *Naturwiss.* **16**, 245.
Bohr, N.: 1934, *Atomic Theory and the Description of Nature*, Cambridge University Press, Cambridge.
Bohr, N.: 1961, *Atomic Physics and Human Knowledge*, Science, New York.
Bohr, N., and L. Rosenfeld: 1933, *Kgl. Danske Vidensk. Selsk. Mat.-Fys. Medd.* **12**, No. 8.
Bohr, N., and L. Rosenfeld: 1950, *Phys. Rev.* **78**, 794.
Born, M.: 1926, *Z. Phys.* **37**, 863.
Born, M.: 1938, *Proc. Roy. Soc. London* **A165**, 291.
Born, M.: 1939, *Proc. Roy. Soc. Edinburgh* **59**, 219.
Born, M.: 1949a, *Nature* **163**, 207.
Born, M.: 1949b, *Rev. Mod. Phys.* **21**, 463.
Born, M.: 1953, *Scientific Papers: Presented to Max Born*, Oliver and Boyd, Edinburgh.
Born, M.: 1955a, *Physik. Blätter* **11**, 49.
Born, M.: 1955b, *Science* **122**, 675.
Born, M.: 1955c, *Dan. Mat. Fys. Medd.* **30**, no. 2.
Born, M.: 1956, *Physics in My Generation*, Pergamon Press, London.
Born, M.: 1962, *Physics and Politics*, Oliver and Boyd, Edinburgh.
Born, M.: 1969a, *Atomic Physics* (8th edn), J. Dougall, R. J. Blin-Stoyle and J. M. Radcliffe (trans.), Blackie & Son, London.
Born, M.: 1969b, *Phys. Blätter* **25**, 112.
Born, M.: 1971, *The Born–Einstein Letters*, Macmillan, London.
Born, M., and K. Fuchs: 1940a, *Proc. Roy. Soc. Edinburgh* **60**, 100.
Born, M., and K. Fuchs: 1940b, *Proc. Roy. Soc. Edinburgh* **60**, 141.
Born, M., H. S. Green, K. C. Cheng and A. E. Rodriguez: 1949, *Proc. Roy. Soc. Edinburgh* **62**, 470.
Born, M., W. Heisenberg, and P. Jordan: 1926, *Z. Phys.* **35**, 557.
Born, M., and P. Jordan: 1925, *Z. Phys.* **34**, 858.
Born, M., and W. Ludwig: 1958, *Z. Phys.* **150**, 106.
Breit, G.: 1928, *Proc. Nat. Acad. Sci. U.S.A.* **14**, 553.
Brooke, J. A.: 1978, *J. Math. Phys.* **19**, 952.
Brooke, J. A.: 1980, *J. Math. Phys.* **21**, 617.
Brooke, J. A., and W. Guz: 1982, *Lett. Nuovo Cimento* **35**, 265.
Brooke, J. A., and W. Guz: 1983a, 'The Baryon Mass Spectrum and the Reciprocity Principle of Born', *Nuovo Cimento A* (to appear).
Brooke, J. A., and W. Guz: 1983b, 'Relativistic Canonical Commutation Relations and the Harmonic Oscillator Model of Elementary Particles', *Nuovo Cimento A* (to appear).
Brooke, J. A., W. Guz and E. Prugovečki: 1982, *Hadronic. J.* **5**, 1717.
Brooke, J. A., and E. Prugovečki, 1982, *Lett. Nuovo Cimento* **33**, 171.
Brooke, J. A., and E. Prugovečki: 1983a, 'Relativistic Canonical Commutation Relations and the Geometrization of Quantum Mechanics' (University of Toronto preprint).
Brooke, J. A., and E. Prugovečki: 1983b, 'Hadronic Exciton States and Propagators Derived from Reciprocity Theory on Quantum Spacetime' (University of Toronto preprint).
Brown, L. M.: 1958, *Phys. Rev.* **111**, 957.

Broyles, A. A.: 1970, *Phys. Rev.* **D1**, 979.
Bunch, T. S., and L. Parker: 1979, *Phys. Rev.* **D20**, 2499.
Busch, P.: 1982, *Unbestimmtheitsrelation und simultane Messungen in der Quantentheorie*, Doctoral Thesis, Universität zu Köln, Köln, Germany.
Butkov, E.: 1968, *Mathematical Physics*, Addison-Wesley, Reading, Mass.
Caianiello, E. R. (ed.): 1974, *Renormalization and Invariance in Quantum Field Theory*, Plenum, New York.
Caianiello, E. R.: 1979, *Lett. Nuovo Cimento* **25**, 225.
Caianiello, E. R.: 1980a, *Lett. Nuovo Cimento* **27**, 89.
Caianiello, E. R.: 1980b, *Nuovo Cimento* **59B**, 350.
Caianiello, E. R.: 1981, *Lett. Nuovo Cimento* **32**, 65.
Caianiello, E. R., and G. Vilasi: 1981, *Lett. Nuovo Cimento* **30**, 469.
Caianiello, E. R., S. De Filippo and G. Vilasi: 1982a, *Lett. Nuovo Cimento* **33**, 555.
Caianiello, E. R., S. De Filippo, G. Marmo, and G. Vilasi: 1982b, *Lett. Nuovo Cimento* **34**, 112.
Cameron, R. H.: 1960, *J. Math. Phys.* **39**, 126.
Carmeli, M.: 1977, *Group Theory and General Relativity*, McGraw-Hill, New York.
Castagnino, M.: 1971, *J. Math. Phys.* **12**, 2203.
Castrigiano, D. P. L., and R. W. Heinrichs: 1980, *Lett. Math. Phys.* **4**, 169.
Castrigiano, D. P. L., and U. Mutze: 1982, *Phys. Rev.* **D26**, 3499.
Cattaneo, U.: 1979, *Comment Math. Helvetici* **54**, 629.
Chiu, C. B.: 1972, *Ann. Rev. Nucl. Sci.* **22**, 255.
Chretien, M., and R. Peierls: 1953, *Nuovo Cimento* **10**, 669.
Close, F. E.: 1979, *An Introduction to Quarks and Partons,* Academic Press, New York.
Cohen, L.: 1966, *J. Math. Phys.* **7**, 781.
Collins, P. D. B.: 1977, *An Introduction to Regge Theory and High Energy Physics*, Cambridge University Press, Cambridge.
Corbett, J. V., and C. A. Hurst: 1978, *J. Austral. Math. Soc.* **20B**, 182.
Corinaldesi, F.: 1951, *Nuovo Cimento* **8**, 494.
Corinaldesi, F.: 1953, *Suppl. Nuovo Cimento* **10**, No. 2.
Currie, D. G., T. F. Jordan and E. C. G. Sudarshan: 1963, *Rev. Mod. Phys.* **35**, 350.
Das, A.: 1960, *Nuovo Cimento* **18**, 482.
Davidson, M.: 1979, *Lett. Math. Phys.* **3**, 271. 367.
Davidson, M.: 1980, *Lett. Math. Phys.* **4**, 101, 475.
Davies, E. B., and J. T. Lewis: 1969, *Commun. Math. Phys.* **17**, 239.
Davies, E. B.: 1976, *Quantum Theory of Open Systems*, Academic Press, New York.
Davies, P. C. W.: 1980, in *General Relativity and Gravitation*, Vol. 1, A. Held (ed.), Plenum, New York.
de Broglie, L.: 1964, *Ann. Inst. Henri Poincaré* **1**, 1.
de Broglie, L.: 1979, in *Perspectives in Quantum Theory*, W. Yourgrau and A. van der Merwe (eds), Dover, New York.
de Falco, D., De Martino, S. and De Siena, S.: 1982, *Phys. Rev. Lett.* **49**, 181.
de la Pena-Auerbach, L.: 1969, *J. Math. Phys.* **10**, 1620.
de la Pena-Auerbach, L.: 1971, *J. Math. Phys.* **12**, 453.
de la Pena-Auerbach, L., and A. M. Cetto: 1975, *Found. Phys.* **5**, 355.
Deser, S.: 1957, *Rev. Mod. Phys.* **29**, 417.
Deser, S.: 1980, in *Unification of the Fundamental Particle Interactions*, S. Ferrara, J. Ellis and P. van Nieuwenhuizen (eds), Plenum, New York.
De Sabbata, V., and M. Gasperini: 1982, *Lett. Nuovo Cimento* **34**, 337.
d'Espagnat, B.: 1976, *Conceptual Foundations of Quantum Mechanics* (2nd edn), Benjamin, Reading, Mass.
d'Espagnat, B.: 1981, *Found. Phys.* **11**, 205.
De Witt, B. S.: 1955, *Phys. Rev.* **100**, 905.

De Witt, B. S.: 1962, in *Gravitation: An Introduction to Current Research*, L. Witten (ed.), Wiley, New York.
De Witt, B. S.: 1975, *Phys. Rep.* **19C**, 297.
De Witt, B. S., and N. Graham (eds): 1973, *The Many-World Interpretation of Quantum Mechanics*, Princeton University Press, Princeton, N.J.
De Witt, C., and K. D. Elsworthy (eds): 1981, *Phys. Rep.* **C77**, No. 3.
De Witt-Morette, C., A. Maheshwari and B. Nelson: 1977, *Gen. Rel. Grav.* **8**, 581.
De Witt-Morette, C., A. Maheshwari and B. Nelson: 1979, *Phys. Rep.* **C9**, 255.
Dicke, R. H.: 1964, *The Theoretical Significance of Experimental Relativity*, Gordon and Breach, New York.
Dietrich, C. F.: 1973, *Uncertainty, Calibration and Probability*, Wiley, New York.
Dimock, J.: 1972, *J. Math. Phys.* **13**, 477.
Dirac, P. A. M.: 1927, *Proc. Roy. Soc. London* **114A**, 243.
Dirac, P. A. M.: 1928, *Proc. Roy. Soc. London* **117A**, 610.
Dirac, P. A. M.: 1933, *Physik. Z. Sowjetunion* **3**, 64.
Dirac, P. A. M.: 1945, *Rev. Mod. Phys.* **17**, 195.
Dirac, P. A. M.: 1951, *Proc. Roy. Soc. London* **209A**, 291.
Dirac, P. A. M.: 1962, *Proc. Roy. Soc. London* **268A**, 57.
Dirac, P. A. M.: 1965, *Phys. Rev.* **139B**, 684.
Dirac, P. A. M.: 1973, in *The Past Decade in Particle Theory*, E. C. G. Sudarshan and Y. Ne'eman (eds), Gordon and Breach, London.
Dirac, P. A. M.: 1977, in *Deeper Pathways in High-Energy Physics*, A. Perlmutter and L. F. Scott (eds), Plenum, New York.
Dirac, P. A. M.: 1978a, in *Directions in Physics*, H. Hora and J. R. Shepanski (eds), Wiley, New York.
Dirac, P. A. M.: 1978b, in *Mathematical Foundations of Quantum Mechanics*, A. R. Marlow (ed.), Academic Press, New York.
Dita, P., V. Georgescu, and R. Purice (eds.): 1982, *Gauge Theories: Fundamental Interactions and Rigorous Results*, Birkhäuser, Boston.
Dixon, W. G.: 1965, *Nuovo Cimento* **38**, 1616.
Dothan, Y., M. Gell-Mann, and Y. Ne'eman: 1965, *Phys. Rev. Lett.* **17**, 145.
Drell, S. D. and F. Zachariasen: 1961, *Electromagnetic Structure of Nucleons*, Oxford University Press.
Duane, W.: 1923, *Proc. Nat. Acad. Wash.* **9**, 158.
Dyson, F. J.: 1949, *Phys. Rev.* **75**, 486, 1736.
Dyson, F. J.: 1952, *Phys. Rev.* **85**, 631.
Eddington, A. S.: 1946, *Fundamental Theory*, Cambridge University Press, Cambridge.
Efimov, G. V.: 1972, *Ann. Phys. (N.Y.)* **71**, 466.
Efimov, G. V.: 1977, *Nonlocal Interactions of Quantum Fields* (in Russian), Nauka, Moscow.
Ehlers, J.: 1971, in *General Relativity and Cosmology*, B. K. Sachs (ed.), Academic Press, New York.
Ehlers, J.: 1973, in *Relativity, Astrophysics and Cosmology*, W. Israel (ed.), Reidel, Dordrecht.
Ehlers, J., F. A. E. Pirani and A. Schild: 1972, in *General Relativity*, L. O'Raifeartaigh (ed.), Clarendon Press, Oxford.
Ehlers, J., and A. Schild: 1973, *Comm. Math. Phys.* **32**, 119.
Einstein, A.: 1905, *Ann. Phys.* **17**, 891.
Einstein, A.: 1916, *Ann. Phys.* **49**, 769.
Einstein, A.: 1918, *Ann. Phys.* **55**, 241.
Einstein, A.: 1923, *Sidelights of Relativity*, E. P. Dutton & Co., New York.
Einstein, A.: 1923, 1953, 'Geometry and Experience' (from *Sidelights of Relativity*) reprinted in *Readings on the Philosophy of Science*, H. Feigl and M. Brodbeck (eds), Appleton-Century-Crofts, New York, pp. 189–194.

Einstein, A.: 1949, in *Albert Einstein: Philosopher-Scientist*, A. Schilpp (ed.), The Library of Living Philosophers, Evanston, Illinois.
Einstein, A., H. A. Lorentz, H. Minkowski and H. Weyl: 1923, *The Principle of Relativity*, W. Perrett and G. B. Jeffery (trans.), Methuen, London; reprinted by Dover, New York.
Ellis, J., M. K. Gaillard, L. Maiami and B. Zumino: 1980, in *Unification of the Fundamental Particle Interactions*, S. Ferrara, J. Ellis and P. van Nieuwenhuizen (eds), Plenum, New York.
Erdélyi, A. (ed.): 1954, *Tables of Integral Transforms*, vol. 1, McGraw-Hill, New York.
Eu, B. C.: 1975, *J. Chem. Phys.* **63**, 303.
Everett III, H.: 1957, *Rev. Mod. Phys.* **29**, 454.
Faddeev, L. D.: 1969, *Theor. Math. Phys.* **1**, 3.
Faddeev, L. D., and V. E. Korepin: 1978, *Phys. Rep.* **C42**, 1.
Faddeev, L. D., and A. A. Slavnov: 1980, *Gauge Fields*, D. B. Pontecorvo (trans.), Addison-Wesley, Reading, Mass.
Ferrara, S., J. Ellis and P. van Nieuwenhuizen (eds): 1980, *Unification of the Fundamental Particle Interactions*, Plenum, New York.
Ferrari, R., L. E. Picasso and F. Strocchi: 1974, *Comm. Math. Phys.* **35**, 25.
Ferretti, A.: 1963, *Nuovo Cimento* **27**, 1503.
Feynman, R. P.: 1948a, *Rev. Mod. Phys.* **20**, 367.
Feynman, R. P.: 1948b, *Phys. Rev.* **74**, 939.
Feynman, R. P.: 1949a, *Phys. Rev.* **76**, 749.
Feynman, R. P.: 1949b, *Phys. Rev.* **76**, 769.
Feynman, R. P.: 1950, *Phys. Rev.* **80**, 440.
Feynman, R. P.: 1951, *Phys. Rev.* **84**, 108.
Feynman, R. P., and M. Gell-Man: 1958, *Phys. Rev.* **109**, 193.
Feynman, R. P., and A. R. Hibbs: 1965, *Quantum Mechanics and Path Integrals*, McGraw-Hill, New York.
Feynman, R. P., M. Kislinger and F. Ravndal: 1971, *Phys. Rev.* **D3**, 2706.
Fierz, M.: 1950, *Nuovo Cimento* **27**, 1503.
Fine, T. L.: 1973, *Theories of Probability: An Examination of Foundations*, Academic Press, New York.
Fleming, G. N.: 1965, *Phys. Rev.* **139**, B963.
Flint, H. T.: 1937, *Proc. Roy. Soc. London* **159A**, 45.
Flint, H. T.: 1948, *Phys. Rev.* **74**, 209.
Flint, H. T., and O. W. Richardson: 1928, *Proc. Roy. Soc. London* **117A**, 637.
Foldy, L. L., and S. A. Wouthuysen: 1950, *Phys. Rev.* **78**, 29.
Frampton, P. H., S. L. Glashow and A. Yildiz: 1980, *First Workshop on Grand Unification*, Math Sci Press, Brookline, Mass.
Frank, M. J.: 1971, *J. Math. Anal. Appl.* **34**, 67.
Frederick, C.: 1976, *Phys. Rev.* **D13**, 3183.
Fröhlich, J.: 1980, in *Recent Developments in Gauge Theories*, G.'t Hooft *et al.* (eds), Plenum, New York.
Fröhlich, J., G. Marchio and F. Strocchi: 1979, *Ann. Phys. (NY)* **119**, 241.
Fujimura, K., T. Kobayashi and M. Namiki: 1970, *Prog. Theor. Phys.* **43**, 73.
Fujimura, K., T. Kobayashi and M. Namiki: 1971, *Prog. Theor. Phys.* **44**, 193.
Fulling, S. A.: 1973, *Phys. Rev.* **D7**, 2850.
Fürth, R.: 1929, *Z. Phys.* **57**, 429.
Gagnon, R.: 1983, *Path Integrals and Scattering Theory in Nonrelativistic Stochastic Quantum Mechanics*, Ph.D. Thesis, University of Toronto.
Gallardo, J. A., A. J. Kalnay, A. J. Stec and B. P. Toledo: 1967, *Nuovo Cimento* **48**, 1008; **49**, 393.
Gel'fand, I. M., and G. E. Shilov: 1964, *Generalized Functions*, vol. I, E. Saletan (trans.), Academic Press, New York.

Gel'fand, I. M., and N. Vilenkin: 1968, *Generalized Functions*, vol. IV, A. Feinstein (trans.), Academic Press, New York.
Gel'fand, I. M., and A. M. Yaglom: 1960, *J. Math. Phys.* **1**, 48.
Georgi, H., and C. Jarlskog: 1979, *Phys. Lett.* **86B**, 297.
Geronimus, Yu. V., and M. Y. Tseyllin: 1965, *Tables of Integrals, Series and Products*, Academic Press, New York, pp. 307, 930.
Giles, R.: 1970, *J. Math. Phys.* **11**, 2139.
Giles, R.: 1979, *Studia Logica* **38**, 337.
Giovannini, N.: 1981a, *Lett. Math. Phys.* **5**, 161.
Giovannini, N.: 1981b, *J. Math. Phys.* **22**, 2389, 2397.
Giovannini, N., and C. Piron: 1979, *Helv. Phys. Acta* **52**, 518.
Glaser, W., and K. Sitte: 1934, *Z. Phys.* **87**, 674.
Glimm, J., and A. Jaffe: 1981, *Quantum Physics*, Springer, New York.
Gnadig, P., Z. Kunst, D. Hasenfratz and J. Koti: 1978, *Ann. Phys.* (*NY*) **116**, 380.
Goldstone, J., and R. Jackiw: 1975, *Phys. Rev.* **D11**, 1486.
Gol'fand, Yu. A.: 1960, *Sov. Phys. JETP* **10**, 356.
Gol'fand, Yu. A.: 1963a, *Sov. Phys. JETP* **16**, 184.
Gol'fand, Yu. A.: 1963b, *Sov. Phys. JETP* **17**, 842.
Gordon, W.: 1926, *Z. Phys.* **40**, 117.
Greenwood, D., and E. Prugovečki: 1984, 'Stochastic Microcausality in Relativistic Quantum Mechanics', *Found. Phys.* **14** (to appear).
Gudder, S. P.: 1979, *Stochastic Methods in Quantum Mechanics*, North Holland, New York.
Guerra, F.: 1981, *Phys. Rep.* **77**, 263.
Guerra, F., and P. Ruggiero: 1978, *Lett. Nuovo Cimento* **23**, 529.
Gupta, S. N.: 1950, *Proc. Phys. Soc. London* **63A**, 681.
Guz, W.: 1983a, 'Foundations of the Phase-Space Representability of Quantum Mechanics', *Int. J. Theor. Phys.* (to appear).
Guz, W.: 1983b, 'Stochastic Phase Spaces, Fuzzy Sets, and Statistical Metric Spaces', *Found. Phys.* (to appear).
Haag, R., and D. Kastler: 1964, *J. Math. Phys.* **5**, 848.
Hagedorn, G. A.: 1980, *Comm. Math. Phys.* **71**, 77.
Hagedorn, G. A.: 1981, *Ann. Phys.* (*N.Y.*) **135**, 58.
Halbwachs, F., F. Piperno, and J.-P. Vigier: 1982, *Lett. Nuovo Cimento* **33**, 311.
Halmos, P. R.: 1950, *Measure Theory*, Van Nostrand, Princeton, N.J.
Harari, H.: 1979, *Phys. Lett.* **B86**, 83.
Harvey, A.: 1976, *Gen. Rel. Grav.* **7**, 891.
Hawking, S. W., and G. F. R. Ellis: 1973, *The Large-Scale Structure of Space-Time*, Cambridge University Press, Cambridge.
Hegerfeldt, G. C.: 1974, *Phys. Rev.* **D10**, 3320.
Hegerfeldt, G. C., and J. Henning: 1968, *Fortschr. Physik* **16**, 491.
Hegerfeldt, G. C., K. Kraus and E. P. Wigner: 1968, *J. Math. Phys.* **9**, 2029.
Hegerfeldt, G. C., and S. N. M. Ruijsenaars: 1980, *Phys. Rev.* **D22**, 377.
Heisenberg, W.: 1927, *Z. Phys.* **43**, 172.
Heisenberg, W.: 1930, *Die Physikalischen Prinzipien der Quantentheorie*, S. Hirzel, Leipzig.
Heisenberg, W.: 1938a, *Ann. Phys. Lpz.* **32**, 20.
Heisenberg, W.: 1938b, *Z. Phys.* **110**, 251.
Heisenberg, W.: 1942, *Z. Phys.* **120**, 513.
Heisenberg, W.: 1969, *Phys. Blätter* **25**, 112.
Heisenberg, W.: 1971, *Physics and Beyond*, A. J. Pomerance (trans.), Harper and Row, New York.
Heisenberg, W.: 1976, *Phys. Today* **29**(3), 32.
Heisenberg, W., and W. Pauli: 1929, *Z. Physik* **56**, 1.
Hess, S.: 1967, *Z. Naturforsch.* **22a**, 1871.

Hill, E. L.: 1950, *Phys. Rev.* **100**, 1780.
Hill, R. N.: 1967, *J. Math. Phys.* **8**, 1756.
Hoegh-Krohn, R.: 1970, *Comm. Math. Phys.* **18**, 109.
Hofstadter, R.: 1956, *Rev. Mod. Phys.* **28**, 214.
Hofstadter, R.: 1957, *Ann. Rev. Nucl. Sci.* **7**, 231.
Hofstadter, R., F. Bumiller, and M. Yearian: 1958, *Rev. Mod. Phys.* **30**, 482.
Holstein, D., and F. E. Schroeck, Jr: 1983, 'Experimental Evidence for the Stochastic Quantum Mechanical Mass Formula' (Atlantic Florida University preprint).
Horwitz, L. P., and C. Piron: 1973, *Helv. Phys. Acta* **46**, 316.
Hu, B. L., S. A. Fulling and L. Parker: 1973, *Phys. Rev.* **D8**, 2377.
Huang, K.: 1982, *Quarks, Leptons and Gauge Fields*, World Scientific, Singapore.
Husimi, K.: 1978, *Proc. Phys. Math. Soc. Japan* **22**, 264.
Hwa, R. C.: 1980, *Phys. Rev.* **D22**, 759, 1593.
Hwa, R. C., and M. S. Zahir: 1981, *Phys. Rev.* **D23**, 2539.
Hynes, J. T., and J. M. Deutch: 1969, *J. Chem. Phys.* **50**, 3015.
Infeld, L.: 1957, *Bull. Pol. Akad. Sci.* **5**, 491.
Ingraham, R. L.: 1962, *Nuovo Cimento* **24**, 1117.
Ingraham, R. L.: 1963, *Nuovo Cimento* **27**, 303.
Ingraham, R. L.: 1964, *Nuovo Cimento* **34**, 182.
Ingraham, R. L.: 1967, *Renormalization Theory of Quantum Fields with a Cut-Off*, Gordon and Breach, New York.
Ingraham, R. L.: 1968, *Int. J. Theor. Phys.* **1**, 191.
Ingraham, R. L.: 1972, *Int. J. Theor. Phys.* **6**, 175.
Inönü, E., and E. P. Wigner: 1952, *Nuovo Cimento* **9**, 705.
Inönü, E., and E. P. Wigner: 1953, *Proc. Natl. Acad. Sci. USA* **39**, 510.
Isham, C. J.: 1975, in *Quantum Gravity*, C. J. Isham, R. Penrose and D. W. Sciama (eds), Clarendon Press, Oxford.
Itzykson, C., and J.-B., Zuber: 1980, *Quantum Field Theory*, McGraw-Hill, New York.
Jackiw, R.: 1977, *Rev. Mod. Phys.* **49**, 681.
Jackson, J. D.: 1975, *Classical Electrodynamics* (2nd edn), Wiley, New York.
Jammer, M.: 1974, *The Philosophy of Quantum Mechanics*, Wiley, New York.
Jancewicz, B.: 1975, *Rep. Math. Phys.* **8**, 181.
Jancewicz, B.: 1977, *J. Math. Phys.* **18**, 2487.
Jannussis, A., N. Patargias, A. Leodaris, P. Filippakis, T. Filippakis, A. Streclas and V. Papatheou: 1982, *Lett. Nuovo Cimento* **34**, 553.
Jauch, J. M., and C. Piron: 1967, *Helv. Phys. Acta* **40**, 559.
Jauch, J. M., and F. Rohrlich: 1955, *The Theory of Photons and Electrons*, Addison-Wesley, Cambridge.
Johnson, J. E.: 1969, *Phys. Rev.* **181**, 1755.
Johnson, J. E.: 1971, *Phys. Rev.* **D3**, 1735.
Johnson, K., R. Willey and M. Baker: 1967, *Phys. Rev.* **163**, 1699.
Jordan, T. F. and N. Mukunda: 1963, *Phys. Rev.* **132**, 1842.
Jost, R.: 1965, *The General Theory of Quantum Fields*, American Mathematical Society, Providence, R. I.
Kadyshevskiĭ, V. G.: 1962, *Sov. Phys. JETP* **14**, 1340.
Kadyshevskiĭ, V. G.: 1963, *Sov. Phys. Doklady* **7**, 1031.
Kálnay, A. J.: 1971, in *Problems in the Foundations of Physics*, M. Bunge (ed.), Springer, New York.
Kálnay, A. J., and B. P. Toledo: 1969, *Nuovo Cimento* **48**, 997.
Katayama, H., I. Umemura and H. Yukawa: 1968, *Prog. Theor. Phys. Suppl.* **41**, 22.
Katayama, Y., and H. Yukawa: 1968, *Prog. Theor. Phys. Suppl.* **41**, 1.
Kato, Y., and N. Mugibayashi: 1963, *Prog. Theor. Phys.* **30**, 103.
Kato, Y., and N. Mugibayashi: 1971, *Prog. Theor. Phys.* **38**, 682.

Kaufmann, A.: 1975, *Introduction to the Theory of Fuzzy Subsets*, Academic Press, New York.
Kelly, R. L., *et al*.: 1982, *Phys. Lett.* **111B**, 1.
Kerner, E. H.: 1965, *J. Math. Phys.* **6**, 1218.
Kershaw, D.: 1964, *Phys. Rev.* **136B**, 1850.
Kim, Y. S., and M. Z. Noz: 1973, *Phys. Rev.* **D8**, 3521.
Kim, Y. S., and M. Z. Noz: 1975, *Phys. Rev.* **D12**, 122, 129.
Kim. Y. S., and M. Z. Noz: 1977, *Phys. Rev.* **D15**, 355.
Kim, Y. S., and M. E. Noz: 1979, *Found. Phys.* **9**, 375.
Kirzhintz, D. A.: 1969, *Sov. Phys. Usp.* **9**, 692.
Klauder, J. R., and E. C. G. Sudarshan: 1968, *Fundamentals of Quantum Optics*, Benjamin, New York.
Klauder, J. R.: 1963, *J. Math. Phys.* **4**, 1058.
Klein, O.: 1926, *Z. Phys.* **37**, 895.
Klein, O.: 1929, *Z. Phys.* **53**, 157.
Kobayashi, S., and K. Nomizu: 1963, *Foundations of Differential Geometry*, Vol. 1, Wiley, New York.
Kokkedee, J. J. J.: 1969, *The Quark Model*, Benjamin, New York.
Konuma, M., and T. Maskawa (eds): 1981, *Grand Unified Theories and Related Topics*, World Scientific, Singapore.
Kraus, K.: 1971, *Ann. Phys. (NY)* **64**, 311.
Kretschmann, E.: 1917, *Ann. Physik* **53**, 575.
Kronheimer, E. H., and R. Penrose: 1967, *Proc. Camb. Phil. Soc.* **63**, 481.
Ktoridis, C. N., H. C. Myung and R. M. Santilli: 1980, *Phys. Rev.* **D22**, 892.
Kurşunoglu, B.: 1979, in *On the Path of Albert Einstein*, A. Perlmutter and L. F. Scott (eds), Plenum, New York.
Lai, C. H. (ed.): 1981, *Gauge Theory of Weak and Electromagnetic Interactions*, World Scientific, Singapore.
Landau, L. D.: 1955, in *Niels Bohr and the Development of Physics*, W. Pauli (ed.), Pergamon, New York.
Landau, L. D., and E. Lifshitz: 1971, *The Classical Theory of Fields* (3rd edn), Benjamin, Reading, Mass.
Landau, L., and R. Peierls: 1931, *Z. Phys.* **69**, 56.
Landau, L. D., and I. Pomeranchuk: 1955, *Dokl. Akad. Nauk SSSR* **102**, 489.
Landé, A.; 1939, *Phys. Rev.* **56**, 482, 486.
Landé, A.: 1960, *From Dualism to Unity in Quantum Physics*, Cambridge University Press, Cambridge.
Landé, A.: 1965a, *New Foundations of Quantum Mechanics*, Cambridge University Press, Cambridge.
Landé, A.: 1965b, *Amer. J. Phys.* **33**, 123.
Landé, A.: 1969, *Phys. Blätter* **25**, 105.
Landé, A.: 1971, *Found. Phys.* **1**, 191.
La Rue, G. S., W. M. Fairbank and A. F. Hebard: 1977, *Phys. Rev. Lett.* **38**, 1011.
La Rue, G. S., W. M. Fairbank and J. D. Phillips: 1979, *Phys. Rev. Lett.* **42**, 142.
Latzin, H.: 1927, *Naturwiss.* **15**, 161.
Lehr, W. J., and J. L. Park: 1977, *J. Math. Phys.* **18**, 1235.
Lehmann, H., K. Symanzik and W. Zimmermann: 1955, *Nuovo Cimento* **1**, 205.
Lehmann, H., K. Symanzik and W. Zimmermann: 1957, *Nuovo Cimento* **6**, 319.
Levy-Leblond, J. M.: 1971, in *Group Theory and Its Applications*, vol. II, E. M. Loebl (ed.), Academic Press, New York.
Lewis, J. T.: 1981, *Phys. Rep.* **77**, 339.
Lifshitz, E. M., and I. M. Khalatnikov: 1963, *Adv. Phys.* **12**, 185.
Lindemann, F. A.: 1932, *The Physical Significance of Quantum Theory*, Oxford University Press, London.

Mackey, G. W.: 1963, *Bull. Amer. Math. Soc.* **69**, 628.
Mackey, G. W.: 1968, *Induced Representations of Groups and Quantum Mechanics*, Bejamin, New York.
Malkin, I. A., and V. I. Man'ko: 1965, *JETP Lett* **2**, 230.
March, A.: 1936, *Z. Phys.* **104**, 93, 161.
March, A.: 1937a, *Z. Phys.* **105**, 620.
March, A.: 1937b, *Z. Phys.* **106**, 49.
March, A.: 1937c, *Z. Phys.* **108**, 128.
Margenau, H.: 1949, in *Albert Einstein: Philosopher-Scientist*, A. Schilpp (ed.), The Library of Living Philosophers, Evanston, Illinois.
Marinelli, M., and G. Marpurgo: 1980, *Phys. Lett.* **89B**, 427, 433.
Markov, M.: 1940, *Zh. Eksperim. Teor. Fiz.* **10**, 1311.
Markov, M.: 1956, *Nuovo Cimento Suppl.* **3**, 760.
Markov, M. A.: 1959, *Nucl. Phys.* **10**, 140.
Markov, M. A.: 1960, *Hyperonen und K-Mesonen*, VEB Deutscher Verlag der Wissenschaften, Berlin
Markov, M.A.: 1966, *Zh. Eksperim. Teor. Fiz.* **51**, 878.
Marzke, R. F., and J. A. Wheeler: 1964, in *Gravitation and Relativity*, H. Y. Chiu and W. F. Hoffmann (eds.), Benjamin, New York.
McKenna, J., and H. L. Frisch: 1966, *Phys. Rev.* **145**, 93.
Menger, K.: 1942, *Proc. Nat. Acad. Sci. USA* **28**, 535.
Menger, K.: 1951, *Proc. Nat. Acad. Sci. USA* **37**, 178, 226.
Messiah, A.: 1962, *Quantum Mechanics*, J. Potter (trans.), Wiley, New York.
Mignani, R.: 1981. *Hadronic J.* **4**, 785.
Miles, J. R. N., and J. S. Dahler: 1970, *J. Chem. Phys.* **52**, 616.
Miller, W., Jr: 1972, *Symmetry Groups and Their Applications*, Academic Press, New York.
Misner, C. W., K. S. Thorne and J. A. Wheeler: 1973, *Gravitation*, Freeman, San Francisco.
Misra, B., and E. C. G. Sudarshan: 1977, *J. Math. Phys.* **18**, 756.
Misra, B., I. Prigogine and M. Courbage: 1979a, *Proc. Natl. Acad. Sci. USA* **76**, 4768.
Misra, B., I. Prigogine and M. Courbage: 1979b, *Physica* **98A**, 1.
Mitroff, I. I.: 1974, *The Subjective Side of Science*, Elsevier, Amsterdam.
Mohapatra, R. N., and C. H. Lai (eds): 1981, *Gauge Theories of Fundamental Interactions*, World Scientific, Singapore.
Møller, C.: 1962, in *Royaumont Conference Proceedings*, CNRS, Paris.
Montroll, E. W.: 1952, *Commun. Pure. Appl. Math.* **5**, 415.
Moravetz, C. S., and W. A. Strauss: 1972, *Comm. Pure Appl. Math.* **25**, 1.
Mori, H., and S. Ono: 1952, *Prog. Theor. Phys.* **8**, 327.
Moyer, D. F.: 1979, in *On the Path of Albert Einstein*, A. Perlmutter and L. G. Scott (eds), Plenum, New York.
Mukunda, N., L. O'Raifeartaigh and E. C. G. Sudarshan: 1965, *Phys. Rev. Lett.* **15**, 1041.
Mukunda, N., H. van Dam, and L. C. Biedenharn: 1980, *Phys. Rev.* **D22**, 1938.
Namsrai, Kh.: 1980, *Found. Phys.* 10, 353, 731.
Namsrai, Kh.: 1981a, *Phys. Lett.* **82A**, 103.
Namsrai, Kh.: 1981b, Intern. *J. Theor. Phys.* **20**, 365, 749.
Nanopoulos, D. V.: 1981, in *Grand Unified Field Theories and Related Topics*, M. Konuma and T. Maskawa (eds.), World Scientific, Singapore.
Nelson, E.: 1964, *J. Math. Phys.* **5**, 332.
Nelson, E.: 1959, *Ann. Math.* **70**, 572.
Nelson, E.: 1966, *Phys. Rev.* **150**, 1079.
Nelson, E.: 1967, *Dynamical Theories of Brownian Motion*, Princeton University Press, Princeton, New Jersey.
Neumann, H.: 1972, *Helv. Phys. Acta* **45**, 811.
Newton, R. G.: 1979, *Found. Phys.* **9**, 929.

Newton, T. D., and E. P. Wigner: 1949, *Rev. Mod. Phys.* **21**, 400.
Nordheim, L.: 1928, *Proc. Roy. Soc. London* **A119**, 689.
O'Connell, R. F.: 1983, *Found. Phys.* **13**, 83.
O'Connell, R. F., and A. K. Rajagopal: 1982, *Phys. Rev. Lett.* **48**, 525.
O'Connell, R. F., and E. P. Wigner: 1981, *Phys. Lett.* **85A**, 121.
Ono, S.: 1954, *Prog. Theor. Phys.* **12**, 113.
Osterwalder, K.: 1982, in *Gauge Theories: Fundamental Interactions and Rigorous Results*, P. Dita, V. Georgescu and R. Purice (eds), Birkhäuser, Boston.
Pais, A.: 1972, in *Aspects of Quantum Theory*, A. Salam and E. P. Wigner (eds), Cambridge University Press, Cambridge.
Pais, A., and G. E. Uhlenbeck: 1950, *Phys. Rev.* **79**, 145.
Pandit, L. K.: 1959, *Suppl. Nuovo Cimento* **11**, 157.
Papadopoulos, G. J., and J. T. Devreese (eds): 1978, *Path Integrals*, Plenum, New York.
Papp, E. W. R.: 1974a, *Int. J. Theor. Phys.* **9**, 101.
Papp, E. W. R.: 1974b, *Int. J. Theor. Phys.* **10**, 123, 385.
Papp, E. W. R.: 1975, *Ann. Phys. Lpz.* **32**, 285.
Papp, E. W. R.: 1977, in *The Uncertainty Principle and Foundations of Quantum Mechanics*, W. C. Price and S. S. Chissick (eds), Wiley, New York.
Parker, L.: 1973, *Phys. Rev.* **D7**, 976.
Parker, L.: 1977, in *Asymptotic Structure of Space-Time*, F. P. Esposito and L. Witten (eds), Academic Press, New York.
Parker, L., and S. A. Fulling: 1973, *Phys. Rev.* **D7**, 2377.
Parker, L., and S. A. Fulling: 1974, *Phys. Rev.* **D9**, 341.
Pasupathy, J.: 1976, *Phys. Rev. Lett.* **37**, 1336.
Pathria, R. K.: 1972, *Statistical Mechanics*, Pergamon Press, Oxford.
Pati, J. C., A. Salam and J. Strathdee: 1975, *Phys. Lett.* **B59**, 265.
Pauli, W.: 1953, *Nuovo Cimento* **10**, 648.
Pauli, W.: 1956, *Helv. Phys. Acta Suppl.* **4**, 68.
Peierls, R.: 1979, *Surprises in Theoretical Physics*, Princeton University Press, Princeton, N.J.
Penrose, R.: 1968, in *Battelle Rencontres*, C. M. DeWitt and J. A. Wheeler (eds), Benjamin, New York.
Penrose, R.: 1974, in *Confrontation of Cosmological Theories with Observational Data*, M. S. Longair (ed), Reidel, Dordrecht.
Penrose, R., and M. A. H. MacCallum: 1973, *Phys. Rep.* **C6**, 243.
Perez, J. F., and I. F. Wilde: 1977, *Phys. Rev.* **D16**, 315.
Poincaré, H.: 1905, *Science and Hypothesis*, J. Larmor (trans.), Walter Scott Publ. Co., New York.
Pokrowski, G. I.: 1928, *Z. Phys.* **51**, 737.
Pool, J. C. T.: 1966, *J. Math. Phys.* **7**, 66.
Prigogine, I.: 1980, *From Being to Becoming*, Freeman, San Francisco.
Prugovečki, E.: 1966, *J. Math. Phys.* **7**, 1054, 1070.
Prugovečki, E.: 1967, *Can. J. Phys.* **45**, 2173.
Prugovečki, E.: 1972, *J. Math. Phys.* **13**, 969.
Prugovečki, E.: 1973a, *Found. Phys.* **3**, 3.
Prugovečki, E.: 1973b, *J. Math. Phys.* **14**, 1410.
Prugovečki, E.: 1974, *Found Phys.* **4**, 9.
Prugovečki, E.: 1975, *Found. Phys.* **5**, 557.
Prugovečki, E.: 1976a, *J. Math. Phys.* **17**, 517.
Prugovečki, E.: 1976b, *J. Math. Phys.* **17**, 1673.
Prugovečki, E.: 1976c, *J. Phys. A.: Math. Gen.* **9**, 1851.
Prugovečki, E.: 1977a, *Int. J. Theor. Phys.* **16**, 321.
Prugovečki, E.: 1977b, *J. Phys. A.: Math. Gen.* **10**, 543.
Prugovečki, E.: 1978a, *Ann. Phys.* (*N.Y.*) **110**, 102.

Prugovečki, E.: 1978b, *Physica* **91A**, 202.
Prugovečki, E.: 1978c, *Physica* **91A**, 229.
Prugovečki, E.: 1978d, *J. Math. Phys.* **19**, 2260.
Prugovečki, E.: 1978e, *J. Math. Phys.* **19**, 2271.
Prugovečki, E.: 1978f, *Phys. Rev.* **D18**, 3655.
Prugovečki, E.: 1979, *Found. Phys.* **9**, 575.
Prugovečki, E.: 1980, *Rep. Math. Phys.* **17**, 401.
Prugovečki, E.: 1981a, *Quantum Mechanics in Hilbert Space* (2nd edn), Academic Press, New York.
Prugovečki, E.: 1981b, *Nuovo Cimento* **61A**, 85.
Prugovečki, E.: 1981c, *Found. Phys.* **11**, 355.
Prugovečki, E.: 1981d, *Found. Phys.* **11**, 501.
Prugovečki, E.: 1981e, *Nuovo Cimento* **62B**, 17.
Prugovečki, E.: 1981f, *Hadronic J.* **4**, 1018.
Prugovečki, E.: 1981g, *Lett. Nuovo Cimento* **32**, 277.
Prugovečki, E.: 1981h, *Lett. Nuovo Cimento* **32**, 481; **33**, 480.
Prugovečki, E.: 1982a, *Found. Phys.* **12**, 555.
Prugovečki, E.: 1982b, *Nuovo Cimento* **68B**, 261.
Prugovečki, E.: 1982c, *Phys. Rev. Lett.* **49**, 1065.
Prugovečki, E.: 1983, 'Finite Charge and Field Renormalizations in Reciprocally Invariant Quantum Electrodynamics', *Lett. Nuovo Cimento* (to appear).
Prugovečki, E., and E. B. Manoukian: 1972a, *Nuovo Cimento* **10B**, 421.
Prugovečki, E., and E. B. Manoukian: 1972b, *Comm. Math. Phys.* **24**, 133.
Prugovečki, E., and A. Tip: 1974, *J. Math. Phys.* **15**, 275.
Prugovečki, E., and A. Tip: 1975, *Comp. Math.* **30**, 113.
Pryce, M. H. L.: 1948, *Proc. Roy. Soc. London* **A195**, 62.
Putnam, C. R.: 1967, *Commutation Properties of Hilbert Space Operators and Related Topics*, Springer, New York.
Rączka, R.: 1975, *J. Math. Phys.* **16**, 173.
Raine, D. J., and C. P. Winlove: 1975, *Phys. Rev.* **D12**, 946.
Rajaraman, R.: 1975, *Phys. Rep.* **C21**, 227.
Rajasekaran, G.: 1980, in *Gravitation, Quanta and the Universe*, A. R. Prasanna, J. V. Narlikar and C. V. Vishveshwara (eds), Wiley, New York.
Rañada, A. F., and G. S. Rodero: 1980, *Phys. Rev.* **D22**, 385.
Rayski, J., and J. M. Rayski, Jr: 1977, in *The Uncertainty Principle and Foundations of Quantum Mechanics*, W. C. Price and S. S. Chissick (eds), Wiley, New York.
Recami, E.: 1977, in *The Uncertainty Principle and Foundation of Quantum Mechanics*, W. C. Price and S. S. Chissick (eds), Wiley, New York.
Reuse, F.: 1978, *Helv. Phys. Acta* **51**, 157.
Reuse, F.: 1979, *Found. Phys.* **9**, 865.
Reuse, F.: 1980, *Helv. Phys. Acta* **53**, 416, 522.
Rohrlich, F.: 1974, in *Physical Reality and Mathematical Description*, C. P. Enz and J. Mehra (eds), Reidel, Dordrecht.
Rohrlich, F.: 1980, in *Foundations of Radiation Theory and Quantum Electrodynamics*, A. O. Barut (ed), Plenum, New York.
Roman, P.: 1969, *Introduction to Quantum Field Theory*, Wiley, New York.
Rosen, G.: 1972, *Phys. Rev.* **D5**, 1100.
Rosen, G.: 1982, *Lett. Nuovo Cimento* **34**, 71.
Ross, J., and J. G. Kirkwood: 1954, *J. Chem. Phys.* **22**, 1094.
Roy, S.: 1980, *J. Math. Phys.* **21**, 71.
Roy, S.: 1981, *Nuovo Cimento* **64B**, 81.
Ruark, A. E.: 1928, *Proc. Nat. Acad. Sci. Wash.* **14**, 322.
Ruijsenaars, S. N. M.: 1981, *Ann. Phys.* (*NY*) **137**, 33.

Rühl, W. (ed): 1980, *Field Theoretical Methods in Particle Physics*, Plenum, New York.
Russell, B.: 1945, *A History of Western Philosophy*, Simon and Schuster, New York.
Russell, B.: 1948, *Human Knowledge*, Simon and Schuster, New York.
Sachs, M.: 1979, *Ann. Fond. Louis de Broglie* **4**, 85, 175.
Sachs, M.: 1980, *Nuovo Cimento* **58A**, 1.
Sachs, M.: 1982, *Hadronic J.* **5**, 1781.
Saenz, A. W.: 1957, *Phys. Rev.* **105**, 546.
Salam, A.: 1979, *Nobel Prize Talk*, reprinted by Lai (1981).
Salecker, H., and E. P. Wigner: 1958, *Phys. Rev.* **109**, 571.
Santilli, R. M.: 1978, *Hadronic J.* **1**, 574.
Santilli, R. M.: 1979, *Hadronic J.* **2**, 1460.
Santilli, R. M.: 1981, *Found. Phys.* **11**, 383.
Santilli, R. M.: 1982, *Lett. Nuovo Cimento* **33**, 145.
Schames, L.: 1933, *Z. Phys.* **81**, 270.
Schechter, M.: 1981, *Operator Methods in Quantum Mechanics*, North-Holland, New York.
Schild, A.: 1948, *Phys. Rev.* **73**, 414.
Schlegel, R.: 1980, *Found. Phys.* **10**, 345.
Schönberg, M.: 1953, *Nuovo Cimento* **10**, 419.
Schrödinger, E.: 1930, *Sitzber. Preuss. Acad. Wiss. Berlin* **24**, 418.
Schroeck, F. E., Jr: 1971, *J. Math. Phys.* **12**, 1849.
Schroeck, F. E., Jr: 1973, *J. Math. Phys.* **14**, 130.
Schroeck, F. E., Jr: 1975, *J. Math. Phys.* **16**, 729, 2112.
Schroeck, F. E., Jr: 1978, in *Mathematical Foundations of Quantum Mechanics*, A. R. Marlow (ed.), Academic Press, New York.
Schroeck, F. E., Jr: 1981a, in *Quantum Mechanics in Mathematics, Chemistry, and Physics*, K. E. Gustafson and W. D. Reinhardt (eds), Plenum, New York.
Schroeck, F. E., Jr: 1981b, *J. Math. Phys.* **22**, 2562.
Schroeck, F. E., Jr: 1982a, *Found. Phys.* **12**, 479.
Schroeck, F. E., Jr: 1982b, *Found. Phys.* **12**, 825.
Schroeck, F. E., Jr: 1983a, 'On the Nonoccurrence of Two Paradoxes in the Measurement Scheme of Stochastic Quantum Mechanics' (Florida Atlantic University preprint).
Schroeck, F. E., Jr: 1983b, 'A Note on the Fields in Quantum Spacetime' (Florida Atlantic University preprint).
Schulman, L. S.: 1981, *Techniques and Application of Path Integration*, Wiley, New York.
Schupe, M.: 1979, *Phys. Lett.* **B86**, 87.
Schwartz, L.: 1973, *Radon Measures on Arbitrary Topological Spaces and Cylindrical Measures*, Oxford University Press, London.
Schwebel, S. L.: 1978, *Int. J. Theor. Phys.* **17**, 931.
Schweber, S. S.: 1961, *An Introduction to Relativistic Quantum Field Theory*, Row-Peterson, Evanston, Illinois.
Schweber, S. S.: 1962, *J. Math. Phys.* **3**, 831.
Schweizer, B., A. Sklar and E. Thorpe: 1960, *Pac. J. Math.* **10**, 673.
Schweizer, B., and A. Sklar: 1962, *Theory Prob. and Appl.* 7, 447.
Schwinger, J.: 1948, *Phys. Rev.* **73**, 416.
Schwinger, J.: 1949a, *Phys. Rev.* **75**, 898.
Schwinger, J.: 1949b, *Phys. Rev.* **76**, 790.
Schwinger, J. (ed.): 1958, *Quantum Electrodynamics*, Dover, New York.
Schwinger, J.: 1962, *Phys. Rev.* **125**, 397; **128**, 2425.
Scutaru, H.: 1979, *Lett. Math. Phys.* **2**, 101.
Segal, E.: 1976, in *Mathematical Physics and Physical Mathematics*, K. Maurin and R. Rączka (eds), Reidel, Dordrecht.
Sen, P.: 1958, *Nuovo Cimento* 8, 407.

Shimony, A.: 1966, *Phys. Today* **19**, 85.
Simms, D. J., and N. M. J. Woodhouse: 1976, *Lectures on Geometric Quantization*, Springer, Berlin.
Skagerstam, B. K.: 1976, *Int. J. Theor. Phys.* **15**, 213.
Śniatycki, J.: 1979, *Geometric Quantization and Quantum Mechanics*, Springer, Berlin.
Snider, R. F., and B. C. Sanctuary: 1971, *J. Chem. Phys.* **55**, 1555.
Snyder, H. S.: 1947a, *Phys. Rev.* **71**, 38.
Snyder, H. S.: 1947b, *Phys. Rev.* **72**, 68.
Spivak, M.: 1970, *A Comprehensive Introduction to Differential Geometry*, Vol. 1, Publish or Perish, Boston.
Srinivas, M. D., and E. Wolf: 1975, *Phys. Rev.* **D11**, 1477.
Stachel, J.: 1980, in *General Relativity and Gravitation*, vol. 1, A. Held (ed.), Plenum, New York.
Stapp, H. P.: 1972, *Amer. J. Phys.* **40**, 1098.
Steigman, G.: 1976, *Ann. Rev. Astrophys. Astr.* **14**, 339.
Sternberg, S.: 1977, *Proc. Nat. Acad. Sci. USA* **74**, 5253.
Sternberg, S.: 1978, in *Differential Geometric Methods in Mathematical Physics II*, K. Bleuler, H. R. Petry, and A. Reetz (eds), Springer, New York.
Streater, R. F., and A. S. Wightman: 1964, *PCT, Spin and Statistics and All That*, Benjamin, New York.
Strocchi, F., and A. S. Wightman: 1974, *J. Math. Phys.* **12**, 2198.
Stueckelberg, E. C. G.: 1941, *Helv. Phys. Acta* **14**, 372, 588.
Stueckelberg, E., and G. Vanders: 1954: *Helv. Phys. Acta* **27**, 667.
Synge, J. L.: 1970, *Proc. Roy. Soc. London* **A319**, 307.
Takabayashi, T.: 1967, *Prog. Theor. Phys.* **38**, 966.
Takabayashi, T.: 1970, *Prog. Theor. Phys.* **44**, 1429.
Takabayashi, T.: 1971, *Prog. Theor. Phys.* **46**, 1528, 1924.
Takabayashi, T.: 1979, *Prog. Theor. Phys. Suppl.* **67**, 1.
Taylor, J. C.: 1976, *Gauge Theories of Weak Interactions*, Cambridge University Press, Cambridge.
Taylor, J. G.: 1978, *Phys. Rev.* **D18**, 3544.
Taylor, J. G.: 1979, *Phys. Rev.* **D19**, 2336.
Taylor, J. R.: 1972, *Scattering Theory: The Quantum Theory of Nonrelativistic Collisions*, Wiley, New York.
Tip, A.: 1971, *Physica* **52**, 493.
Tomonaga, S.: 1946, *Prog. Theor. Phys.* **1**, 27.
Trautman, A.: 1965, *Lectures on General Relativity*, Prentice-Hall, Englewood Cliffs, N.J.
Trotter, H.: 1959, *Proc. Amer. Math. Soc.* **10**, 545.
Turner, R. E., and R. F. Snider: 1980, *Can. J. Phys.* **58**, 1171.
Uhlenbeck, G. E.: 1973, in *The Boltzmann Equation*, E. G. D. Cohen and W. Thirring (eds), Springer, Wien and New York.
Uhling, E. A., and G. E. Uhlenbeck: 1933, *Phys. Rev.* **43**, 552.
Unruh, W. G.: 1976, *Phys. Rev.* **D14**, 870.
van Dam, H., and L. C. Biedenharn: 1976a, *Phys. Rev.* **14**, 405.
van Dam, H., and L. C. Biedenharn: 1976b, *Phys. Lett.* **B62**, 190.
van Dam, H., L. C. Biedenharn and N. Mukunda: 1981, *Phys. Rev.* **D23**, 1451.
van Nieuwenhuizen, P.: 1977, in *Marcel Grossmann Meeting on General Relativity*, R. Ruffini (ed.), North-Holland, Amsterdam.
Velo, G., and A. S. Wightman (eds): 1978, *Invariant Wave Equations*, Lecture Notes in Physics, vol. 73, Springer, Berlin.
Velo, G., and D. Zwanziger: 1969a, *Phys. Rev.* **186**, 1337.
Velo, G., and D. Zwanziger: 1969b, *Phys. Rev.* **188**, 2218.

Vigier, J.-P.: 1979, *Lett. Nuovo Cimento* **24**, 258.

Vogt, A.: 1978, in *Mathematical Foundations of Quantum Theory*, A. R. Marlow (ed.), Academic Press, New York.

von Neumann, J.: 1931, *Math. Ann.* **104**, 570.

von Neumann, J.: 1955, *Mathematical Foundations of Quantum Mechanics*, R. T. Beyer (trans.), Princeton University Press, Princeton, N.J.

Vyaltsev, A. N.: 1965, *Discrete Space-Time* (in Russian), Nauka, Moscow.

Wald, A.: 1943, *Proc. Nat. Acad. Sci. U.S.A.* **29**, 196.

Wald, A.: 1955, *Selected Papers in Statistics and Probability*, McGraw-Hill, New York.

Weidlich, W., and A. K. Mitra: 1966, *Nuovo Cimento* **30**, 385.

Weinberg, S.: 1972, *Gravitation and Cosmology: Principles and Application of the General Theory of Relativity*, Wiley, New York.

Weinberg, S.: 1981, in *Gauge Theory of Weak and Electromagnetic Interactions*, C. H. Lai (ed.), World Scientific, Singapore.

Welton, T. A.: 1948, *Phys. Rev.* **74**, 1157.

Wheeler, J. A.: 1962, *Geometrodynamics*, Academic Press, New York.

Wheeler, J. A.: 1967, in *Battelle Rencontres*, C. De Witt and J. A. Wheeler (eds), Benjamin, New York.

Wheeler, J. A.: 1979, in *Problems in the Foundations of Physics*, G. T. Di Francia (ed.), North-Holland, Amsterdam.

Whitham, G. B.: 1974, *Linear and Nonlinear Waves*, Wiley, New York.

Whittaker, E. T.: 1951, *History of the Theories of Aether and Electricity* (2nd edn), Nelson, London.

Wick, G. C., E. P. Wigner and A. S. Wightman: 1952, *Phys. Rev.* **88**, 101.

Wightman, A. S.: 1962, *Rev. Mod. Phys.* **34**, 845.

Wightman, A. S.: 1971, *Troubles in the External Field Problem for Invariant Wave Equations*, Gordon and Breach, New York.

Wightman, A. S.: 1972, in *Aspects of Quantum Theory*, A. Salam and E. P. Wigner (eds), Cambridge University Press, Cambridge.

Wightman, A. S., and S. S. Schweber: 1955, *Phys. Rev.* **98**, 812.

Wigner, E. P.: 1932, *Phys. Rev.* **40**, 749.

Wigner, E. P.: 1939, *Ann. Math.* **40**, 149.

Wigner, E. P.: 1952, *Z. Phys.* **131**, 101.

Wigner, E. P.: 1955, *Helv. Phys. Acta Suppl.* **4**, 210.

Wigner, E. P.: 1963, *Amer. J. Phys.* **31**, 6.

Wigner, E. P.: 1972, in *Aspects of Quantum Theory*, A. Salam and E. P. Wigner (eds), Cambridge University Press, Cambridge.

Woodhouse, N. M. J.: 1973, *J. Math. Phys.* **14**, 495.

Woolf, H. (ed.): 1980, *Some Strangeness in the Proportion*, Addison-Wesley, Reading, Mass.

Yaes, R.: 1974, *New Scientist* **63**, 462.

Yang, C. N.: 1947, *Phys. Rev.* **72**, 874.

Yang, C. N., and R. L. Mills: 1954, *Phys. Rev.* **96**, 191.

Yosida, K.: 1974, *Functional Analysis* (4th edn), Springer, Berlin.

Yourgrau W., and A. van der Merwe (eds): 1979, *Perspectives in Quantum Theory*, Dover, New York.

Yuen, H. P., and J. H. Shapiro: 1980, *IEEE Trans. Inform. Theory* **26**, 78.

Yuen, H. P.: 1982, *Phys. Lett.* **91A**, 101.

Yukawa, H.: 1949, *Phys. Rev.* **76**, 300.

Yukawa, H.: 1950a, *Phys. Rev.* **77**, 219, 849.

Yukawa, H.: 1950b, *Phys. Rev.* **80**, 1047.

Yukawa, H.: 1953, *Phys. Rev.* **91**, 416.

Zadeh, L. A.: 1965, *Inform. Control* **8**, 338.

Zadeh, L. A.: 1968, *J. Math. Anal. Appl.* **23**, 421.
Zaslansky, G. M.: 1981, *Phys. Rep.* **C80**, 157.
Zinn-Justin, J.: 1982, in *Non-Perturbative Aspects of Quantum Field Theory*, J. Julve and M. Ramon-Medrano (eds.), World Scientific, Singapore.
Zwanziger, D.: 1979, *Phys. Rev.* **D19**, 473.

Index